SCHAUM'S OUTLINE SERIES

MATHEMATICAL HANDBOOK

of

Formulas and Tables

•

BY

MURRAY R. SPIEGEL, Ph.D.

Professor of Mathematics

Rensselaer Polytechnic Institute

•

SCHAUM'S OUTLINE SERIES

McGRAW-HILL BOOK COMPANY

New York, St. Louis, San Francisco, Toronto, Sydney

60224

7 8 9 10 11 12 13 14 15 SH SH 7 5 4 3

Preface

The purpose of this handbook is to supply a collection of mathematical formulas and tables which will prove to be valuable to students and research workers in the fields of mathematics, physics, engineering and other sciences. To accomplish this, care has been taken to include those formulas and tables which are most likely to be needed in practice rather than highly specialized results which are rarely used. Every effort has been made to present results concisely as well as precisely so that they may be referred to with a maximum of ease as well as confidence.

Topics covered range from elementary to advanced. Elementary topics include those from algebra, geometry, trigonometry, analytic geometry and calculus. Advanced topics include those from differential equations, vector analysis, Fourier series, gamma and beta functions, Bessel and Legendre functions, Fourier and Laplace transforms, elliptic functions and various other special functions of importance. This wide coverage of topics has been adopted so as to provide within a single volume most of the important mathematical results needed by the student or research worker regardless of his particular field of interest or level of attainment.

The book is divided into two main parts. Part I presents mathematical formulas together with other material, such as definitions, theorems, graphs, diagrams, etc., essential for proper understanding and application of the formulas. Included in this first part are extensive tables of integrals and Laplace transforms which should be extremely useful to the student and research worker. Part II presents numerical tables such as the values of elementary functions (trigonometric, logarithmic, exponential, hyperbolic, etc.) as well as advanced functions (Bessel, Legendre, elliptic, etc.). In order to eliminate confusion, especially to the beginner in mathematics, the numerical tables for each function are separated. Thus, for example, the sine and cosine functions for angles in degrees and minutes are given in separate tables rather than in one table so that there is no need to be concerned about the possibility of error due to looking in the wrong column or row.

I wish to thank the various authors and publishers who gave me permission to adapt data from their books for use in several tables of this handbook. Appropriate references to such sources are given next to the corresponding tables. In particular I am indebted to the Literary Executor of the late Sir Ronald A. Fisher, F.R.S., to Dr. Frank Yates, F.R.S., and to Oliver and Boyd Ltd., Edinburgh, for permission to use data from Table III of their book *Statistical Tables for Biological, Agricultural and Medical Research.*

I also wish to express my gratitude to Nicola Monti, Henry Hayden and Jack Margolin for their excellent editorial cooperation.

M. R. Spiegel

Rensselaer Polytechnic Institute
September, 1968

CONTENTS

Part I	**FORMULAS**

Part II TABLES

CONTENTS

Part I

FORMULAS

THE GREEK ALPHABET

Greek name	Greek letter	
	Lower case	Capital
Alpha	α	A
Beta	β	B
Gamma	γ	Γ
Delta	δ	Δ
Epsilon	ϵ	E
Zeta	ζ	Z
Eta	η	H
Theta	θ	Θ
Iota	ι	I
Kappa	κ	K
Lambda	λ	Λ
Mu	μ	M

Greek name	Greek letter	
	Lower case	Capital
Nu	ν	N
Xi	ξ	Ξ
Omicron	o	O
Pi	π	Π
Rho	ρ	P
Sigma	σ	Σ
Tau	τ	T
Upsilon	υ	Υ
Phi	ϕ	Φ
Chi	χ	X
Psi	ψ	Ψ
Omega	ω	Ω

1 SPECIAL CONSTANTS

1.1 $\pi = 3.14159\ 26535\ 89793\ 23846\ 2643\ldots$

1.2 $e = 2.71828\ 18284\ 59045\ 23536\ 0287\ldots = \lim_{n\to\infty}\left(1+\frac{1}{n}\right)^n$
= natural base of logarithms

1.3 $\sqrt{2} = 1.41421\ 35623\ 73095\ 0488\ldots$

1.4 $\sqrt{3} = 1.73205\ 08075\ 68877\ 2935\ldots$

1.5 $\sqrt{5} = 2.23606\ 79774\ 99789\ 6964\ldots$

1.6 $\sqrt[3]{2} = 1.25992\ 1050\ldots$

1.7 $\sqrt[3]{3} = 1.44224\ 9570\ldots$

1.8 $\sqrt[5]{2} = 1.14869\ 8355\ldots$

1.9 $\sqrt[5]{3} = 1.24573\ 0940\ldots$

1.10 $e^{\pi} = 23.14069\ 26327\ 79269\ 006\ldots$

1.11 $\pi^{e} = 22.45915\ 77183\ 61045\ 47342\ 715\ldots$

1.12 $e^{e} = 15.15426\ 22414\ 79264\ 190\ldots$

1.13 $\log_{10} 2 = 0.30102\ 99956\ 63981\ 19521\ 37389\ldots$

1.14 $\log_{10} 3 = 0.47712\ 12547\ 19662\ 43729\ 50279\ldots$

1.15 $\log_{10} e = 0.43429\ 44819\ 03251\ 82765\ldots$

1.16 $\log_{10} \pi = 0.49714\ 98726\ 94133\ 85435\ 12683\ldots$

1.17 $\log_e 10 = \ln 10 = 2.30258\ 50929\ 94045\ 68401\ 7991\ldots$

1.18 $\log_e 2 = \ln 2 = 0.69314\ 71805\ 59945\ 30941\ 7232\ldots$

1.19 $\log_e 3 = \ln 3 = 1.09861\ 22886\ 68109\ 69139\ 5245\ldots$

1.20 $\gamma = 0.57721\ 56649\ 01532\ 86060\ 6512\ldots$ = *Euler's constant*
$= \lim_{n\to\infty}\left(1+\frac{1}{2}+\frac{1}{3}+\cdots+\frac{1}{n}-\ln n\right)$

1.21 $e^{\gamma} = 1.78107\ 24179\ 90197\ 9852\ldots$ [see 1.20]

1.22 $\sqrt{e} = 1.64872\ 12707\ 00128\ 1468\ldots$

1.23 $\sqrt{\pi} = \Gamma(\frac{1}{2}) = 1.77245\ 38509\ 05516\ 02729\ 8167\ldots$
where Γ is the *gamma function* [see pages 101-102].

1.24 $\Gamma(\frac{1}{3}) = 2.67893\ 85347\ 07748\ldots$

1.25 $\Gamma(\frac{1}{4}) = 3.62560\ 99082\ 21908\ldots$

1.26 $1 \text{ radian} = 180°/\pi = 57.29577\ 95130\ 8232\ldots°$

1.27 $1° = \pi/180 \text{ radians} = 0.01745\ 32925\ 19943\ 29576\ 92\ldots \text{ radians}$

2 SPECIAL PRODUCTS and FACTORS

2.1 $(x+y)^2 = x^2 + 2xy + y^2$

2.2 $(x-y)^2 = x^2 - 2xy + y^2$

2.3 $(x+y)^3 = x^3 + 3x^2y + 3xy^2 + y^3$

2.4 $(x-y)^3 = x^3 - 3x^2y + 3xy^2 - y^3$

2.5 $(x+y)^4 = x^4 + 4x^3y + 6x^2y^2 + 4xy^3 + y^4$

2.6 $(x-y)^4 = x^4 - 4x^3y + 6x^2y^2 - 4xy^3 + y^4$

2.7 $(x+y)^5 = x^5 + 5x^4y + 10x^3y^2 + 10x^2y^3 + 5xy^4 + y^5$

2.8 $(x-y)^5 = x^5 - 5x^4y + 10x^3y^2 - 10x^2y^3 + 5xy^4 - y^5$

2.9 $(x+y)^6 = x^6 + 6x^5y + 15x^4y^2 + 20x^3y^3 + 15x^2y^4 + 6xy^5 + y^6$

2.10 $(x-y)^6 = x^6 - 6x^5y + 15x^4y^2 - 20x^3y^3 + 15x^2y^4 - 6xy^5 + y^6$

The results 2.1 to 2.10 above are special cases of the *binomial formula* [see page 3].

2.11 $x^2 - y^2 = (x-y)(x+y)$

2.12 $x^3 - y^3 = (x-y)(x^2+xy+y^2)$

2.13 $x^3 + y^3 = (x+y)(x^2-xy+y^2)$

2.14 $x^4 - y^4 = (x-y)(x+y)(x^2+y^2)$

2.15 $x^5 - y^5 = (x-y)(x^4+x^3y+x^2y^2+xy^3+y^4)$

2.16 $x^5 + y^5 = (x+y)(x^4-x^3y+x^2y^2-xy^3+y^4)$

2.17 $x^6 - y^6 = (x-y)(x+y)(x^2+xy+y^2)(x^2-xy+y^2)$

2.18 $x^4 + x^2y^2 + y^4 = (x^2+xy+y^2)(x^2-xy+y^2)$

2.19 $x^4 + 4y^4 = (x^2+2xy+2y^2)(x^2-2xy+2y^2)$

Some generalizations of the above are given by the following results where n is a positive integer.

2.20
$$\begin{aligned} x^{2n+1} - y^{2n+1} &= (x-y)(x^{2n} + x^{2n-1}y + x^{2n-2}y^2 + \cdots + y^{2n}) \\ &= (x-y)\left(x^2 - 2xy\cos\frac{2\pi}{2n+1} + y^2\right)\left(x^2 - 2xy\cos\frac{4\pi}{2n+1} + y^2\right) \\ &\quad \cdots \left(x^2 - 2xy\cos\frac{2n\pi}{2n+1} + y^2\right) \end{aligned}$$

2.21
$$\begin{aligned} x^{2n+1} + y^{2n+1} &= (x+y)(x^{2n} - x^{2n-1}y + x^{2n-2}y^2 - \cdots + y^{2n}) \\ &= (x+y)\left(x^2 + 2xy\cos\frac{2\pi}{2n+1} + y^2\right)\left(x^2 + 2xy\cos\frac{4\pi}{2n+1} + y^2\right) \\ &\quad \cdots \left(x^2 + 2xy\cos\frac{2n\pi}{2n+1} + y^2\right) \end{aligned}$$

2.22
$$\begin{aligned} x^{2n} - y^{2n} &= (x-y)(x+y)(x^{n-1} + x^{n-2}y + x^{n-3}y^2 + \cdots)(x^{n-1} - x^{n-2}y + x^{n-3}y^2 - \cdots) \\ &= (x-y)(x+y)\left(x^2 - 2xy\cos\frac{\pi}{n} + y^2\right)\left(x^2 - 2xy\cos\frac{2\pi}{n} + y^2\right) \\ &\quad \cdots \left(x^2 - 2xy\cos\frac{(n-1)\pi}{n} + y^2\right) \end{aligned}$$

2.23
$$\begin{aligned} x^{2n} + y^{2n} &= \left(x^2 + 2xy\cos\frac{\pi}{2n} + y^2\right)\left(x^2 + 2xy\cos\frac{3\pi}{2n} + y^2\right) \\ &\quad \cdots \left(x^2 + 2xy\cos\frac{(2n-1)\pi}{2n} + y^2\right) \end{aligned}$$

3 The BINOMIAL FORMULA and BINOMIAL COEFFICIENTS

FACTORIAL n

If $n = 1, 2, 3, \ldots$ *factorial n* or *n factorial* is defined as

3.1 $$n! = 1 \cdot 2 \cdot 3 \cdot \cdots \cdot n$$

We also define *zero factorial* as

3.2 $$0! = 1$$

BINOMIAL FORMULA FOR POSITIVE INTEGRAL n

If $n = 1, 2, 3, \ldots$ then

3.3 $$(x+y)^n = x^n + nx^{n-1}y + \frac{n(n-1)}{2!}x^{n-2}y^2 + \frac{n(n-1)(n-2)}{3!}x^{n-3}y^3 + \cdots + y^n$$

This is called the *binomial formula*. It can be extended to other values of n and then is an infinite series [see *Binomial Series*, page 110].

BINOMIAL COEFFICIENTS

The result 3.3 can also be written

3.4 $$(x+y)^n = x^n + \binom{n}{1}x^{n-1}y + \binom{n}{2}x^{n-2}y^2 + \binom{n}{3}x^{n-3}y^3 + \cdots + \binom{n}{n}y^n$$

where the coefficients, called *binomial coefficients*, are given by

3.5 $$\binom{n}{k} = \frac{n(n-1)(n-2)\cdots(n-k+1)}{k!} = \frac{n!}{k!\,(n-k)!} = \binom{n}{n-k}$$

PROPERTIES OF BINOMIAL COEFFICIENTS

3.6 $$\binom{n}{k} + \binom{n}{k+1} = \binom{n+1}{k+1}$$

This leads to *Pascal's triangle* [see page 236].

3.7 $$\binom{n}{0} + \binom{n}{1} + \binom{n}{2} + \cdots + \binom{n}{n} = 2^n$$

3.8 $$\binom{n}{0} - \binom{n}{1} + \binom{n}{2} - \cdots (-1)^n \binom{n}{n} = 0$$

3.9 $$\binom{n}{n} + \binom{n+1}{n} + \binom{n+2}{n} + \cdots + \binom{n+m}{n} = \binom{n+m+1}{n+1}$$

3.10 $$\binom{n}{0} + \binom{n}{2} + \binom{n}{4} + \cdots = 2^{n-1}$$

3.11 $$\binom{n}{1} + \binom{n}{3} + \binom{n}{5} + \cdots = 2^{n-1}$$

3.12 $$\binom{n}{0}^2 + \binom{n}{1}^2 + \binom{n}{2}^2 + \cdots + \binom{n}{n}^2 = \binom{2n}{n}$$

3.13 $$\binom{m}{0}\binom{n}{p} + \binom{m}{1}\binom{n}{p-1} + \cdots + \binom{m}{p}\binom{n}{0} = \binom{m+n}{p}$$

3.14 $$(1)\binom{n}{1} + (2)\binom{n}{2} + (3)\binom{n}{3} + \cdots + (n)\binom{n}{n} = n2^{n-1}$$

3.15 $$(1)\binom{n}{1} - (2)\binom{n}{2} + (3)\binom{n}{3} - \cdots (-1)^{n+1}(n)\binom{n}{n} = 0$$

MULTINOMIAL FORMULA

3.16 $$(x_1 + x_2 + \cdots + x_p)^n = \sum \frac{n!}{n_1!\, n_2! \cdots n_p!} x_1^{n_1} x_2^{n_2} \cdots x_p^{n_p}$$

where the sum, denoted by Σ, is taken over all nonnegative integers $n_1, n_2, \ldots, n_p$ for which $n_1 + n_2 + \cdots + n_p = n$.

4 GEOMETRIC FORMULAS

RECTANGLE OF LENGTH b AND WIDTH a

4.1 Area $= ab$

4.2 Perimeter $= 2a + 2b$

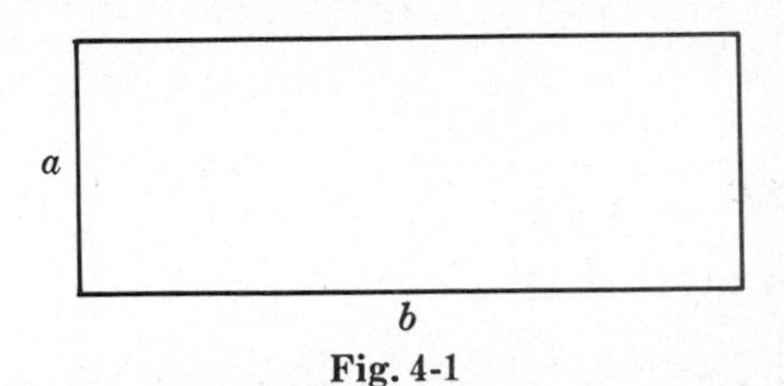

Fig. 4-1

PARALLELOGRAM OF ALTITUDE h AND BASE b

4.3 Area $= bh = ab \sin\theta$

4.4 Perimeter $= 2a + 2b$

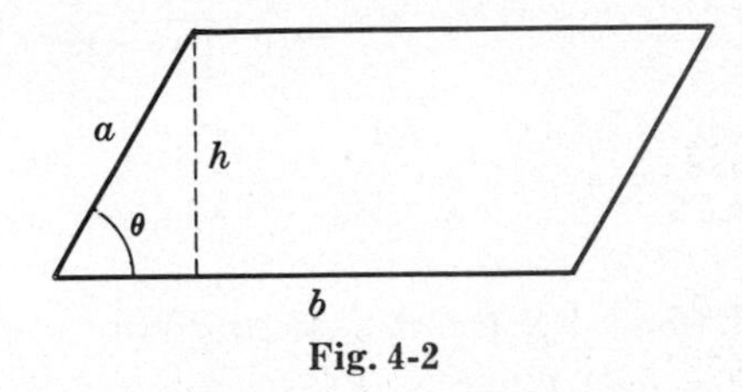

Fig. 4-2

TRIANGLE OF ALTITUDE h AND BASE b

4.5 Area $= \frac{1}{2}bh = \frac{1}{2}ab \sin\theta$

$$= \sqrt{s(s-a)(s-b)(s-c)}$$

where $s = \frac{1}{2}(a+b+c) =$ semiperimeter

4.6 Perimeter $= a + b + c$

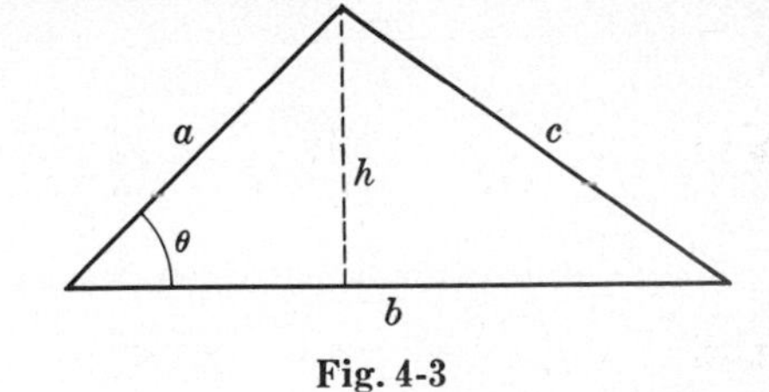

Fig. 4-3

TRAPEZOID OF ALTITUDE h AND PARALLEL SIDES a AND b

4.7 Area $= \frac{1}{2}h(a+b)$

4.8 Perimeter $= a + b + h\left(\frac{1}{\sin\theta} + \frac{1}{\sin\phi}\right)$

$$= a + b + h(\csc\theta + \csc\phi)$$

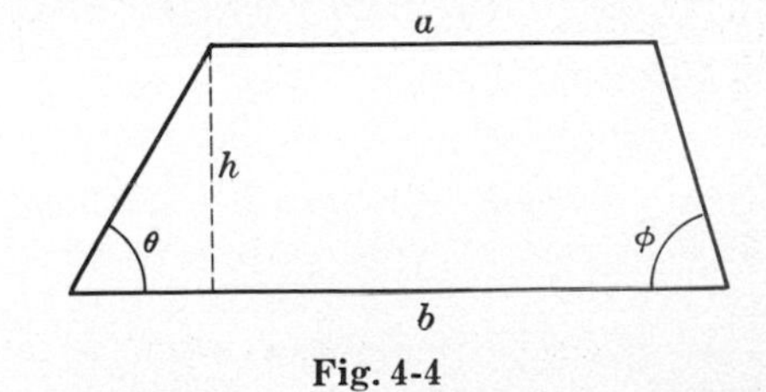

Fig. 4-4

REGULAR POLYGON OF n SIDES EACH OF LENGTH b

4.9 Area $= \frac{1}{4}nb^2 \cot\frac{\pi}{n} = \frac{1}{4}nb^2 \frac{\cos(\pi/n)}{\sin(\pi/n)}$

4.10 Perimeter $= nb$

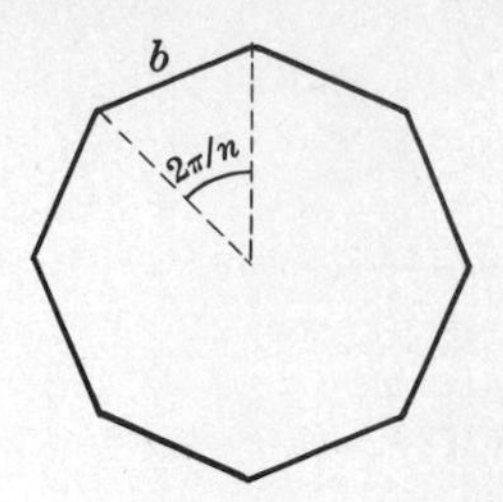

Fig. 4-5

CIRCLE OF RADIUS r

4.11 Area $= \pi r^2$

4.12 Perimeter $= 2\pi r$

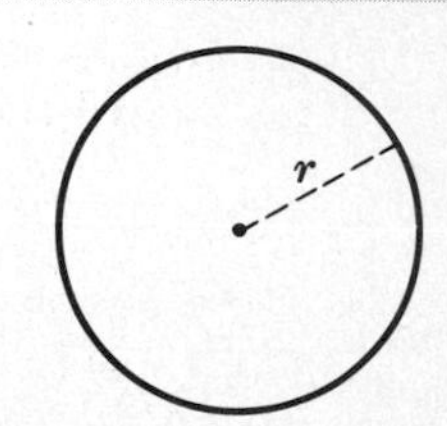

Fig. 4-6

SECTOR OF CIRCLE OF RADIUS r

4.13 Area $= \frac{1}{2}r^2\theta$ [θ in radians]

4.14 Arc length $s = r\theta$

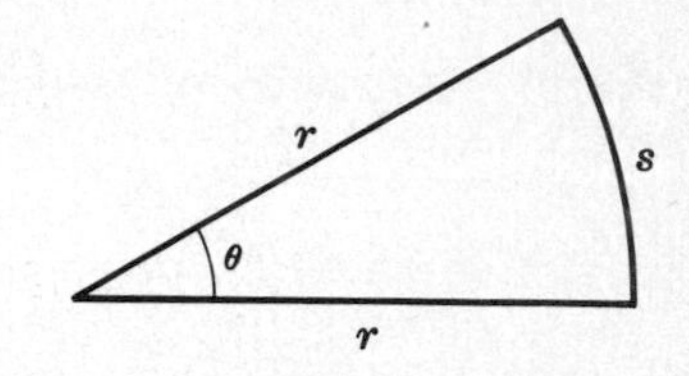

Fig. 4-7

RADIUS OF CIRCLE INSCRIBED IN A TRIANGLE OF SIDES a, b, c

4.15
$$r = \frac{\sqrt{s(s-a)(s-b)(s-c)}}{s}$$

where $s = \frac{1}{2}(a+b+c)$ = semiperimeter

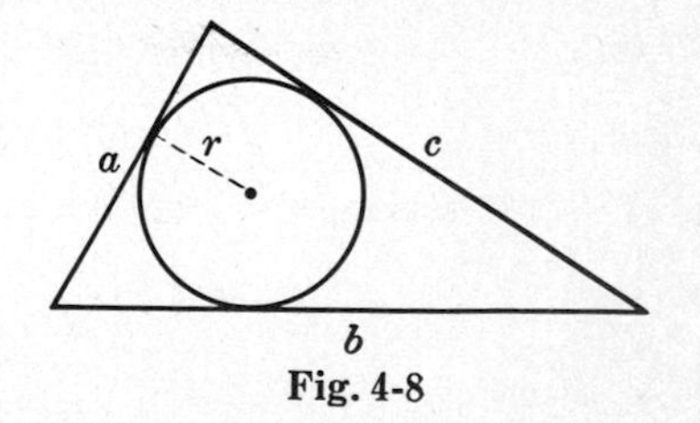

Fig. 4-8

RADIUS OF CIRCLE CIRCUMSCRIBING A TRIANGLE OF SIDES a, b, c

4.16
$$R = \frac{abc}{4\sqrt{s(s-a)(s-b)(s-c)}}$$

where $s = \frac{1}{2}(a+b+c)$ = semiperimeter

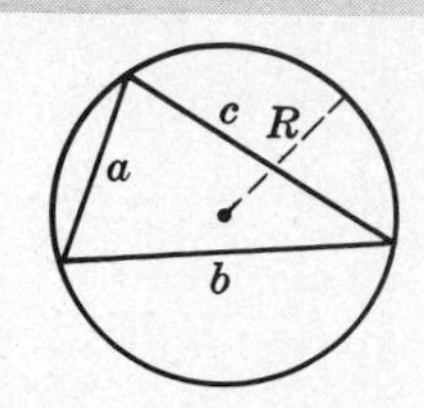

Fig. 4-9

REGULAR POLYGON OF n SIDES INSCRIBED IN CIRCLE OF RADIUS r

4.17 Area $= \frac{1}{2}nr^2 \sin\frac{2\pi}{n} = \frac{1}{2}nr^2 \sin\frac{360°}{n}$

4.18 Perimeter $= 2nr \sin\frac{\pi}{n} = 2nr \sin\frac{180°}{n}$

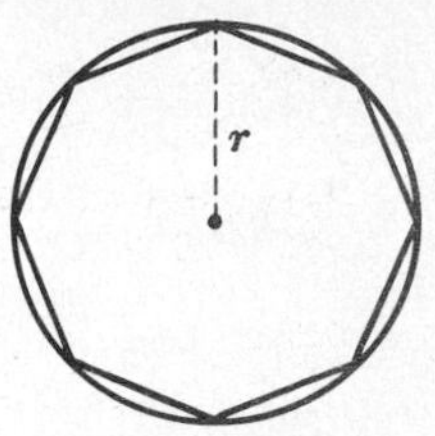

Fig. 4-10

REGULAR POLYGON OF n SIDES CIRCUMSCRIBING A CIRCLE OF RADIUS r

4.19 Area $= nr^2 \tan\frac{\pi}{n} = nr^2 \tan\frac{180°}{n}$

4.20 Perimeter $= 2nr \tan\frac{\pi}{n} = 2nr \tan\frac{180°}{n}$

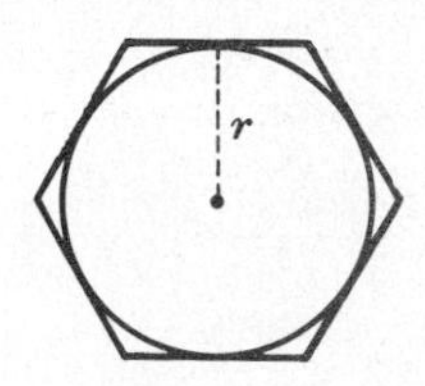

Fig. 4-11

SEGMENT OF CIRCLE OF RADIUS r

4.21 Area of shaded part $= \frac{1}{2}r^2(\theta - \sin\theta)$

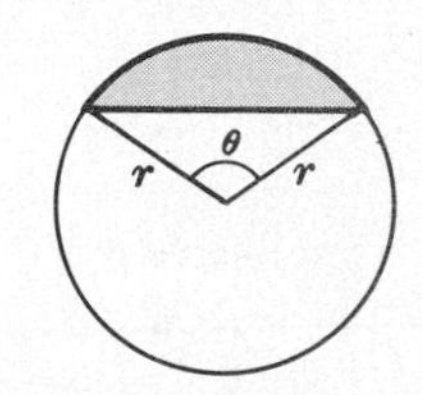

Fig. 4-12

ELLIPSE OF SEMI-MAJOR AXIS a AND SEMI-MINOR AXIS b

4.22 Area $= \pi ab$

4.23 Perimeter $= 4a\int_0^{\pi/2}\sqrt{1 - k^2\sin^2\theta}\,d\theta$

$= 2\pi\sqrt{\tfrac{1}{2}(a^2 + b^2)}$ [approximately]

where $k = \sqrt{a^2 - b^2}/a$. See page 254 for numerical tables.

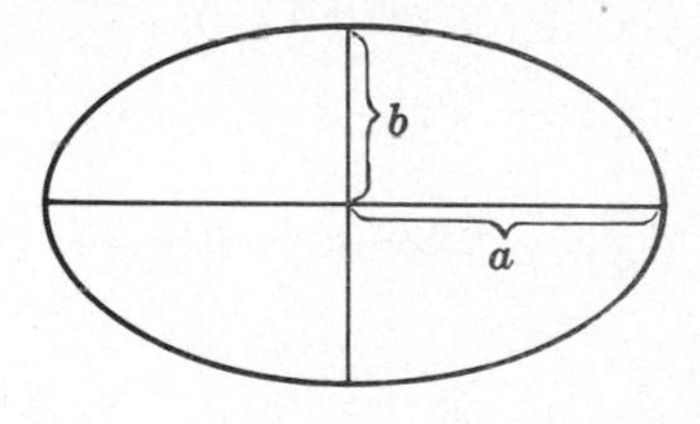

Fig. 4-13

SEGMENT OF A PARABOLA

4.24 Area $= \frac{2}{3}ab$

4.25 Arc length $ABC = \frac{1}{2}\sqrt{b^2 + 16a^2} + \frac{b^2}{8a}\ln\left(\frac{4a + \sqrt{b^2 + 16a^2}}{b}\right)$

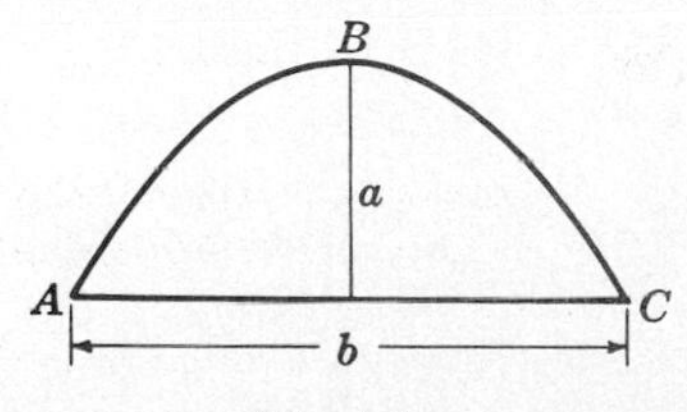

Fig. 4-14

RECTANGULAR PARALLELEPIPED OF LENGTH a, HEIGHT l, WIDTH c

4.26 Volume $= abc$

4.27 Surface area $= 2(ab + ac + bc)$

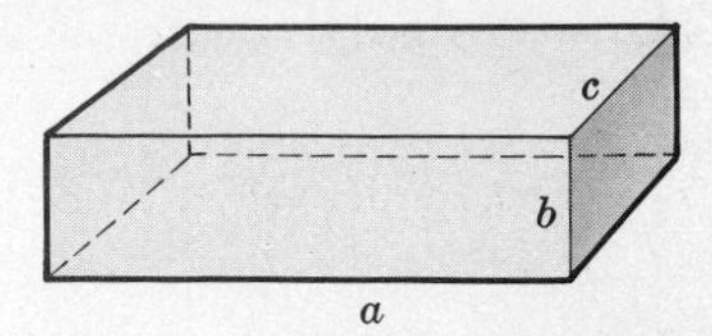

Fig. 4-15

PARALLELEPIPED OF CROSS-SECTIONAL AREA A AND HEIGHT h

4.28 Volume $= Ah = abc \sin\theta$

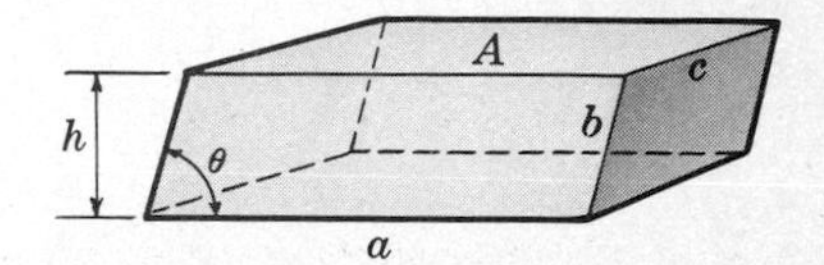

Fig. 4-16

SPHERE OF RADIUS r

4.29 Volume $= \frac{4}{3}\pi r^3$

4.30 Surface area $= 4\pi r^2$

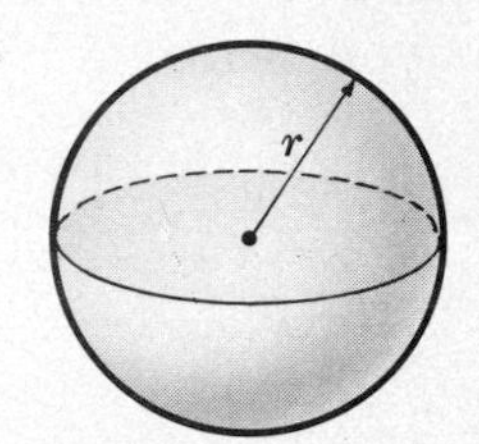

Fig. 4-17

RIGHT CIRCULAR CYLINDER OF RADIUS r AND HEIGHT h

4.31 Volume $= \pi r^2 h$

4.32 Lateral surface area $= 2\pi rh$

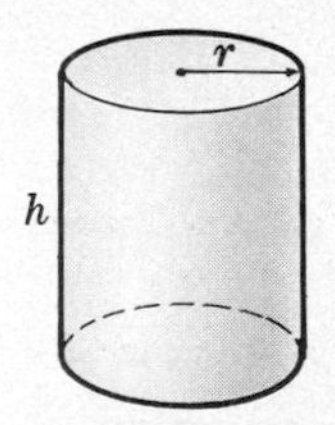

Fig. 4-18

CIRCULAR CYLINDER OF RADIUS r AND SLANT HEIGHT l

4.33 Volume $= \pi r^2 h = \pi r^2 l \sin\theta$

4.34 Lateral surface area $= 2\pi rl = \dfrac{2\pi rh}{\sin\theta} = 2\pi rh \csc\theta$

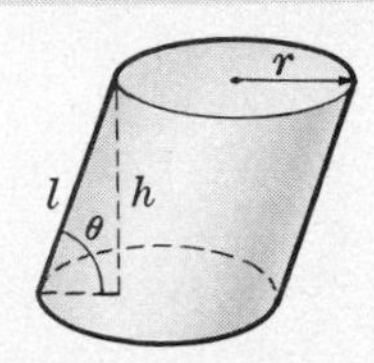

Fig. 4-19

CYLINDER OF CROSS-SECTIONAL AREA A AND SLANT HEIGHT l

4.35 Volume $= Ah = Al \sin \theta$

4.36 Lateral surface area $= pl = \dfrac{ph}{\sin \theta} = ph \csc \theta$

Note that formulas 4.31 to 4.34 are special cases.

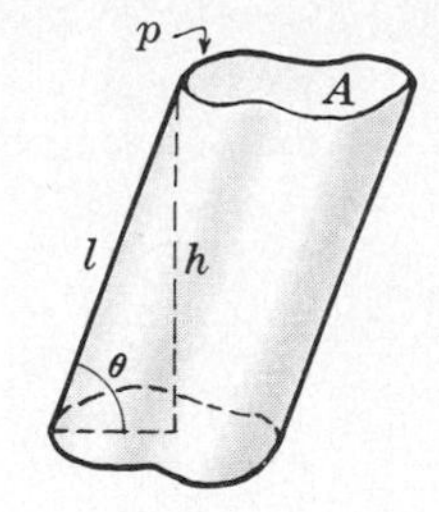

Fig. 4-20

RIGHT CIRCULAR CONE OF RADIUS r AND HEIGHT h

4.37 Volume $= \frac{1}{3}\pi r^2 h$

4.38 Lateral surface area $= \pi r \sqrt{r^2 + h^2} = \pi r l$

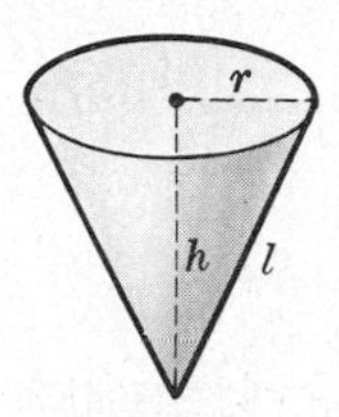

Fig. 4-21

PYRAMID OF BASE AREA A AND HEIGHT h

4.39 Volume $= \frac{1}{3}Ah$

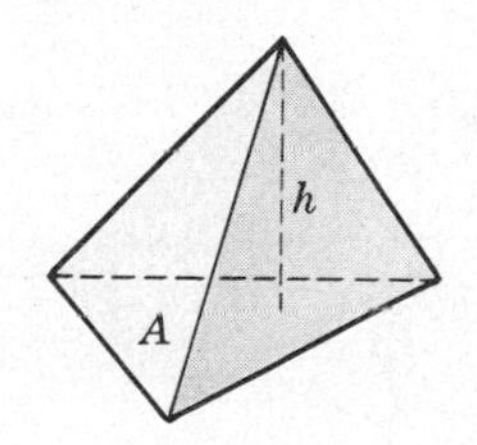

Fig. 4-22

SPHERICAL CAP OF RADIUS r AND HEIGHT h

4.40 Volume (shaded in figure) $= \frac{1}{3}\pi h^2(3r - h)$

4.41 Surface area $= 2\pi r h$

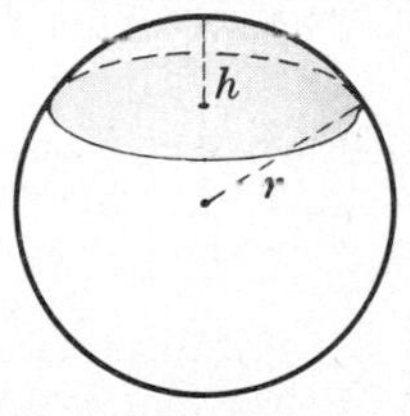

Fig. 4-23

FRUSTRUM OF RIGHT CIRCULAR CONE OF RADII a, b AND HEIGHT h

4.42 Volume $= \frac{1}{3}\pi h(a^2 + ab + b^2)$

4.43 Lateral surface area $= \pi(a + b)\sqrt{h^2 + (b - a)^2}$

$= \pi(a + b)l$

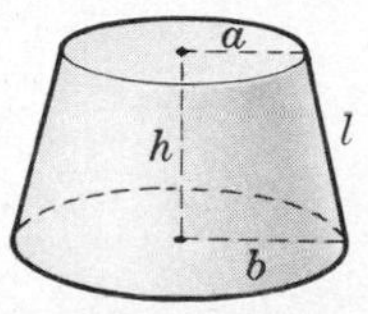

Fig. 4-24

SPHERICAL TRIANGLE OF ANGLES A, B, C ON SPHERE OF RADIUS r

4.44 Area of triangle $ABC = (A + B + C - \pi)r^2$

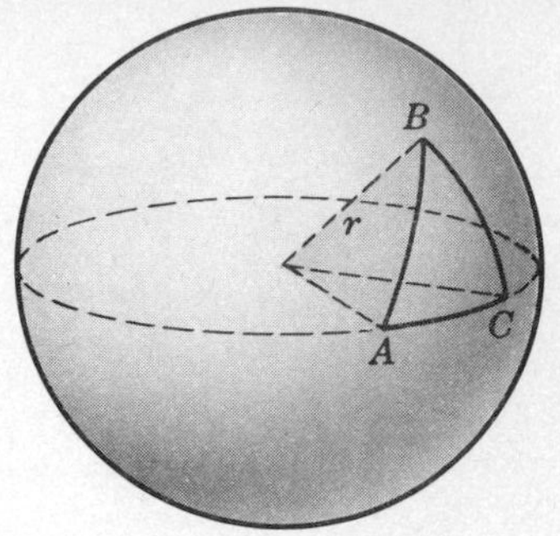

Fig. 4-25

TORUS OF INNER RADIUS a AND OUTER RADIUS b

4.45 Volume $= \frac{1}{4}\pi^2(a+b)(b-a)^2$

4.46 Surface area $= \pi^2(b^2 - a^2)$

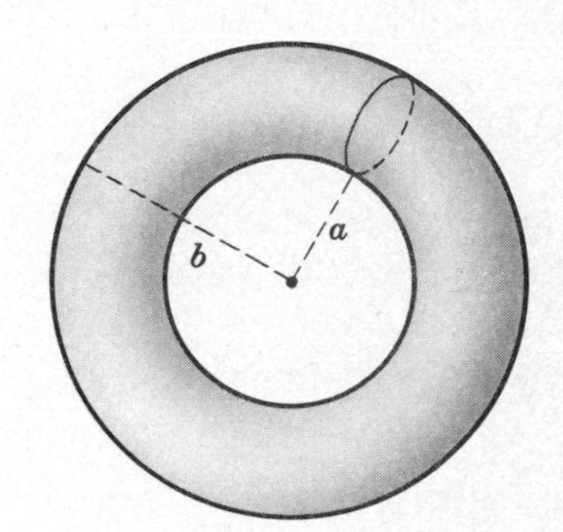

Fig. 4-26

ELLIPSOID OF SEMI-AXES a, b, c

4.47 Volume $= \frac{4}{3}\pi abc$

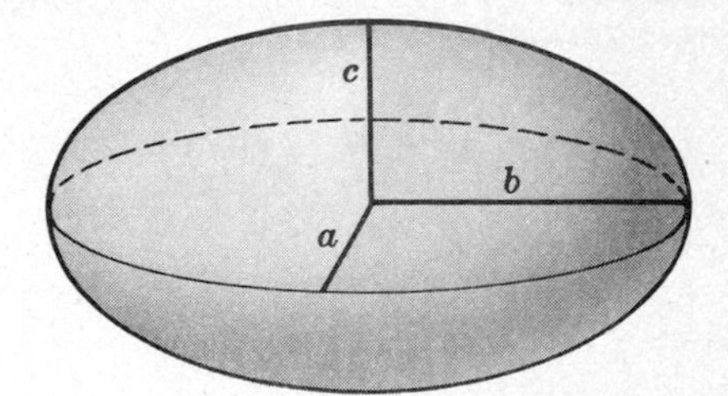

Fig. 4-27

PARABOLOID OF REVOLUTION

4.48 Volume $= \frac{1}{2}\pi b^2 a$

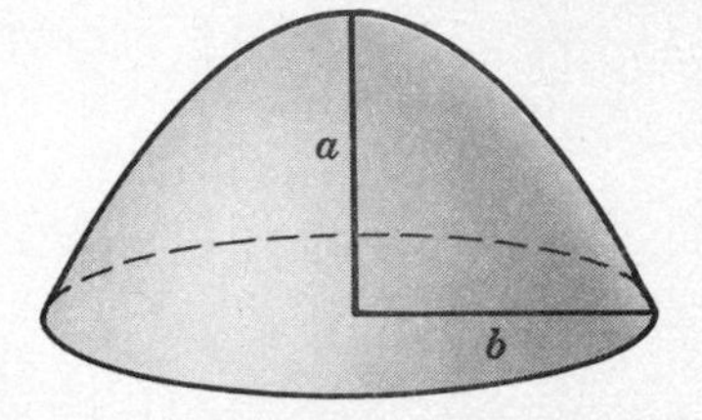

Fig. 4-28

5 TRIGONOMETRIC FUNCTIONS

DEFINITION OF TRIGONOMETRIC FUNCTIONS FOR A RIGHT TRIANGLE

Triangle ABC has a right angle (90°) at C and sides of length a, b, c. The trigonometric functions of angle A are defined as follows.

5.1 $$\textit{sine} \text{ of } A = \sin A = \frac{a}{c} = \frac{\text{opposite}}{\text{hypotenuse}}$$

5.2 $$\textit{cosine} \text{ of } A = \cos A = \frac{b}{c} = \frac{\text{adjacent}}{\text{hypotenuse}}$$

5.3 $$\textit{tangent} \text{ of } A = \tan A = \frac{a}{b} = \frac{\text{opposite}}{\text{adjacent}}$$

5.4 $$\textit{cotangent} \text{ of } A = \cot A = \frac{b}{a} = \frac{\text{adjacent}}{\text{opposite}}$$

5.5 $$\textit{secant} \text{ of } A = \sec A = \frac{c}{b} = \frac{\text{hypotenuse}}{\text{adjacent}}$$

5.6 $$\textit{cosecant} \text{ of } A = \csc A = \frac{c}{a} = \frac{\text{hypotenuse}}{\text{opposite}}$$

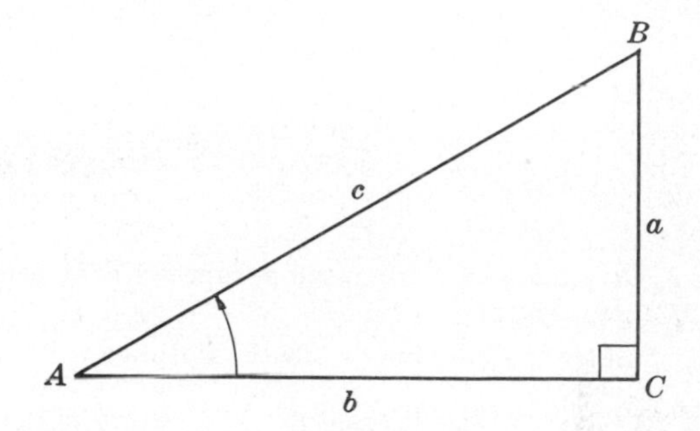

Fig. 5-1

EXTENSIONS TO ANGLES WHICH MAY BE GREATER THAN 90°

Consider an xy coordinate system [see Fig. 5-2 and 5-3 below]. A point P in the xy plane has coordinates (x, y) where x is considered as positive along OX and negative along OX' while y is positive along OY and negative along OY'. The distance from origin O to point P is positive and denoted by $r = \sqrt{x^2 + y^2}$. The angle A described *counterclockwise* from OX is considered *positive*. If it is described *clockwise* from OX it is considered *negative*. We call $X'OX$ and $Y'OY$ the x and y axis respectively.

The various quadrants are denoted by I, II, III and IV called the first, second, third and fourth quadrants respectively. In Fig. 5-2, for example, angle A is in the second quadrant while in Fig. 5-3 angle A is in the third quadrant.

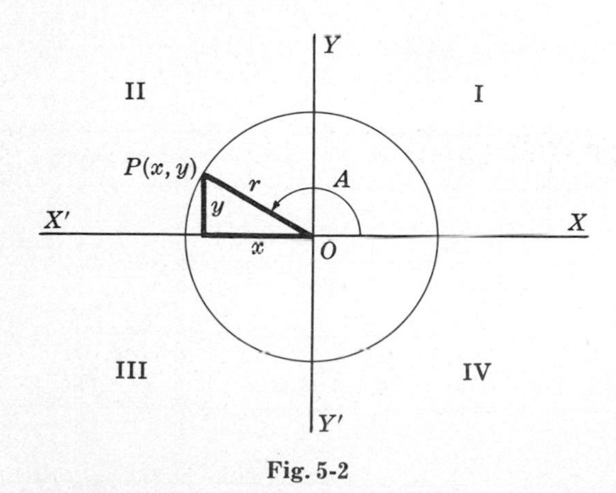

Fig. 5-2

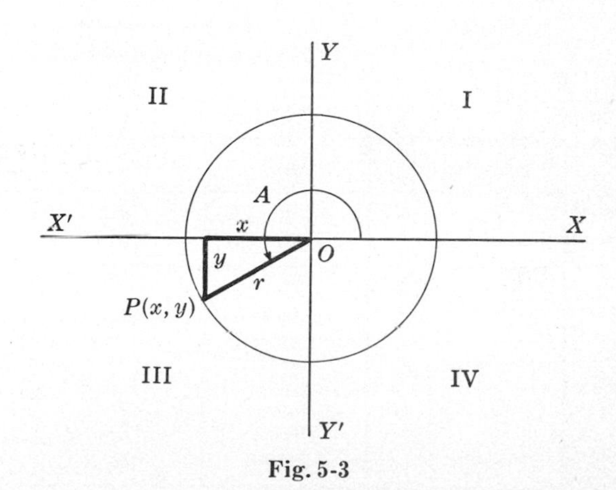

Fig. 5-3

For an angle A in any quadrant the trigonometric functions of A are defined as follows.

5.7 $$\sin A \;=\; y/r$$

5.8 $$\cos A \;=\; x/r$$

5.9 $$\tan A \;=\; y/x$$

5.10 $$\cot A \;=\; x/y$$

5.11 $$\sec A \;=\; r/x$$

5.12 $$\csc A \;=\; r/y$$

RELATIONSHIP BETWEEN DEGREES AND RADIANS

A *radian* is that angle θ subtended at center O of a circle by an arc MN equal to the radius r.

Since 2π radians $=$ $360°$ we have

5.13 $$1 \text{ radian} \;=\; 180°/\pi \;=\; 57.29577\ 95130\ 8232\ldots°$$

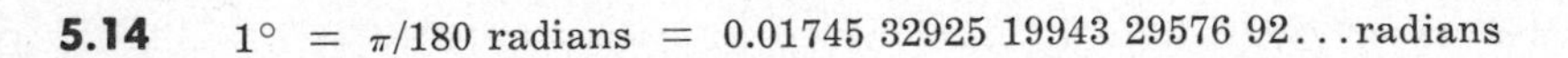

5.14 $$1° \;=\; \pi/180 \text{ radians} \;=\; 0.01745\ 32925\ 19943\ 29576\ 92\ldots\text{radians}$$

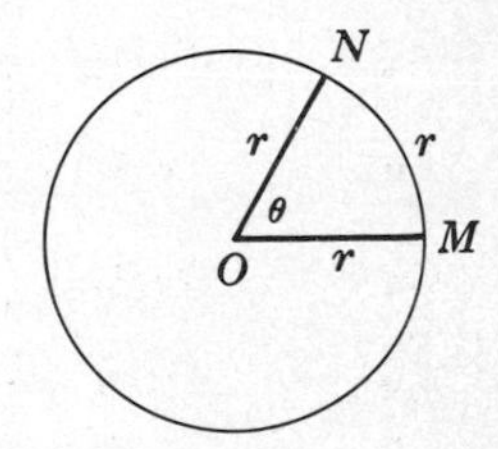

Fig. 5-4

RELATIONSHIPS AMONG TRIGONOMETRIC FUNCTIONS

5.15 $$\tan A \;=\; \frac{\sin A}{\cos A}$$

5.16 $$\cot A \;=\; \frac{1}{\tan A} \;=\; \frac{\cos A}{\sin A}$$

5.17 $$\sec A \;=\; \frac{1}{\cos A}$$

5.18 $$\csc A \;=\; \frac{1}{\sin A}$$

5.19 $$\sin^2 A + \cos^2 A \;=\; 1$$

5.20 $$\sec^2 A - \tan^2 A \;=\; 1$$

5.21 $$\csc^2 A - \cot^2 A \;=\; 1$$

SIGNS AND VARIATIONS OF TRIGONOMETRIC FUNCTIONS

Quadrant	$\sin A$	$\cos A$	$\tan A$	$\cot A$	$\sec A$	$\csc A$
I	$+$ 0 to 1	$+$ 1 to 0	$+$ 0 to ∞	$+$ ∞ to 0	$+$ 1 to ∞	$+$ ∞ to 1
II	$+$ 1 to 0	$-$ 0 to -1	$-$ $-\infty$ to 0	$-$ 0 to $-\infty$	$-$ $-\infty$ to -1	$+$ 1 to ∞
III	$-$ 0 to -1	$-$ -1 to 0	$+$ 0 to ∞	$+$ ∞ to 0	$-$ -1 to $-\infty$	$-$ $-\infty$ to -1
IV	$-$ -1 to 0	$+$ 0 to 1	$-$ $-\infty$ to 0	$-$ 0 to $-\infty$	$+$ ∞ to 1	$-$ -1 to $-\infty$

EXACT VALUES FOR TRIGONOMETRIC FUNCTIONS OF VARIOUS ANGLES

Angle A in degrees	Angle A in radians	$\sin A$	$\cos A$	$\tan A$	$\cot A$	$\sec A$	$\csc A$
$0°$	0	0	1	0	∞	1	∞
$15°$	$\pi/12$	$\frac{1}{4}(\sqrt{6}-\sqrt{2})$	$\frac{1}{4}(\sqrt{6}+\sqrt{2})$	$2-\sqrt{3}$	$2+\sqrt{3}$	$\sqrt{6}-\sqrt{2}$	$\sqrt{6}+\sqrt{2}$
$30°$	$\pi/6$	$\frac{1}{2}$	$\frac{1}{2}\sqrt{3}$	$\frac{1}{3}\sqrt{3}$	$\sqrt{3}$	$\frac{2}{3}\sqrt{3}$	2
$45°$	$\pi/4$	$\frac{1}{2}\sqrt{2}$	$\frac{1}{2}\sqrt{2}$	1	1	$\sqrt{2}$	$\sqrt{2}$
$60°$	$\pi/3$	$\frac{1}{2}\sqrt{3}$	$\frac{1}{2}$	$\sqrt{3}$	$\frac{1}{3}\sqrt{3}$	2	$\frac{2}{3}\sqrt{3}$
$75°$	$5\pi/12$	$\frac{1}{4}(\sqrt{6}+\sqrt{2})$	$\frac{1}{4}(\sqrt{6}-\sqrt{2})$	$2+\sqrt{3}$	$2-\sqrt{3}$	$\sqrt{6}+\sqrt{2}$	$\sqrt{6}-\sqrt{2}$
$90°$	$\pi/2$	1	0	$\pm\infty$	0	$\pm\infty$	1
$105°$	$7\pi/12$	$\frac{1}{4}(\sqrt{6}+\sqrt{2})$	$-\frac{1}{4}(\sqrt{6}-\sqrt{2})$	$-(2+\sqrt{3})$	$-(2-\sqrt{3})$	$-(\sqrt{6}+\sqrt{2})$	$\sqrt{6}-\sqrt{2}$
$120°$	$2\pi/3$	$\frac{1}{2}\sqrt{3}$	$-\frac{1}{2}$	$-\sqrt{3}$	$-\frac{1}{3}\sqrt{3}$	-2	$\frac{2}{3}\sqrt{3}$
$135°$	$3\pi/4$	$\frac{1}{2}\sqrt{2}$	$-\frac{1}{2}\sqrt{2}$	-1	-1	$-\sqrt{2}$	$\sqrt{2}$
$150°$	$5\pi/6$	$\frac{1}{2}$	$-\frac{1}{2}\sqrt{3}$	$-\frac{1}{3}\sqrt{3}$	$-\sqrt{3}$	$-\frac{2}{3}\sqrt{3}$	2
$165°$	$11\pi/12$	$\frac{1}{4}(\sqrt{6}-\sqrt{2})$	$-\frac{1}{4}(\sqrt{6}+\sqrt{2})$	$-(2-\sqrt{3})$	$-(2+\sqrt{3})$	$-(\sqrt{6}-\sqrt{2})$	$\sqrt{6}+\sqrt{2}$
$180°$	π	0	-1	0	$\mp\infty$	-1	$\pm\infty$
$195°$	$13\pi/12$	$-\frac{1}{4}(\sqrt{6}-\sqrt{2})$	$-\frac{1}{4}(\sqrt{6}+\sqrt{2})$	$2-\sqrt{3}$	$2+\sqrt{3}$	$-(\sqrt{6}-\sqrt{2})$	$-(\sqrt{6}+\sqrt{2})$
$210°$	$7\pi/6$	$-\frac{1}{2}$	$-\frac{1}{2}\sqrt{3}$	$\frac{1}{3}\sqrt{3}$	$\sqrt{3}$	$-\frac{2}{3}\sqrt{3}$	-2
$225°$	$5\pi/4$	$-\frac{1}{2}\sqrt{2}$	$-\frac{1}{2}\sqrt{2}$	1	1	$-\sqrt{2}$	$-\sqrt{2}$
$240°$	$4\pi/3$	$-\frac{1}{2}\sqrt{3}$	$-\frac{1}{2}$	$\sqrt{3}$	$\frac{1}{3}\sqrt{3}$	-2	$-\frac{2}{3}\sqrt{3}$
$255°$	$17\pi/12$	$-\frac{1}{4}(\sqrt{6}+\sqrt{2})$	$-\frac{1}{4}(\sqrt{6}-\sqrt{2})$	$2+\sqrt{3}$	$2-\sqrt{3}$	$-(\sqrt{6}+\sqrt{2})$	$-(\sqrt{6}-\sqrt{2})$
$270°$	$3\pi/2$	-1	0	$\pm\infty$	0	$\mp\infty$	-1
$285°$	$19\pi/12$	$-\frac{1}{4}(\sqrt{6}+\sqrt{2})$	$\frac{1}{4}(\sqrt{6}-\sqrt{2})$	$-(2+\sqrt{3})$	$-(2-\sqrt{3})$	$\sqrt{6}+\sqrt{2}$	$-(\sqrt{6}-\sqrt{2})$
$300°$	$5\pi/3$	$-\frac{1}{2}\sqrt{3}$	$\frac{1}{2}$	$-\sqrt{3}$	$-\frac{1}{3}\sqrt{3}$	2	$-\frac{2}{3}\sqrt{3}$
$315°$	$7\pi/4$	$-\frac{1}{2}\sqrt{2}$	$\frac{1}{2}\sqrt{2}$	-1	-1	$\sqrt{2}$	$-\sqrt{2}$
$330°$	$11\pi/6$	$-\frac{1}{2}$	$\frac{1}{2}\sqrt{3}$	$-\frac{1}{3}\sqrt{3}$	$-\sqrt{3}$	$\frac{2}{3}\sqrt{3}$	-2
$345°$	$23\pi/12$	$-\frac{1}{4}(\sqrt{6}-\sqrt{2})$	$\frac{1}{4}(\sqrt{6}+\sqrt{2})$	$-(2-\sqrt{3})$	$-(2+\sqrt{3})$	$\sqrt{6}-\sqrt{2}$	$-(\sqrt{6}+\sqrt{2})$
$360°$	2π	0	1	0	$\mp\infty$	1	$\mp\infty$

For tables involving other angles see pages 206-211 and 212-215.

GRAPHS OF TRIGONOMETRIC FUNCTIONS

In each graph x is in radians.

5.22 $y = \sin x$

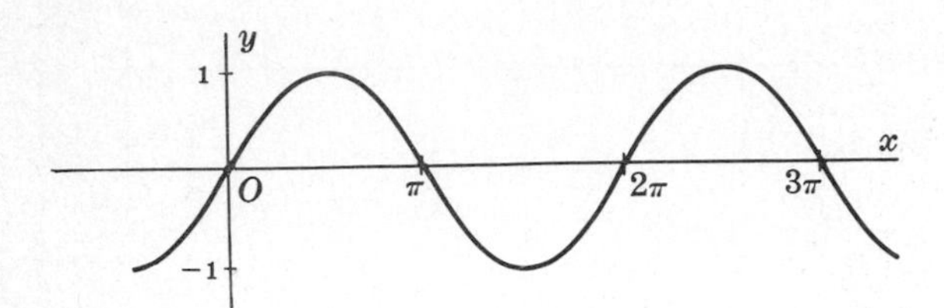

Fig. 5-5

5.23 $y = \cos x$

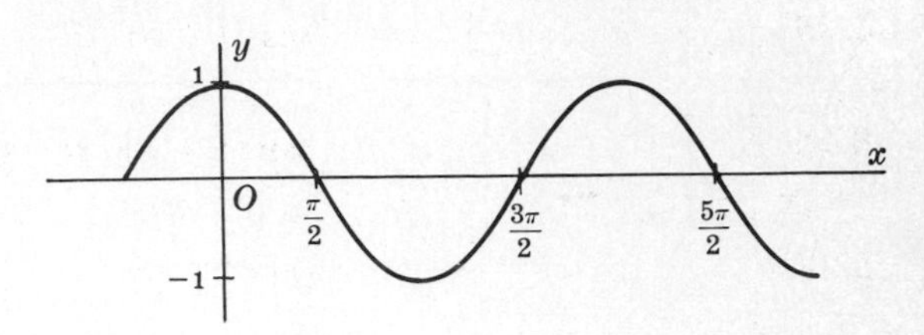

Fig. 5-6

5.24 $y = \tan x$

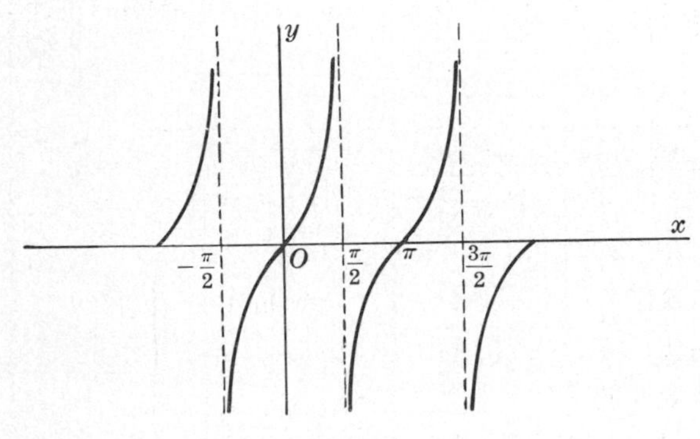

Fig. 5-7

5.25 $y = \cot x$

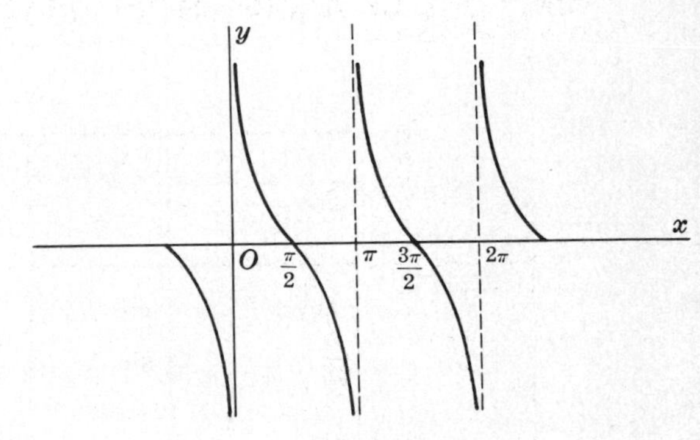

Fig. 5-8

5.26 $y = \sec x$

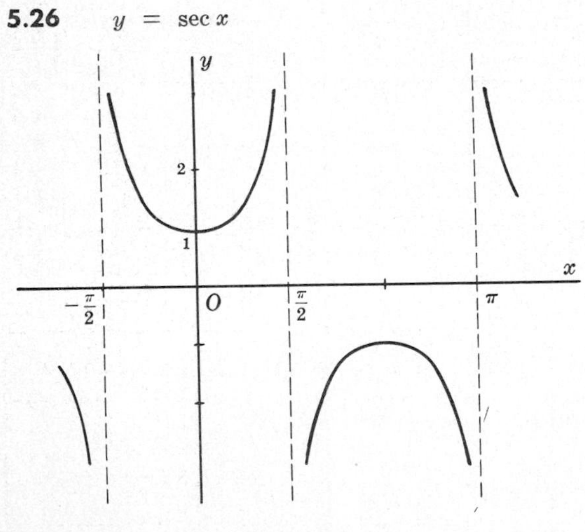

Fig. 5-9

5.27 $y = \csc x$

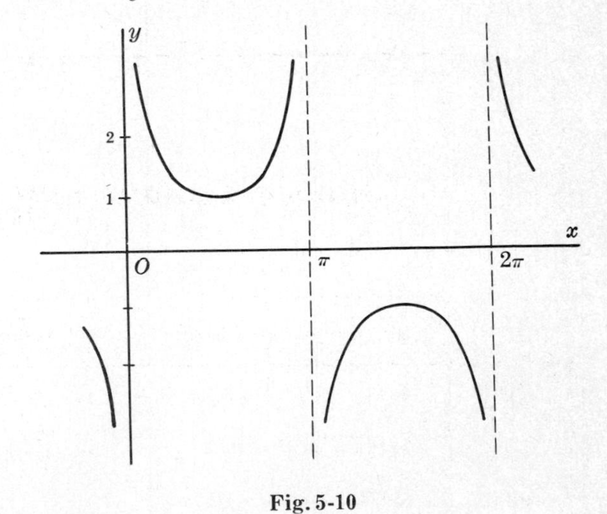

Fig. 5-10

FUNCTIONS OF NEGATIVE ANGLES

5.28 $\sin(-A) = -\sin A$ **5.29** $\cos(-A) = \cos A$ **5.30** $\tan(-A) = -\tan A$

5.31 $\csc(-A) = -\csc A$ **5.32** $\sec(-A) = \sec A$ **5.33** $\cot(-A) = -\cot A$

ADDITION FORMULAS

5.34 $$\sin(A \pm B) = \sin A \cos B \pm \cos A \sin B$$

5.35 $$\cos(A \pm B) = \cos A \cos B \mp \sin A \sin B$$

5.36 $$\tan(A \pm B) = \frac{\tan A \pm \tan B}{1 \mp \tan A \tan B}$$

5.37 $$\cot(A \pm B) = \frac{\cot A \cot B \mp 1}{\cot B \pm \cot A}$$

FUNCTIONS OF ANGLES IN ALL QUADRANTS IN TERMS OF THOSE IN QUADRANT I

	$-A$	$90^\circ \pm A$ $\frac{\pi}{2} \pm A$	$180^\circ \pm A$ $\pi \pm A$	$270^\circ \pm A$ $\frac{3\pi}{2} \pm A$	$k(360^\circ) \pm A$ $2k\pi \pm A$ k = integer
sin	$-\sin A$	$\cos A$	$\mp \sin A$	$-\cos A$	$\pm \sin A$
cos	$\cos A$	$\mp \sin A$	$-\cos A$	$\pm \sin A$	$\cos A$
tan	$-\tan A$	$\mp \cot A$	$\pm \tan A$	$\mp \cot A$	$\pm \tan A$
csc	$-\csc A$	$\sec A$	$\mp \csc A$	$-\sec A$	$\pm \csc A$
sec	$\sec A$	$\mp \csc A$	$-\sec A$	$\pm \csc A$	$\sec A$
cot	$-\cot A$	$\mp \tan A$	$\pm \cot A$	$\mp \tan A$	$\pm \cot A$

RELATIONSHIPS AMONG FUNCTIONS OF ANGLES IN QUADRANT I

	$\sin A = u$	$\cos A = u$	$\tan A = u$	$\cot A = u$	$\sec A = u$	$\csc A = u$
$\sin A$	u	$\sqrt{1-u^2}$	$u/\sqrt{1+u^2}$	$1/\sqrt{1+u^2}$	$\sqrt{u^2-1}/u$	$1/u$
$\cos A$	$\sqrt{1-u^2}$	u	$1/\sqrt{1+u^2}$	$u/\sqrt{1+u^2}$	$1/u$	$\sqrt{u^2-1}/u$
$\tan A$	$u/\sqrt{1-u^2}$	$\sqrt{1-u^2}/u$	u	$1/u$	$\sqrt{u^2-1}$	$1/\sqrt{u^2-1}$
$\cot A$	$\sqrt{1-u^2}/u$	$u/\sqrt{1-u^2}$	$1/u$	u	$1/\sqrt{u^2-1}$	$\sqrt{u^2-1}$
$\sec A$	$1/\sqrt{1-u^2}$	$1/u$	$\sqrt{1+u^2}$	$\sqrt{1+u^2}/u$	u	$u/\sqrt{u^2-1}$
$\csc A$	$1/u$	$1/\sqrt{1-u^2}$	$\sqrt{1+u^2}/u$	$\sqrt{1+u^2}$	$u/\sqrt{u^2-1}$	$\sqrt{1+u^2}$

For extensions to other quadrants use appropriate signs as given in the preceding table.

DOUBLE ANGLE FORMULAS

5.38 $\sin 2A = 2 \sin A \cos A$

5.39 $\cos 2A = \cos^2 A - \sin^2 A = 1 - 2\sin^2 A = 2\cos^2 A - 1$

5.40 $\tan 2A = \dfrac{2 \tan A}{1 - \tan^2 A}$

HALF ANGLE FORMULAS

5.41 $\sin \dfrac{A}{2} = \pm\sqrt{\dfrac{1 - \cos A}{2}}$ $\left[\begin{array}{l} + \text{ if } A/2 \text{ is in quadrant I or II} \\ - \text{ if } A/2 \text{ is in quadrant III or IV} \end{array}\right]$

5.42 $\cos \dfrac{A}{2} = \pm\sqrt{\dfrac{1 + \cos A}{2}}$ $\left[\begin{array}{l} + \text{ if } A/2 \text{ is in quadrant I or IV} \\ - \text{ if } A/2 \text{ is in quadrant II or III} \end{array}\right]$

5.43 $\tan \dfrac{A}{2} = \pm\sqrt{\dfrac{1 - \cos A}{1 + \cos A}}$ $\left[\begin{array}{l} + \text{ if } A/2 \text{ is in quadrant I or III} \\ - \text{ if } A/2 \text{ is in quadrant II or IV} \end{array}\right]$

$$= \frac{\sin A}{1 + \cos A} = \frac{1 - \cos A}{\sin A} = \csc A - \cot A$$

MULTIPLE ANGLE FORMULAS

5.44 $\sin 3A = 3 \sin A - 4 \sin^3 A$

5.45 $\cos 3A = 4 \cos^3 A - 3 \cos A$

5.46 $\tan 3A = \dfrac{3 \tan A - \tan^3 A}{1 - 3 \tan^2 A}$

5.47 $\sin 4A = 4 \sin A \cos A - 8 \sin^3 A \cos A$

5.48 $\cos 4A = 8 \cos^4 A - 8 \cos^2 A + 1$

5.49 $\tan 4A = \dfrac{4 \tan A - 4 \tan^3 A}{1 - 6 \tan^2 A + \tan^4 A}$

5.50 $\sin 5A = 5 \sin A - 20 \sin^3 A + 16 \sin^5 A$

5.51 $\cos 5A = 16 \cos^5 A - 20 \cos^3 A + 5 \cos A$

5.52 $\tan 5A = \dfrac{\tan^5 A - 10 \tan^3 A + 5 \tan A}{1 - 10 \tan^2 A + 5 \tan^4 A}$

See also formulas 5.68 and 5.69.

POWERS OF TRIGONOMETRIC FUNCTIONS

5.53 $\sin^2 A = \frac{1}{2} - \frac{1}{2} \cos 2A$

5.54 $\cos^2 A = \frac{1}{2} + \frac{1}{2} \cos 2A$

5.55 $\sin^3 A = \frac{3}{4} \sin A - \frac{1}{4} \sin 3A$

5.56 $\cos^3 A = \frac{3}{4} \cos A + \frac{1}{4} \cos 3A$

5.57 $\sin^4 A = \frac{3}{8} - \frac{1}{2} \cos 2A + \frac{1}{8} \cos 4A$

5.58 $\cos^4 A = \frac{3}{8} + \frac{1}{2} \cos 2A + \frac{1}{8} \cos 4A$

5.59 $\sin^5 A = \frac{5}{8} \sin A - \frac{5}{16} \sin 3A + \frac{1}{16} \sin 5A$

5.60 $\cos^5 A = \frac{5}{8} \cos A + \frac{5}{16} \cos 3A + \frac{1}{16} \cos 5A$

See also formulas 5.70 through 5.73.

SUM, DIFFERENCE AND PRODUCT OF TRIGONOMETRIC FUNCTIONS

5.61 $$\sin A + \sin B = 2\sin\tfrac{1}{2}(A+B)\cos\tfrac{1}{2}(A-B)$$

5.62 $$\sin A - \sin B = 2\cos\tfrac{1}{2}(A+B)\sin\tfrac{1}{2}(A-B)$$

5.63 $$\cos A + \cos B = 2\cos\tfrac{1}{2}(A+B)\cos\tfrac{1}{2}(A-B)$$

5.64 $$\cos A - \cos B = 2\sin\tfrac{1}{2}(A+B)\sin\tfrac{1}{2}(B-A)$$

5.65 $$\sin A\sin B = \tfrac{1}{2}\{\cos(A-B) - \cos(A+B)\}$$

5.66 $$\cos A\cos B = \tfrac{1}{2}\{\cos(A-B) + \cos(A+B)\}$$

5.67 $$\sin A\cos B = \tfrac{1}{2}\{\sin(A-B) + \sin(A+B)\}$$

GENERAL FORMULAS

5.68 $$\sin nA = \sin A\left\{(2\cos A)^{n-1} - \binom{n-2}{1}(2\cos A)^{n-3} + \binom{n-3}{2}(2\cos A)^{n-5} - \cdots\right\}$$

5.69 $$\cos nA = \frac{1}{2}\left\{(2\cos A)^n - \frac{n}{1}(2\cos A)^{n-2} + \frac{n}{2}\binom{n-3}{1}(2\cos A)^{n-4} - \frac{n}{3}\binom{n-4}{2}(2\cos A)^{n-6} + \cdots\right\}$$

5.70 $$\sin^{2n-1}A = \frac{(-1)^{n-1}}{2^{2n-2}}\left\{\sin(2n-1)A - \binom{2n-1}{1}\sin(2n-3)A + \cdots (-1)^{n-1}\binom{2n-1}{n-1}\sin A\right\}$$

5.71 $$\cos^{2n-1}A = \frac{1}{2^{2n-2}}\left\{\cos(2n-1)A + \binom{2n-1}{1}\cos(2n-3)A + \cdots + \binom{2n-1}{n-1}\cos A\right\}$$

5.72 $$\sin^{2n}A = \frac{1}{2^{2n}}\binom{2n}{n} + \frac{(-1)^n}{2^{2n-1}}\left\{\cos 2nA - \binom{2n}{1}\cos(2n-2)A + \cdots (-1)^{n-1}\binom{2n}{n-1}\cos 2A\right\}$$

5.73 $$\cos^{2n}A = \frac{1}{2^{2n}}\binom{2n}{n} + \frac{1}{2^{2n-1}}\left\{\cos 2nA + \binom{2n}{1}\cos(2n-2)A + \cdots + \binom{2n}{n-1}\cos 2A\right\}$$

INVERSE TRIGONOMETRIC FUNCTIONS

If $x = \sin y$ then $y = \sin^{-1} x$, i.e. *the angle whose sine is x* or *inverse sine of x*, is a many-valued function of x which is a collection of single-valued functions called *branches*. Similarly the other inverse trigonometric functions are multiple-valued.

For many purposes a particular branch is required. This is called the *principal branch* and the values for this branch are called *principal values*.

PRINCIPAL VALUES FOR INVERSE TRIGONOMETRIC FUNCTIONS

Principal values for $x \geqq 0$	Principal values for $x < 0$
$0 \leqq \sin^{-1} x \leqq \pi/2$	$-\pi/2 \leqq \sin^{-1} x < 0$
$0 \leqq \cos^{-1} x \leqq \pi/2$	$\pi/2 < \cos^{-1} x \leqq \pi$
$0 \leqq \tan^{-1} x < \pi/2$	$-\pi/2 < \tan^{-1} x < 0$
$0 < \cot^{-1} x \leqq \pi/2$	$\pi/2 < \cot^{-1} x < \pi$
$0 \leqq \sec^{-1} x < \pi/2$	$\pi/2 < \sec^{-1} x \leqq \pi$
$0 < \csc^{-1} x \leqq \pi/2$	$-\pi/2 \leqq \csc^{-1} x < 0$

RELATIONS BETWEEN INVERSE TRIGONOMETRIC FUNCTIONS

In all cases it is assumed that principal values are used.

5.74 $\sin^{-1} x + \cos^{-1} x = \pi/2$

5.75 $\tan^{-1} x + \cot^{-1} x = \pi/2$

5.76 $\sec^{-1} x + \csc^{-1} x = \pi/2$

5.77 $\csc^{-1} x = \sin^{-1}(1/x)$

5.78 $\sec^{-1} x = \cos^{-1}(1/x)$

5.79 $\cot^{-1} x = \tan^{-1}(1/x)$

5.80 $\sin^{-1}(-x) = -\sin^{-1} x$

5.81 $\cos^{-1}(-x) = \pi - \cos^{-1} x$

5.82 $\tan^{-1}(-x) = -\tan^{-1} x$

5.83 $\cot^{-1}(-x) = \pi - \cot^{-1} x$

5.84 $\sec^{-1}(-x) = \pi - \sec^{-1} x$

5.85 $\csc^{-1}(-x) = -\csc^{-1} x$

GRAPHS OF INVERSE TRIGONOMETRIC FUNCTIONS

In each graph y is in radians. Solid portions of curves correspond to principal values.

5.86 $y = \sin^{-1} x$

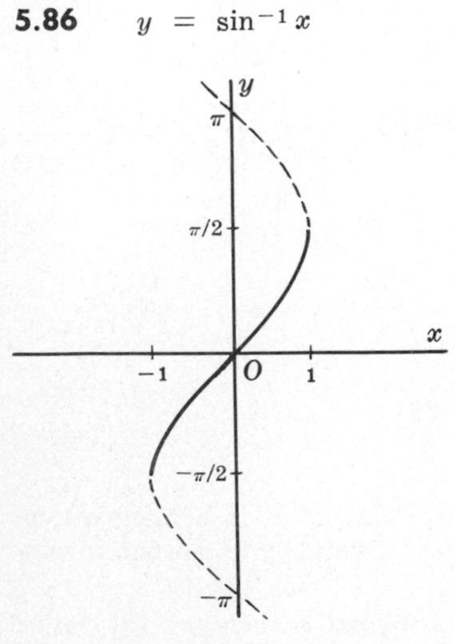

Fig. 5-11

5.87 $y = \cos^{-1} x$

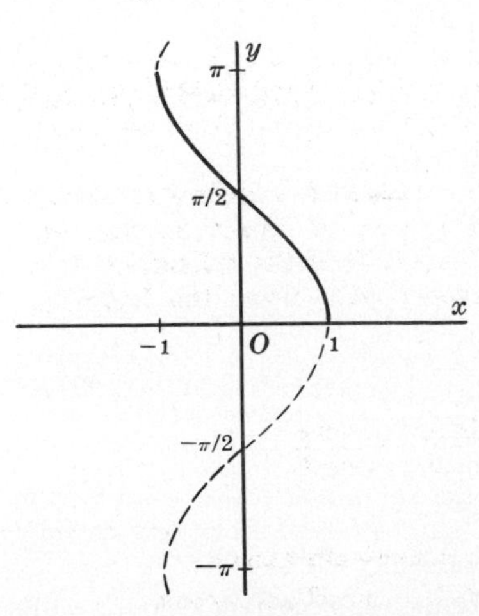

Fig. 5-12

5.88 $y = \tan^{-1} x$

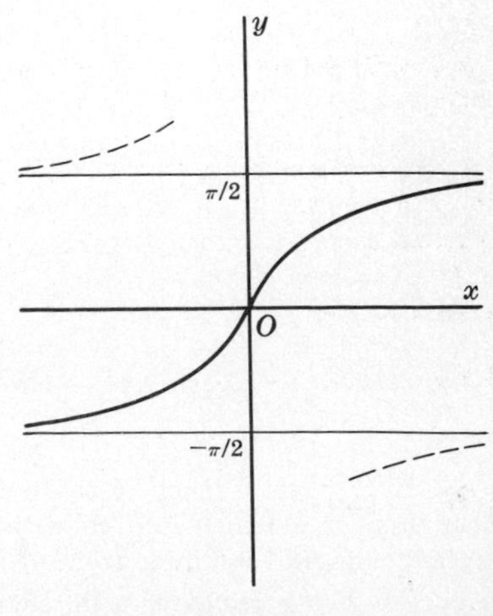

Fig. 5-13

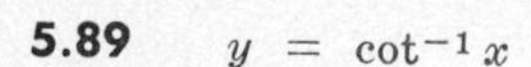

5.89 $y = \cot^{-1} x$

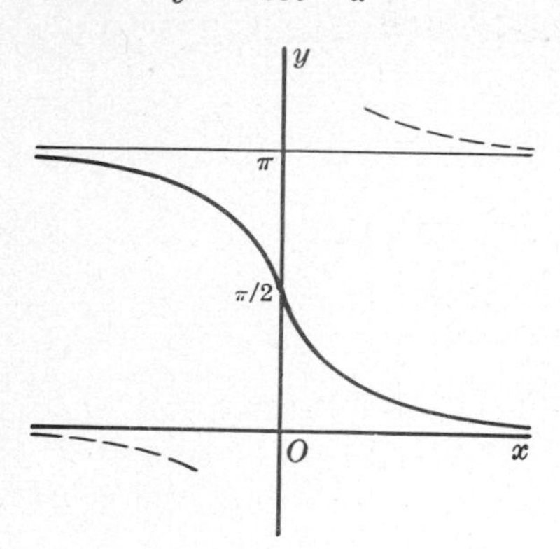

Fig. 5-14

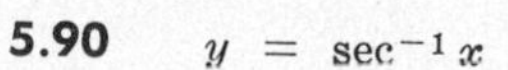

5.90 $y = \sec^{-1} x$

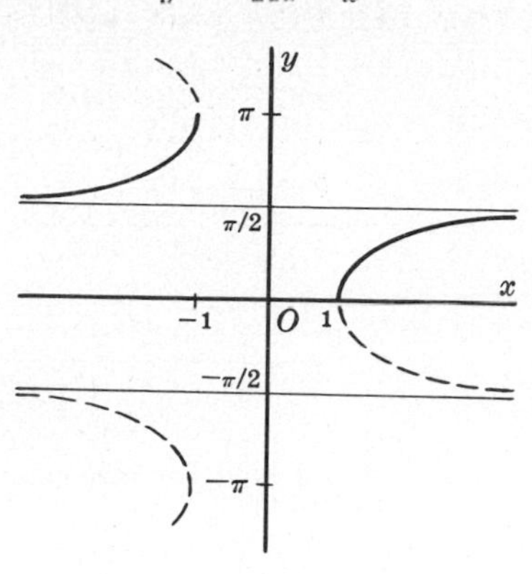

Fig. 5-15

5.91 $y = \csc^{-1} x$

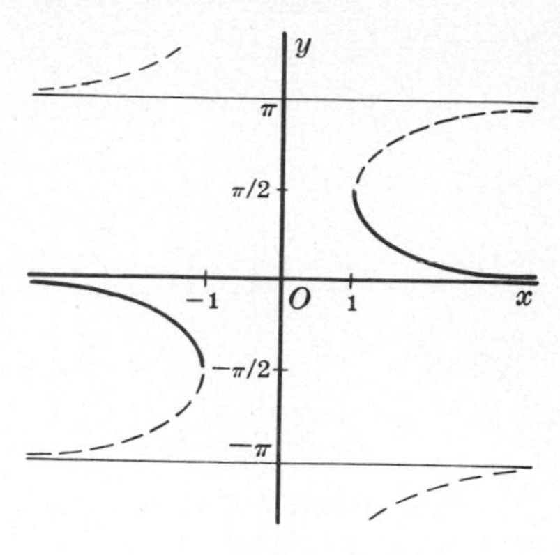

Fig. 5-16

RELATIONSHIPS BETWEEN SIDES AND ANGLES OF A PLANE TRIANGLE

The following results hold for any plane triangle ABC with sides a, b, c and angles A, B, C.

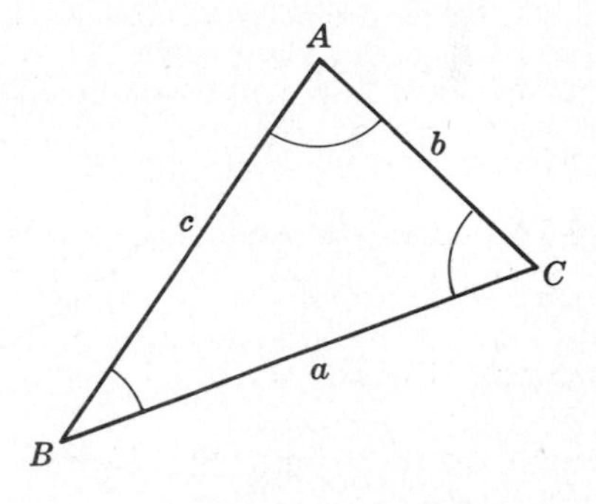

Fig. 5-17

5.92 Law of Sines

$$\frac{a}{\sin A} = \frac{b}{\sin B} = \frac{c}{\sin C}$$

5.93 Law of Cosines

$$c^2 = a^2 + b^2 - 2ab \cos C$$

with similar relations involving the other sides and angles.

5.94 Law of Tangents

$$\frac{a+b}{a-b} = \frac{\tan \frac{1}{2}(A+B)}{\tan \frac{1}{2}(A-B)}$$

with similar relations involving the other sides and angles.

5.95

$$\sin A = \frac{2}{bc}\sqrt{s(s-a)(s-b)(s-c)}$$

where $s = \frac{1}{2}(a+b+c)$ is the semiperimeter of the triangle. Similar relations involving angles B and C can be obtained.

See also formulas 4.5, page 5; 4.15 and 4.16, page 6.

RELATIONSHIPS BETWEEN SIDES AND ANGLES OF A SPHERICAL TRIANGLE

Spherical triangle ABC is on the surface of a sphere as shown in Fig. 5-18. Sides a, b, c [which are arcs of great circles] are measured by their angles subtended at center O of the sphere. A, B, C are the angles opposite sides a, b, c respectively. Then the following results hold.

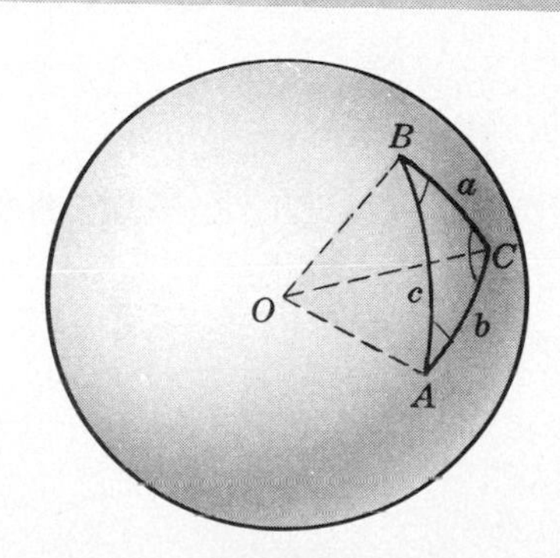

Fig. 5-18

5.96 Law of Sines

$$\frac{\sin a}{\sin A} = \frac{\sin b}{\sin B} = \frac{\sin c}{\sin C}$$

5.97 Law of Cosines

$$\cos a = \cos b \cos c + \sin b \sin c \cos A$$

$$\cos A = -\cos B \cos C + \sin B \sin C \cos a$$

with similar results involving other sides and angles.

5.98 **Law of Tangents**

$$\frac{\tan \frac{1}{2}(A+B)}{\tan \frac{1}{2}(A-B)} = \frac{\tan \frac{1}{2}(a+b)}{\tan \frac{1}{2}(a-b)}$$

with similar results involving other sides and angles.

5.99

$$\cos \frac{A}{2} = \sqrt{\frac{\sin s \sin (s-c)}{\sin b \sin c}}$$

where $s = \frac{1}{2}(a+b+c)$. Similar results hold for other sides and angles.

5.100

$$\cos \frac{a}{2} = \sqrt{\frac{\cos (S-B) \cos (S-C)}{\sin B \sin C}}$$

where $S = \frac{1}{2}(A+B+C)$. Similar results hold for other sides and angles.

See also formula 4.44, page 10.

NAPIER'S RULES FOR RIGHT ANGLED SPHERICAL TRIANGLES

Except for right angle C, there are five parts of spherical triangle ABC which if arranged in the order as given in Fig. 5-19 would be a, b, A, c, B.

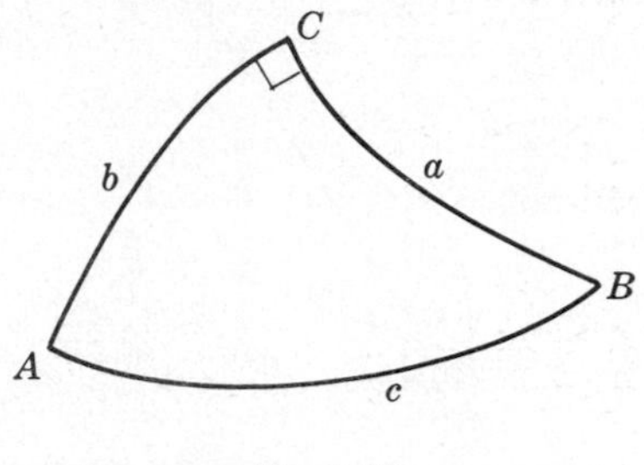

Fig. 5-19

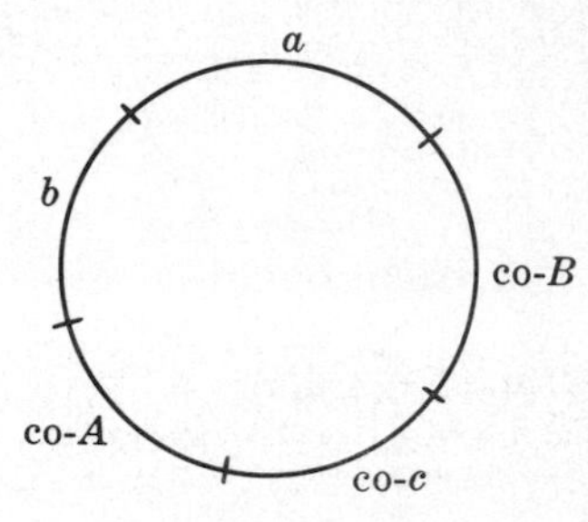

Fig. 5-20

Suppose these quantities are arranged in a circle as in Fig. 5-20 where we attach the prefix co [indicating *complement*] to hypotenuse c and angles A and B.

Any one of the parts of this circle is called a *middle part*, the two neighboring parts are called *adjacent parts* and the two remaining parts are called *opposite parts*. Then Napier's rules are

5.101 The sine of any middle part equals the product of the tangents of the adjacent parts.

5.102 The sine of any middle part equals the product of the cosines of the opposite parts.

Example: Since co-$A = 90° - A$, co-$B = 90° - B$, we have

$$\sin a = \tan b \tan (\text{co-}B) \quad \text{or} \quad \sin a = \tan b \cot B$$

$$\sin (\text{co-}A) = \cos a \cos (\text{co-}B) \quad \text{or} \quad \cos A = \cos a \sin B$$

These can of course be obtained also from the results 5.97 on page 19.

6 COMPLEX NUMBERS

DEFINITIONS INVOLVING COMPLEX NUMBERS

A *complex number* is generally written as $a + bi$ where a and b are real numbers and i, called the *imaginary unit*, has the property that $i^2 = -1$. The real numbers a and b are called the *real* and *imaginary parts* of $a + bi$ respectively.

The complex numbers $a + bi$ and $a - bi$ are called *complex conjugates* of each other.

EQUALITY OF COMPLEX NUMBERS

6.1 $$a + bi = c + di \quad \text{if and only if} \quad a = c \text{ and } b = d$$

ADDITION OF COMPLEX NUMBERS

6.2 $$(a + bi) + (c + di) = (a + c) + (b + d)i$$

SUBTRACTION OF COMPLEX NUMBERS

6.3 $$(a + bi) - (c + di) = (a - c) + (b - d)i$$

MULTIPLICATION OF COMPLEX NUMBERS

6.4 $$(a + bi)(c + di) = (ac - bd) + (ad + bc)i$$

DIVISION OF COMPLEX NUMBERS

6.5 $$\frac{a + bi}{c + di} = \frac{a + bi}{c + di} \cdot \frac{c - di}{c - di} = \frac{ac + bd}{c^2 + d^2} + \left(\frac{bc - ad}{c^2 + d^2}\right)i$$

Note that the above operations are obtained by using the ordinary rules of algebra and replacing i^2 by -1 wherever it occurs.

GRAPH OF A COMPLEX NUMBER

A complex number $a + bi$ can be plotted as a point (a, b) on an xy plane called an *Argand diagram* or *Gaussian plane*. For example in Fig. 6-1 P represents the complex number $-3 + 4i$.

A complex number can also be interpreted as a *vector* from O to P.

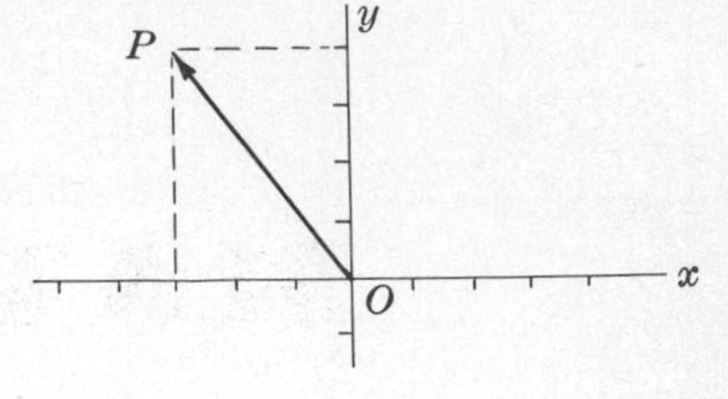

Fig. 6-1

POLAR FORM OF A COMPLEX NUMBER

In Fig. 6-2 point P with coordinates (x, y) represents the complex number $x + iy$. Point P can also be represented by *polar coordinates* (r, θ). Since $x = r\cos\theta,\ y = r\sin\theta$ we have

6.6 $$x + iy = r(\cos\theta + i\sin\theta)$$

called the *polar form* of the complex number. We often call $r = \sqrt{x^2 + y^2}$ the *modulus* and θ the *amplitude* of $x + iy$.

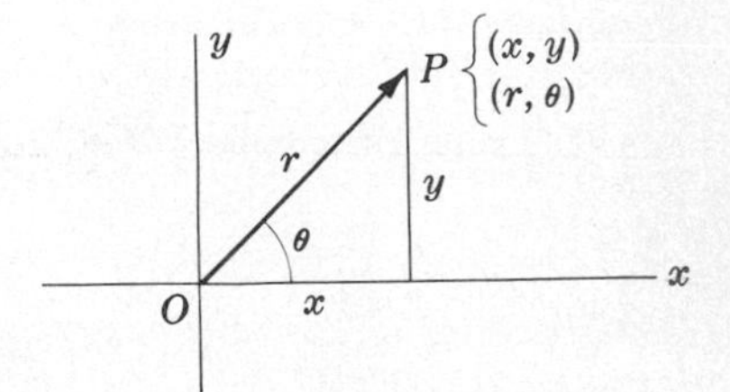

Fig. 6-2

MULTIPLICATION AND DIVISION OF COMPLEX NUMBERS IN POLAR FORM

6.7 $$[r_1(\cos\theta_1 + i\sin\theta_1)][r_2(\cos\theta_2 + i\sin\theta_2)] = r_1r_2[\cos(\theta_1 + \theta_2) + i\sin(\theta_1 + \theta_2)]$$

6.8 $$\frac{r_1(\cos\theta_1 + i\sin\theta_1)}{r_2(\cos\theta_2 + i\sin\theta_2)} = \frac{r_1}{r_2}[\cos(\theta_1 - \theta_2) + i\sin(\theta_1 - \theta_2)]$$

DE MOIVRE'S THEOREM

If p is any real number, De Moivre's theorem states that

6.9 $$[r(\cos\theta + i\sin\theta)]^p = r^p(\cos p\theta + i\sin p\theta)$$

ROOTS OF COMPLEX NUMBERS

If $p = 1/n$ where n is any positive integer, 6.9 can be written

6.10 $$[r(\cos\theta + i\sin\theta)]^{1/n} = r^{1/n}\left[\cos\frac{\theta + 2k\pi}{n} + i\sin\frac{\theta + 2k\pi}{n}\right]$$

where k is any integer. From this the n nth roots of a complex number can be obtained by putting $k = 0, 1, 2, \ldots, n-1$.

7 EXPONENTIAL and LOGARITHMIC FUNCTIONS

LAWS OF EXPONENTS

In the following p, q are real numbers, a, b are positive numbers and m, n are positive integers.

7.1 $a^p \cdot a^q = a^{p+q}$

7.2 $a^p/a^q = a^{p-q}$

7.3 $(a^p)^q = a^{pq}$

7.4 $a^0 = 1, \quad a \neq 0$

7.5 $a^{-p} = 1/a^p$

7.6 $(ab)^p = a^p b^p$

7.7 $\sqrt[n]{a} = a^{1/n}$

7.8 $\sqrt[n]{a^m} = a^{m/n}$

7.9 $\sqrt[n]{a/b} = \sqrt[n]{a}/\sqrt[n]{b}$

In a^p, p is called the *exponent*, a is the *base* and a^p is called the pth *power of* a. The function $y = a^x$ is called an *exponential function.*

LOGARITHMS AND ANTILOGARITHMS

If $a^p = N$ where $a \neq 0$ or 1, then $p = \log_a N$ is called the *logarithm* of N to the base a. The number $N = a^p$ is called the *antilogarithm* of p to the base a, written $\text{antilog}_a\, p$.

Example: Since $3^2 = 9$ we have $\log_3 9 = 2$, $\text{antilog}_3 2 = 9$.

The function $y = \log_a x$ is called a *logarithmic function.*

LAWS OF LOGARITHMS

7.10 $$\log_a MN = \log_a M + \log_a N$$

7.11 $$\log_a \frac{M}{N} = \log_a M - \log_a N$$

7.12 $$\log_a M^p = p \log_a M$$

COMMON LOGARITHMS AND ANTILOGARITHMS

Common logarithms and antilogarithms [also called *Briggsian*] are those in which the base $a = 10$. The common logarithm of N is denoted by $\log_{10} N$ or briefly $\log N$. For tables of common logarithms and antilogarithms, see pages 202-205. For illustrations using these tables see pages 194-196.

NATURAL LOGARITHMS AND ANTILOGARITHMS

Natural logarithms and antilogarithms [also called Napierian] are those in which the base $a = e = 2.71828\,18\ldots$ [see page 1]. The natural logarithm of N is denoted by $\log_e N$ or $\ln N$. For tables of natural logarithms see pages 224-225. For tables of natural antilogarithms [i.e. tables giving e^x for values of x] see pages 226-227. For illustrations using these tables see pages 196 and 200.

CHANGE OF BASE OF LOGARITHMS

The relationship between logarithms of a number N to different bases a and b is given by

7.13 $$\log_a N = \frac{\log_b N}{\log_b a}$$

In particular,

7.14 $$\log_e N = \ln N = 2.30258\,50929\,94\ldots \log_{10} N$$

7.15 $$\log_{10} N = \log N = 0.43429\,44819\,03\ldots \log_e N$$

RELATIONSHIP BETWEEN EXPONENTIAL AND TRIGONOMETRIC FUNCTIONS

7.16 $$e^{i\theta} = \cos\theta + i\sin\theta, \qquad e^{-i\theta} = \cos\theta - i\sin\theta$$

These are called *Euler's identities.* Here i is the imaginary unit [see page 21].

7.17 $$\sin\theta = \frac{e^{i\theta} - e^{-i\theta}}{2i}$$

7.18 $$\cos\theta = \frac{e^{i\theta} + e^{-i\theta}}{2}$$

7.19 $$\tan\theta = \frac{e^{i\theta} - e^{-i\theta}}{i(e^{i\theta} + e^{-i\theta})} = -i\left(\frac{e^{i\theta} - e^{-i\theta}}{e^{i\theta} + e^{-i\theta}}\right)$$

7.20 $$\cot\theta = i\left(\frac{e^{i\theta} + e^{-i\theta}}{e^{i\theta} - e^{-i\theta}}\right)$$

7.21 $$\sec\theta = \frac{2}{e^{i\theta} + e^{-i\theta}}$$

7.22 $$\csc\theta = \frac{2i}{e^{i\theta} - e^{-i\theta}}$$

PERIODICITY OF EXPONENTIAL FUNCTIONS

7.23 $$e^{i(\theta + 2k\pi)} = e^{i\theta} \qquad k = \text{integer}$$

From this it is seen that e^x has period $2\pi i$.

POLAR FORM OF COMPLEX NUMBERS EXPRESSED AS AN EXPONENTIAL

The polar form of a complex number $x + iy$ can be written in terms of exponentials [see 6.6, page 22] as

7.24 $$x + iy = r(\cos\theta + i\sin\theta) = re^{i\theta}$$

OPERATIONS WITH COMPLEX NUMBERS IN POLAR FORM

Formulas 6.7 through 6.10 on page 22 are equivalent to the following.

7.25 $$(r_1e^{i\theta_1})(r_2e^{i\theta_2}) = r_1r_2e^{i(\theta_1+\theta_2)}$$

7.26 $$\frac{r_1e^{i\theta_1}}{r_2e^{i\theta_2}} = \frac{r_1}{r_2}e^{i(\theta_1-\theta_2)}$$

7.27 $$(re^{i\theta})^p = r^pe^{ip\theta} \qquad \text{[De Moivre's theorem]}$$

7.28 $$(re^{i\theta})^{1/n} = [re^{i(\theta+2k\pi)}]^{1/n} = r^{1/n}e^{i(\theta+2k\pi)/n}$$

LOGARITHM OF A COMPLEX NUMBER

7.29 $$\ln(re^{i\theta}) = \ln r + i\theta + 2k\pi i \qquad k = \text{integer}$$

8 HYPERBOLIC FUNCTIONS

DEFINITION OF HYPERBOLIC FUNCTIONS

8.1 *Hyperbolic sine* of $x = \sinh x = \dfrac{e^x - e^{-x}}{2}$

8.2 *Hyperbolic cosine* of $x = \cosh x = \dfrac{e^x + e^{-x}}{2}$

8.3 *Hyperbolic tangent* of $x = \tanh x = \dfrac{e^x - e^{-x}}{e^x + e^{-x}}$

8.4 *Hyperbolic cotangent* of $x = \coth x = \dfrac{e^x + e^{-x}}{e^x - e^{-x}}$

8.5 *Hyperbolic secant* of $x = \operatorname{sech} x = \dfrac{2}{e^x + e^{-x}}$

8.6 *Hyperbolic cosecant* of $x = \operatorname{csch} x = \dfrac{2}{e^x - e^{-x}}$

RELATIONSHIPS AMONG HYPERBOLIC FUNCTIONS

8.7 $\tanh x = \dfrac{\sinh x}{\cosh x}$

8.8 $\coth x = \dfrac{1}{\tanh x} = \dfrac{\cosh x}{\sinh x}$

8.9 $\operatorname{sech} x = \dfrac{1}{\cosh x}$

8.10 $\operatorname{csch} x = \dfrac{1}{\sinh x}$

8.11 $\cosh^2 x - \sinh^2 x = 1$

8.12 $\operatorname{sech}^2 x + \tanh^2 x = 1$

8.13 $\coth^2 x - \operatorname{csch}^2 x = 1$

FUNCTIONS OF NEGATIVE ARGUMENTS

8.14 $\sinh(-x) = -\sinh x$

8.15 $\cosh(-x) = \cosh x$

8.16 $\tanh(-x) = -\tanh x$

8.17 $\operatorname{csch}(-x) = -\operatorname{csch} x$

8.18 $\operatorname{sech}(-x) = \operatorname{sech} x$

8.19 $\coth(-x) = -\coth x$

ADDITION FORMULAS

8.20 $\sinh(x \pm y) = \sinh x \cosh y \pm \cosh x \sinh y$

8.21 $\cosh(x \pm y) = \cosh x \cosh y \pm \sinh x \sinh y$

8.22 $\tanh(x \pm y) = \dfrac{\tanh x \pm \tanh y}{1 \pm \tanh x \tanh y}$

8.23 $\coth(x \pm y) = \dfrac{\coth x \coth y \pm 1}{\coth y \pm \coth x}$

DOUBLE ANGLE FORMULAS

8.24 $\sinh 2x = 2 \sinh x \cosh x$

8.25 $\cosh 2x = \cosh^2 x + \sinh^2 x = 2\cosh^2 x - 1 = 1 + 2\sinh^2 x$

8.26 $\tanh 2x = \dfrac{2 \tanh x}{1 + \tanh^2 x}$

HALF ANGLE FORMULAS

8.27 $\sinh \dfrac{x}{2} = \pm\sqrt{\dfrac{\cosh x - 1}{2}}$ [+ if $x > 0$, − if $x < 0$]

8.28 $\cosh \dfrac{x}{2} = \sqrt{\dfrac{\cosh x + 1}{2}}$

8.29 $\tanh \dfrac{x}{2} = \pm\sqrt{\dfrac{\cosh x - 1}{\cosh x + 1}}$ [+ if $x > 0$, − if $x < 0$]

$$= \frac{\sinh x}{\cosh x + 1} = \frac{\cosh x - 1}{\sinh x}$$

MULTIPLE ANGLE FORMULAS

8.30 $\sinh 3x = 3\sinh x + 4\sinh^3 x$

8.31 $\cosh 3x = 4\cosh^3 x - 3\cosh x$

8.32 $\tanh 3x = \dfrac{3\tanh x + \tanh^3 x}{1 + 3\tanh^2 x}$

8.33 $\sinh 4x = 8\sinh^3 x \cosh x + 4\sinh x \cosh x$

8.34 $\cosh 4x = 8\cosh^4 x - 8\cosh^2 x + 1$

8.35 $\tanh 4x = \dfrac{4\tanh x + 4\tanh^3 x}{1 + 6\tanh^2 x + \tanh^4 x}$

POWERS OF HYPERBOLIC FUNCTIONS

8.36 $\sinh^2 x = \frac{1}{2}\cosh 2x - \frac{1}{2}$

8.37 $\cosh^2 x = \frac{1}{2}\cosh 2x + \frac{1}{2}$

8.38 $\sinh^3 x = \frac{1}{4}\sinh 3x - \frac{3}{4}\sinh x$

8.39 $\cosh^3 x = \frac{1}{4}\cosh 3x + \frac{3}{4}\cosh x$

8.40 $\sinh^4 x = \frac{3}{8} - \frac{1}{2}\cosh 2x + \frac{1}{8}\cosh 4x$

8.41 $\cosh^4 x = \frac{3}{8} + \frac{1}{2}\cosh 2x + \frac{1}{8}\cosh 4x$

SUM, DIFFERENCE AND PRODUCT OF HYPERBOLIC FUNCTIONS

8.42 $\sinh x + \sinh y = 2\sinh\frac{1}{2}(x+y)\cosh\frac{1}{2}(x-y)$

8.43 $\sinh x - \sinh y = 2\cosh\frac{1}{2}(x+y)\sinh\frac{1}{2}(x-y)$

8.44 $\cosh x + \cosh y = 2\cosh\frac{1}{2}(x+y)\cosh\frac{1}{2}(x-y)$

8.45 $\cosh x - \cosh y = 2\sinh\frac{1}{2}(x+y)\sinh\frac{1}{2}(x-y)$

8.46 $\sinh x\sinh y = \frac{1}{2}\{\cosh(x+y) - \cosh(x-y)\}$

8.47 $\cosh x\cosh y = \frac{1}{2}\{\cosh(x+y) + \cosh(x-y)\}$

8.48 $\sinh x\cosh y = \frac{1}{2}\{\sinh(x+y) + \sinh(x-y)\}$

EXPRESSION OF HYPERBOLIC FUNCTIONS IN TERMS OF OTHERS

In the following we assume $x > 0$. If $x < 0$ use the appropriate sign as indicated by formulas 8.14 to 8.19.

	$\sinh x = u$	$\cosh x = u$	$\tanh x = u$	$\coth x = u$	$\operatorname{sech} x = u$	$\operatorname{csch} x = u$
$\sinh x$	u	$\sqrt{u^2-1}$	$u/\sqrt{1-u^2}$	$1/\sqrt{u^2-1}$	$\sqrt{1-u^2}/u$	$1/u$
$\cosh x$	$\sqrt{1+u^2}$	u	$1/\sqrt{1-u^2}$	$u/\sqrt{u^2-1}$	$1/u$	$\sqrt{1+u^2}/u$
$\tanh x$	$u/\sqrt{1+u^2}$	$\sqrt{u^2-1}/u$	u	$1/u$	$\sqrt{1-u^2}$	$1/\sqrt{1+u^2}$
$\coth x$	$\sqrt{u^2+1}/u$	$u/\sqrt{u^2-1}$	$1/u$	u	$1/\sqrt{1-u^2}$	$\sqrt{1+u^2}$
$\operatorname{sech} x$	$1/\sqrt{1+u^2}$	$1/u$	$\sqrt{1-u^2}$	$\sqrt{u^2-1}/u$	u	$u/\sqrt{1+u^2}$
$\operatorname{csch} x$	$1/u$	$1/\sqrt{u^2-1}$	$\sqrt{1-u^2}/u$	$\sqrt{u^2-1}$	$u/\sqrt{1-u^2}$	u

GRAPHS OF HYPERBOLIC FUNCTIONS

8.49 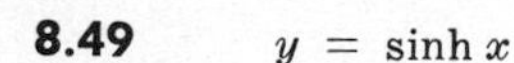 $y = \sinh x$

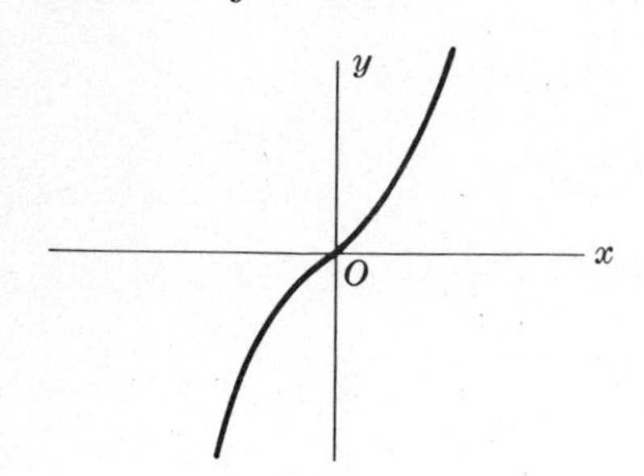

Fig. 8-1

8.50 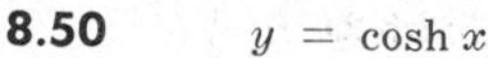 $y = \cosh x$

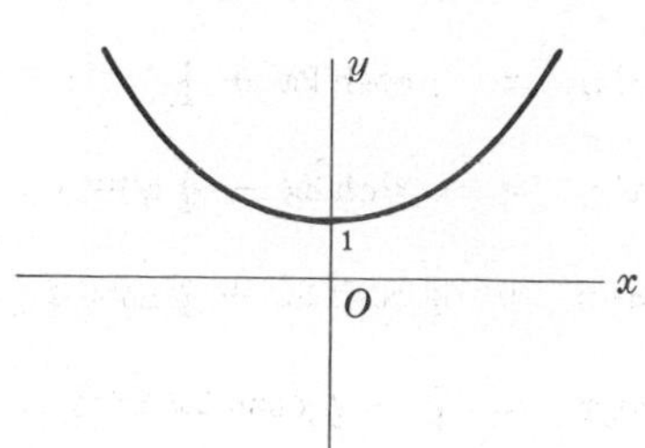

Fig. 8-2

8.51 $y = \tanh x$

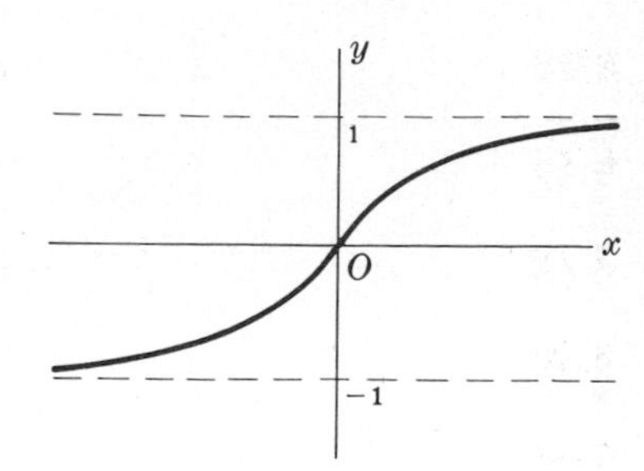

Fig. 8-3

8.52 $y = \coth x$

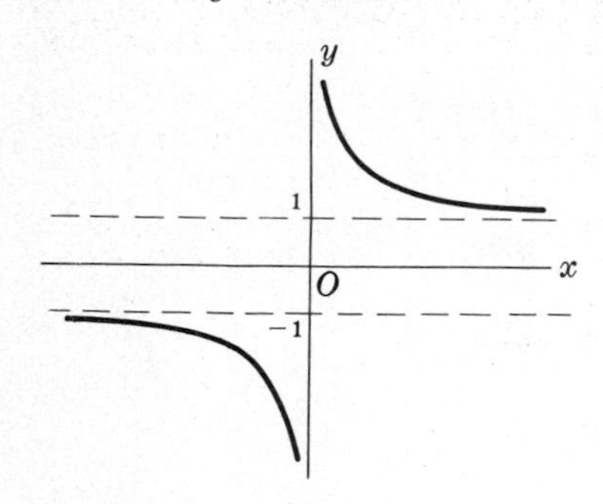

Fig. 8-4

8.53 $y = \operatorname{sech} x$

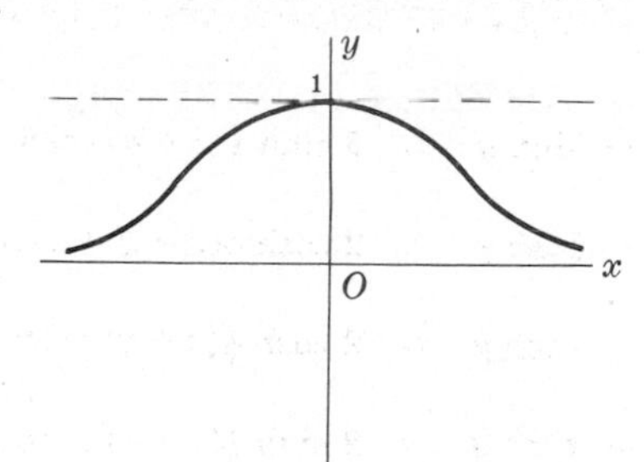

Fig. 8-5

8.54 $y = \operatorname{csch} x$

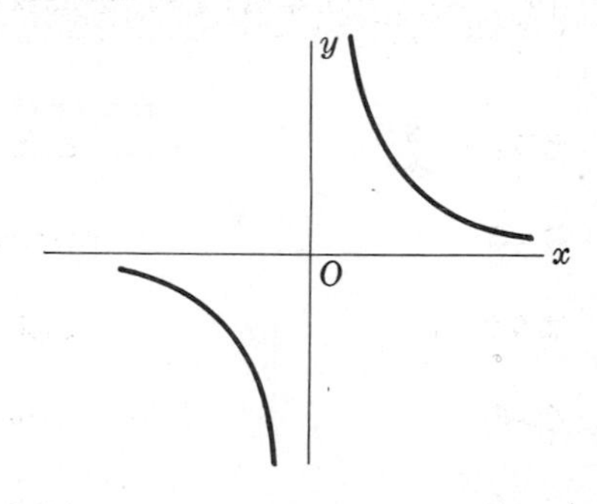

Fig. 8-6

INVERSE HYPERBOLIC FUNCTIONS

If $x = \sinh y$, then $y = \sinh^{-1} x$ is called the *inverse hyperbolic sine* of x. Similarly we define the other inverse hyperbolic functions. The inverse hyperbolic functions are multiple-valued and as in the case of inverse trigonometric functions [see page 17] we restrict ourselves to principal values for which they can be considered as single-valued.

The following list shows the principal values [unless otherwise indicated] of the inverse hyperbolic functions expressed in terms of logarithmic functions which are taken as real valued.

8.55 $\sinh^{-1} x = \ln(x + \sqrt{x^2+1}) \qquad -\infty < x < \infty$

8.56 $\cosh^{-1} x = \ln(x + \sqrt{x^2-1}) \qquad x \geqq 1 \qquad [\cosh^{-1} x > 0 \text{ is principal value}]$

8.57 $\tanh^{-1} x = \dfrac{1}{2} \ln\left(\dfrac{1+x}{1-x}\right) \qquad -1 < x < 1$

8.58 $\coth^{-1} x = \dfrac{1}{2} \ln\left(\dfrac{x+1}{x-1}\right) \qquad x > 1 \text{ or } x < -1$

8.59 $\operatorname{sech}^{-1} x = \ln\left(\dfrac{1}{x} + \sqrt{\dfrac{1}{x^2} - 1}\right) \qquad 0 < x \leqq 1 \qquad [\operatorname{sech}^{-1} x > 0 \text{ is principal value}]$

8.60 $\operatorname{csch}^{-1} x = \ln\left(\dfrac{1}{x} + \sqrt{\dfrac{1}{x^2} + 1}\right) \qquad x \neq 0$

RELATIONS BETWEEN INVERSE HYPERBOLIC FUNCTIONS

8.61 $$\operatorname{csch}^{-1} x = \sinh^{-1}(1/x)$$

8.62 $$\operatorname{sech}^{-1} x = \cosh^{-1}(1/x)$$

8.63 $$\coth^{-1} x = \tanh^{-1}(1/x)$$

8.64 $$\sinh^{-1}(-x) = -\sinh^{-1} x$$

8.65 $$\tanh^{-1}(-x) = -\tanh^{-1} x$$

8.66 $$\coth^{-1}(-x) = -\coth^{-1} x$$

8.67 $$\operatorname{csch}^{-1}(-x) = -\operatorname{csch}^{-1} x$$

GRAPHS OF INVERSE HYPERBOLIC FUNCTIONS

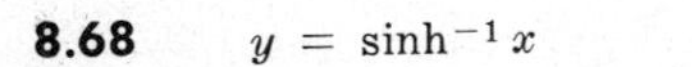

8.68 $y = \sinh^{-1} x$

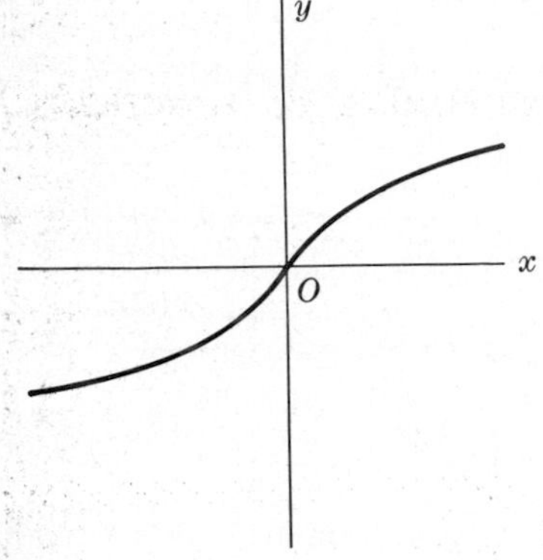

Fig. 8-7

8.69 $y = \cosh^{-1} x$

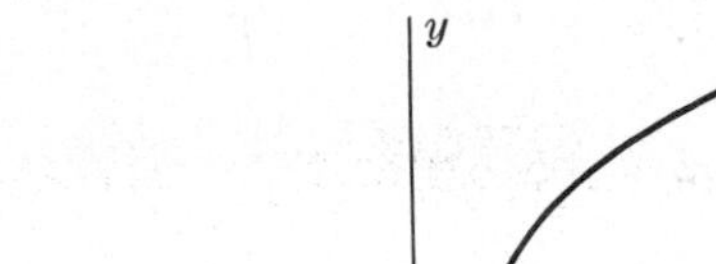

Fig. 8-8

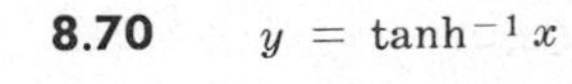

8.70 $y = \tanh^{-1} x$

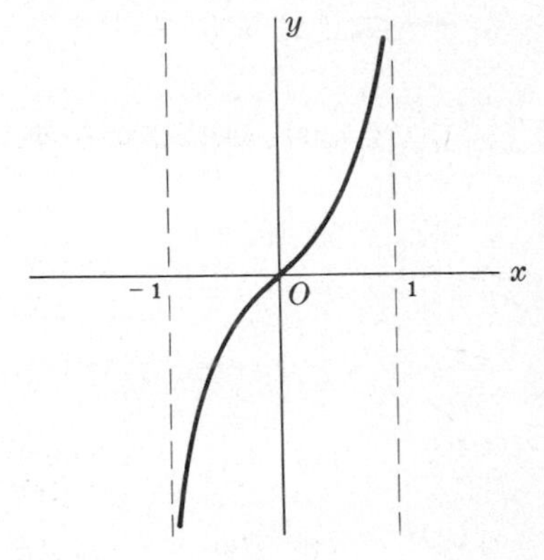

Fig. 8-9

8.71 $y = \coth^{-1} x$

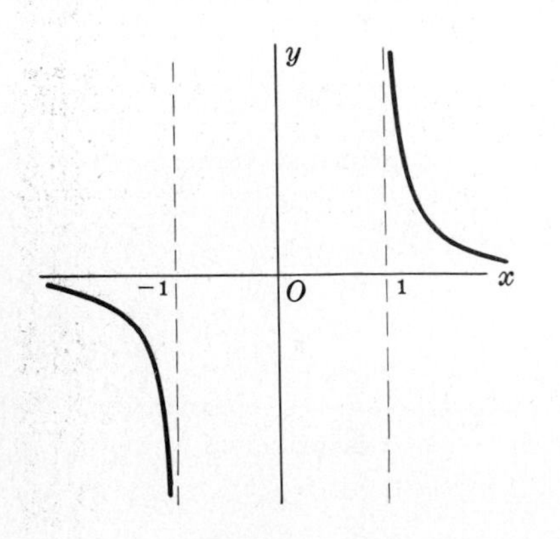

Fig. 8-10

8.72 $y = \operatorname{sech}^{-1} x$

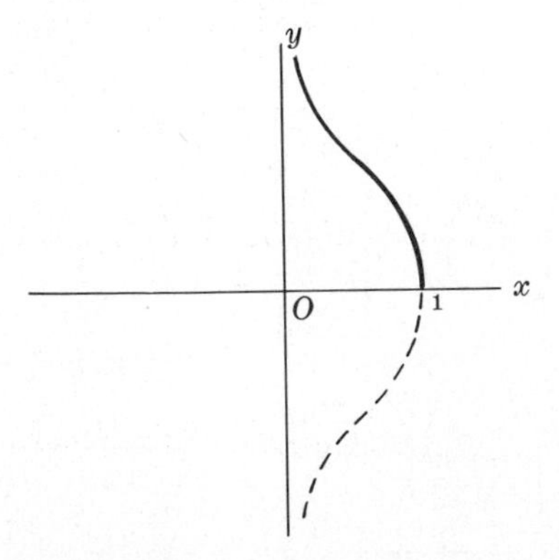

Fig. 8-11

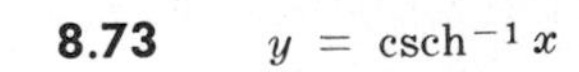

8.73 $y = \operatorname{csch}^{-1} x$

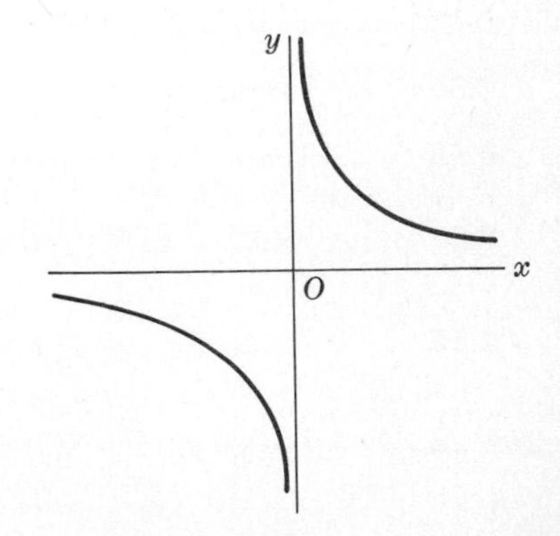

Fig. 8-12

RELATIONSHIP BETWEEN HYPERBOLIC AND TRIGONOMETRIC FUNCTIONS

8.74 $\sin(ix) = i \sinh x$

8.75 $\cos(ix) = \cosh x$

8.76 $\tan(ix) = i \tanh x$

8.77 $\csc(ix) = -i \operatorname{csch} x$

8.78 $\sec(ix) = \operatorname{sech} x$

8.79 $\cot(ix) = -i \coth x$

8.80 $\sinh(ix) = i \sin x$

8.81 $\cosh(ix) = \cos x$

8.82 $\tanh(ix) = i \tan x$

8.83 $\operatorname{csch}(ix) = -i \csc x$

8.84 $\operatorname{sech}(ix) = \sec x$

8.85 $\coth(ix) = -i \cot x$

PERIODICITY OF HYPERBOLIC FUNCTIONS

In the following k is any integer.

8.86 $\sinh(x + 2k\pi i) = \sinh x$

8.87 $\cosh(x + 2k\pi i) = \cosh x$

8.88 $\tanh(x + k\pi i) = \tanh x$

8.89 $\operatorname{csch}(x + 2k\pi i) = \operatorname{csch} x$

8.90 $\operatorname{sech}(x + 2k\pi i) = \operatorname{sech} x$

8.91 $\coth(x + k\pi i) = \coth x$

RELATIONSHIP BETWEEN INVERSE HYPERBOLIC AND INVERSE TRIGONOMETRIC FUNCTIONS

8.92 $\sin^{-1}(ix) = i \sinh^{-1} x$

8.93 $\sinh^{-1}(ix) = i \sin^{-1} x$

8.94 $\cos^{-1} x = \pm i \cosh^{-1} x$

8.95 $\cosh^{-1} x = \pm i \cos^{-1} x$

8.96 $\tan^{-1}(ix) = i \tanh^{-1} x$

8.97 $\tanh^{-1}(ix) = i \tan^{-1} x$

8.98 $\cot^{-1}(ix) = -i \coth^{-1} x$

8.99 $\coth^{-1}(ix) = -i \cot^{-1} x$

8.100 $\sec^{-1} x = \pm i \operatorname{sech}^{-1} x$

8.101 $\operatorname{sech}^{-1} x = \pm i \sec^{-1} x$

8.102 $\csc^{-1}(ix) = -i \operatorname{csch}^{-1} x$

8.103 $\operatorname{csch}^{-1}(ix) = -i \csc^{-1} x$

9 SOLUTIONS of ALGEBRAIC EQUATIONS

QUADRATIC EQUATION: $ax^2 + bx + c = 0$

9.1 Solutions: $$x = \frac{-b \pm \sqrt{b^2 - 4ac}}{2a}$$

If a, b, c are real and if $D = b^2 - 4ac$ is the *discriminant*, then the roots are

(i) real and unequal if $D > 0$

(ii) real and equal if $D = 0$

(iii) complex conjugate if $D < 0$

9.2 If x_1, x_2 are the roots, then $x_1 + x_2 = -b/a$ and $x_1 x_2 = c/a$.

CUBIC EQUATION: $x^3 + a_1 x^2 + a_2 x + a_3 = 0$

Let $$Q = \frac{3a_2 - a_1^2}{9}, \quad R = \frac{9a_1 a_2 - 27a_3 - 2a_1^3}{54},$$
$$S = \sqrt[3]{R + \sqrt{Q^3 + R^2}}, \quad T = \sqrt[3]{R - \sqrt{Q^3 + R^2}}$$

9.3 Solutions: $$\begin{cases} x_1 = S + T - \frac{1}{3}a_1 \\ x_2 = -\frac{1}{2}(S+T) - \frac{1}{3}a_1 + \frac{1}{2}i\sqrt{3}\,(S - T) \\ x_3 = -\frac{1}{2}(S+T) - \frac{1}{3}a_1 - \frac{1}{2}i\sqrt{3}\,(S - T) \end{cases}$$

If a_1, a_2, a_3 are real and if $D = Q^3 + R^2$ is the *discriminant*, then

(i) one root is real and two complex conjugate if $D > 0$

(ii) all roots are real and at least two are equal if $D = 0$

(iii) all roots are real and unequal if $D < 0$.

If $D < 0$, computation is simplified by use of trigonometry.

9.4 Solutions if $D < 0$: $$\begin{cases} x_1 = 2\sqrt{-Q}\cos(\frac{1}{3}\theta) \\ x_2 = 2\sqrt{-Q}\cos(\frac{1}{3}\theta + 120°) \\ x_3 = 2\sqrt{-Q}\cos(\frac{1}{3}\theta + 240°) \end{cases} \qquad \text{where } \cos\theta = -R/\sqrt{-Q^3}$$

9.5 $$x_1 + x_2 + x_3 = -a_1, \quad x_1x_2 + x_2x_3 + x_3x_1 = a_2, \quad x_1x_2x_3 = -a_3$$

where x_1, x_2, x_3 are the three roots.

QUARTIC EQUATION: $x^4 + a_1x^3 + a_2x^2 + a_3x + a_4 = 0$

Let y_1 be a real root of the cubic equation

9.6 $$y^3 - a_2y^2 + (a_1a_3 - 4a_4)y + (4a_2a_4 - a_3^2 - a_1^2a_4) = 0$$

9.7 **Solutions:** The 4 roots of $z^2 + \frac{1}{2}\{a_1 \pm \sqrt{a_1^2 - 4a_2 + 4y_1}\}z + \frac{1}{2}\{y_1 \pm \sqrt{y_1^2 - 4a_4}\} = 0$

If all roots of 9.6 are real, computation is simplified by using that particular real root which produces all real coefficients in the quadratic equation 9.7.

9.8 $$\begin{cases} x_1 + x_2 + x_3 + x_4 = -a_1 \\ x_1x_2 + x_2x_3 + x_3x_4 + x_4x_1 + x_1x_3 + x_2x_4 = a_2 \\ x_1x_2x_3 + x_2x_3x_4 + x_1x_2x_4 + x_1x_3x_4 = -a_3 \\ x_1x_2x_3x_4 = a_4 \end{cases}$$

where x_1, x_2, x_3, x_4 are the four roots.

10 FORMULAS from PLANE ANALYTIC GEOMETRY

DISTANCE d BETWEEN TWO POINTS $P_1(x_1, y_1)$ AND $P_2(x_2, y_2)$

10.1 $$d = \sqrt{(x_2 - x_1)^2 + (y_2 - y_1)^2}$$

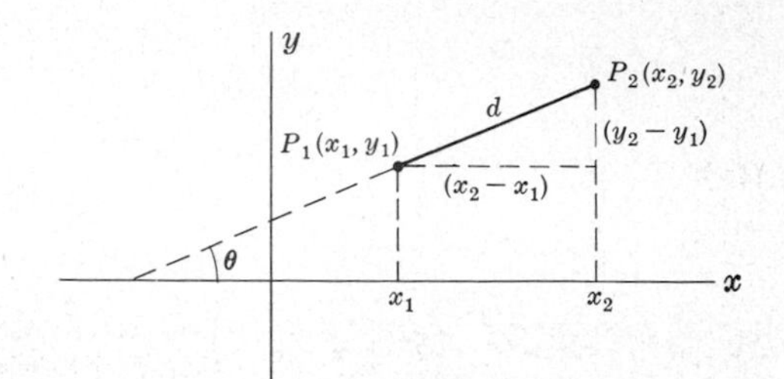

Fig. 10-1

SLOPE m OF LINE JOINING TWO POINTS $P_1(x_1, y_1)$ AND $P_2(x_2, y_2)$

10.2 $$m = \frac{y_2 - y_1}{x_2 - x_1} = \tan\theta$$

EQUATION OF LINE JOINING TWO POINTS $P_1(x_1, y_1)$ AND $P_2(x_2, y_2)$

10.3 $$\frac{y - y_1}{x - x_1} = \frac{y_2 - y_1}{x_2 - x_1} = m \quad \text{or} \quad y - y_1 = m(x - x_1)$$

10.4 $$y = mx + b$$

where $b = y_1 - mx_1 = \dfrac{x_2y_1 - x_1y_2}{x_2 - x_1}$ is the *intercept* on the y axis, i.e. the y *intercept.*

EQUATION OF LINE IN TERMS OF x INTERCEPT $a \neq 0$ AND y INTERCEPT $b \neq 0$

10.5 $$\frac{x}{a} + \frac{y}{b} = 1$$

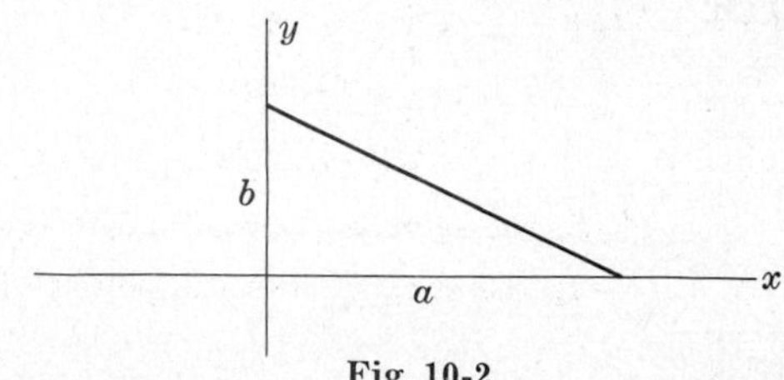

Fig. 10-2

NORMAL FORM FOR EQUATION OF LINE

10.6 $$x \cos\alpha + y \sin\alpha = p$$

where p = perpendicular distance from origin O to line

and α = angle of inclination of perpendicular with positive x axis.

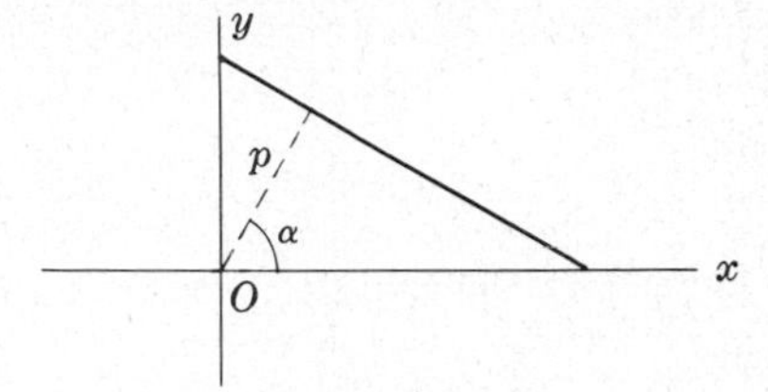

Fig. 10-3

GENERAL EQUATION OF LINE

10.7 $$Ax + By + C = 0$$

DISTANCE FROM POINT (x_1, y_1) TO LINE $Ax + By + C = 0$

10.8 $$\frac{Ax_1 + By_1 + C}{\pm\sqrt{A^2 + B^2}}$$

where the sign is chosen so that the distance is nonnegative.

ANGLE ψ BETWEEN TWO LINES HAVING SLOPES m_1 AND m_2

10.9 $$\tan\psi = \frac{m_2 - m_1}{1 + m_1 m_2}$$

Lines are parallel or coincident if and only if $m_1 = m_2$.

Lines are perpendicular if and only if $m_2 = -1/m_1$.

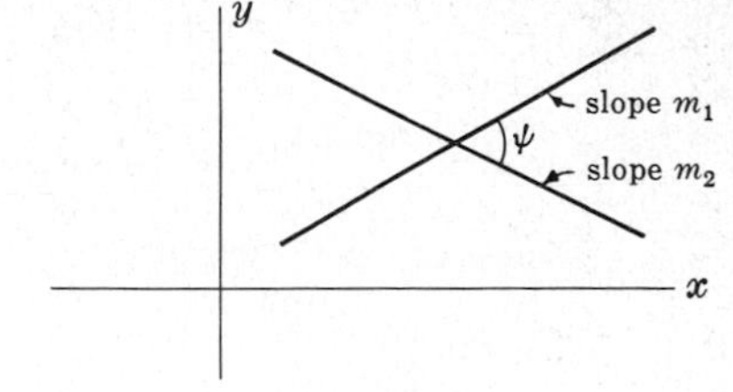

Fig. 10-4

AREA OF TRIANGLE WITH VERTICES AT $(x_1, y_1), (x_2, y_2), (x_3, y_3)$

10.10 $$\text{Area} = \pm\frac{1}{2}\begin{vmatrix} x_1 & y_1 & 1 \\ x_2 & y_2 & 1 \\ x_3 & y_3 & 1 \end{vmatrix}$$

$$= \pm\frac{1}{2}(x_1y_2 + y_1x_3 + y_3x_2 - y_2x_3 - y_1x_2 - x_1y_3)$$

where the sign is chosen so that the area is nonnegative.

If the area is zero the points all lie on a line.

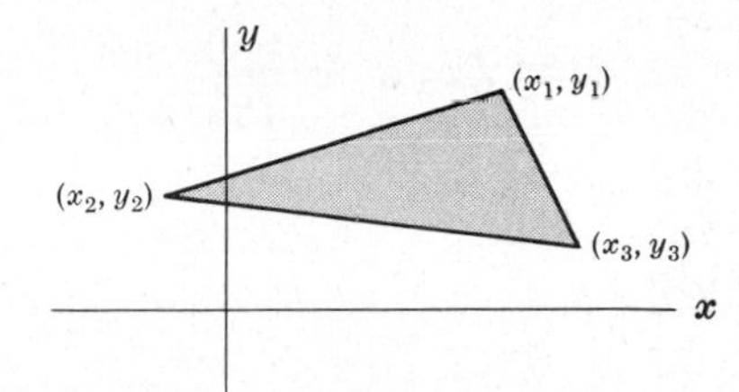

Fig. 10-5

TRANSFORMATION OF COORDINATES INVOLVING PURE TRANSLATION

10.11 $\begin{cases} x = x' + x_0 \\ y = y' + y_0 \end{cases}$ or $\begin{cases} x' = x - x_0 \\ y' = y - y_0 \end{cases}$

where (x, y) are old coordinates [i.e. coordinates relative to xy system], (x', y') are new coordinates [relative to $x'y'$ system] and (x_0, y_0) are the coordinates of the new origin O' relative to the old xy coordinate system.

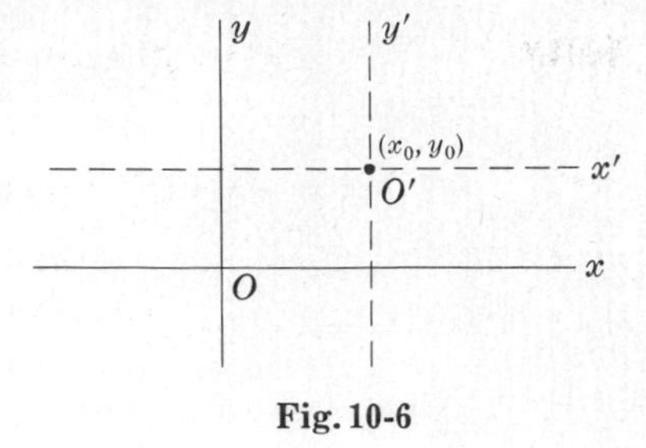

Fig. 10-6

TRANSFORMATION OF COORDINATES INVOLVING PURE ROTATION

10.12 $\begin{cases} x = x' \cos\alpha - y' \sin\alpha \\ y = x' \sin\alpha + y' \cos\alpha \end{cases}$ or $\begin{cases} x' = x \cos\alpha + y \sin\alpha \\ y' = y \cos\alpha - x \sin\alpha \end{cases}$

where the origins of the old $[xy]$ and new $[x'y']$ coordinate systems are the same but the x' axis makes an angle α with the positive x axis.

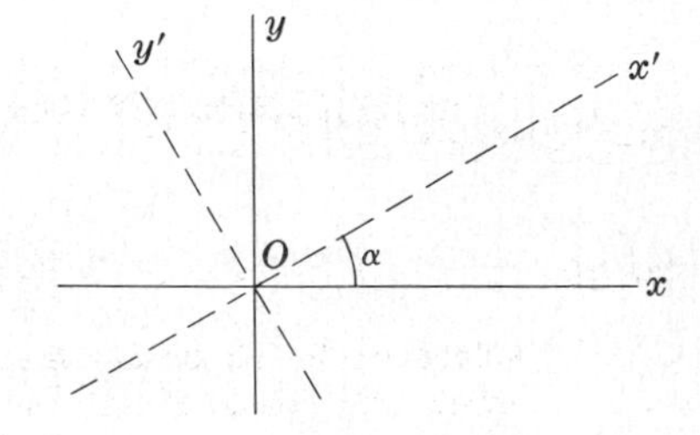

Fig. 10-7

TRANSFORMATION OF COORDINATES INVOLVING TRANSLATION AND ROTATION

10.13 $\begin{cases} x = x' \cos\alpha - y' \sin\alpha + x_0 \\ y = x' \sin\alpha + y' \cos\alpha + y_0 \end{cases}$

or $\begin{cases} x' = (x - x_0) \cos\alpha + (y - y_0) \sin\alpha \\ y' = (y - y_0) \cos\alpha - (x - x_0) \sin\alpha \end{cases}$

where the new origin O' of $x'y'$ coordinate system has coordinates (x_0, y_0) relative to the old xy coordinate system and the x' axis makes an angle α with the positive x axis.

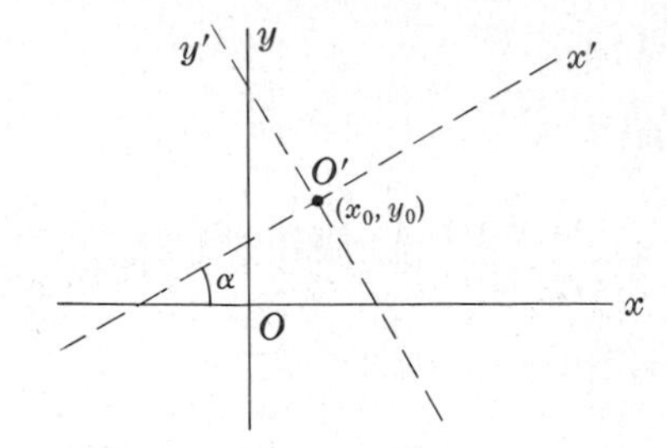

Fig. 10-8

POLAR COORDINATES (r, θ)

A point P can be located by rectangular coordinates (x, y) or polar coordinates (r, θ). The transformation between these coordinates is

10.14 $\begin{cases} x = r \cos\theta \\ y = r \sin\theta \end{cases}$ or $\begin{cases} r = \sqrt{x^2 + y^2} \\ \theta = \tan^{-1}(y/x) \end{cases}$

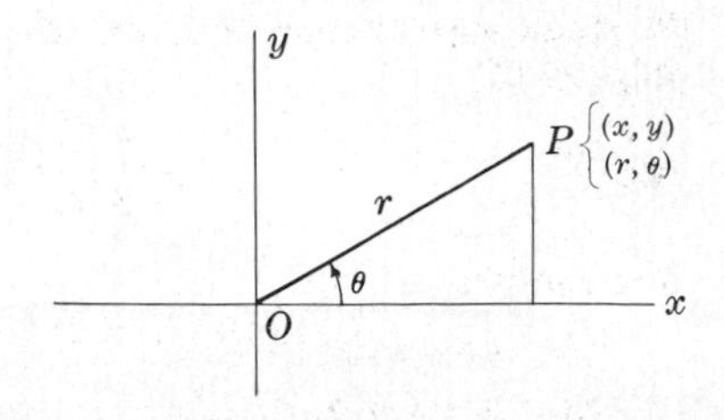

Fig. 10-9

EQUATION OF CIRCLE OF RADIUS R, CENTER AT (x_0, y_0)

10.15 $$(x - x_0)^2 + (y - y_0)^2 = R^2$$

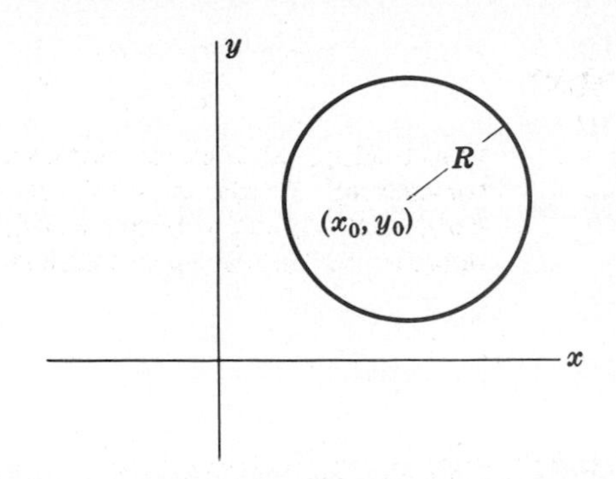

Fig. 10-10

EQUATION OF CIRCLE OF RADIUS R PASSING THROUGH ORIGIN

10.16 $$r = 2R\cos(\theta - \alpha)$$

where (r, θ) are polar coordinates of any point on the circle and (R, α) are polar coordinates of the center of the circle.

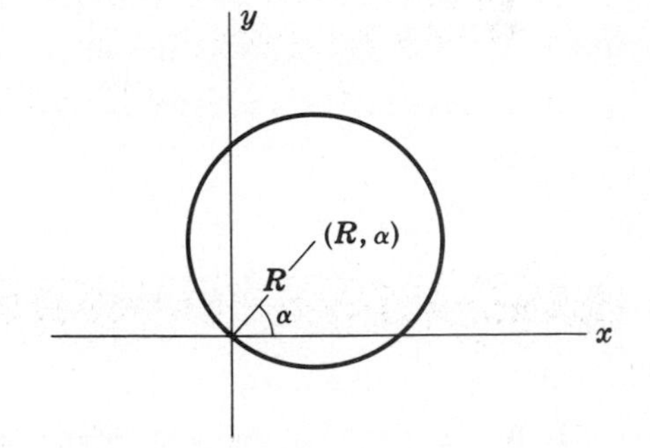

Fig. 10-11

CONICS [ELLIPSE, PARABOLA OR HYPERBOLA]

If a point P moves so that its distance from a fixed point [called the *focus*] divided by its distance from a fixed line [called the *directrix*] is a constant ϵ [called the *eccentricity*], then the curve described by P is called a *conic* [so-called because such curves can be obtained by intersecting a plane and a cone at different angles].

If the focus is chosen at origin O the equation of a conic in polar coordinates (r, θ) is, if $OQ = p$ and $LM = D$, [see Fig. 10-12]

10.17 $$r = \frac{p}{1 - \epsilon\cos\theta} = \frac{\epsilon D}{1 - \epsilon\cos\theta}$$

The conic is

(i) an ellipse if $\epsilon < 1$

(ii) a parabola if $\epsilon = 1$

(iii) a hyperbola if $\epsilon > 1$.

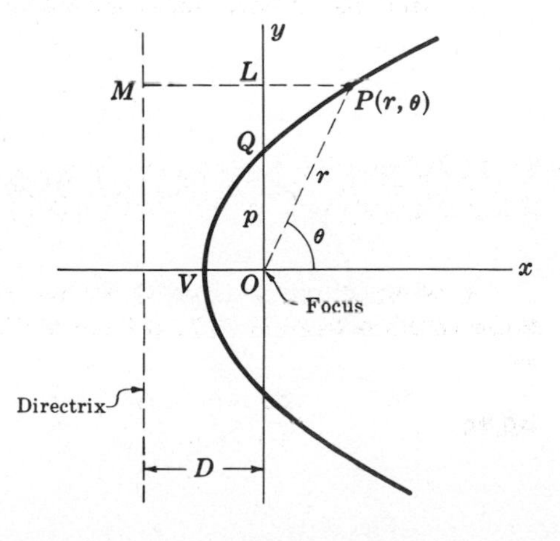

Fig. 10-12

ELLIPSE WITH CENTER $C(x_0, y_0)$ AND MAJOR AXIS PARALLEL TO x AXIS

10.18 Length of major axis $A'A = 2a$

10.19 Length of minor axis $B'B = 2b$

10.20 Distance from center C to focus F or F' is

$$c = \sqrt{a^2 - b^2}$$

10.21 Eccentricity $= \epsilon = \dfrac{c}{a} = \dfrac{\sqrt{a^2 - b^2}}{a}$

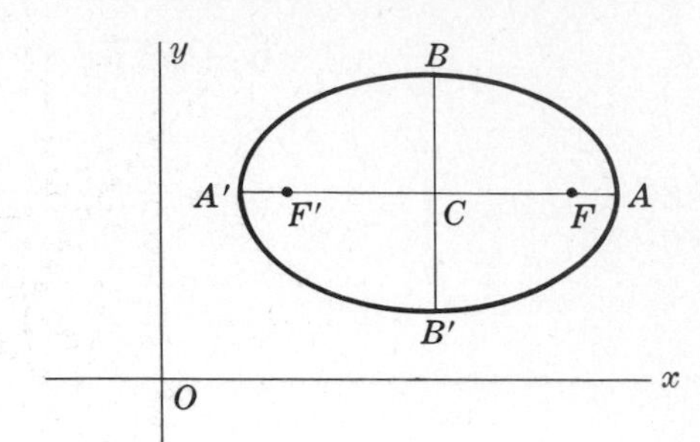

Fig. 10-13

10.22 Equation in rectangular coordinates:

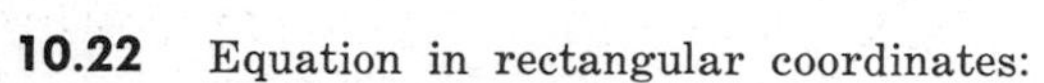

$$\frac{(x - x_0)^2}{a^2} + \frac{(y - y_0)^2}{b^2} = 1$$

10.23 Equation in polar coordinates if C is at O: $\quad r^2 = \dfrac{a^2 b^2}{a^2 \sin^2 \theta + b^2 \cos^2 \theta}$

10.24 Equation in polar coordinates if C is on x axis and F' is at O: $\quad r = \dfrac{a(1 - \epsilon^2)}{1 - \epsilon \cos \theta}$

10.25 If P is any point on the ellipse, $PF + PF' = 2a$

If the major axis is parallel to the y axis, interchange x and y in the above or replace θ by $\frac{1}{2}\pi - \theta$ [or $90° - \theta$].

PARABOLA WITH AXIS PARALLEL TO x AXIS

If vertex is at $A(x_0, y_0)$ and the distance from A to focus F is $a > 0$, the equation of the parabola is

10.26 $\quad (y - y_0)^2 = 4a(x - x_0) \quad$ if parabola opens to right [Fig. 10-14]

10.27 $\quad (y - y_0)^2 = -4a(x - x_0) \quad$ if parabola opens to left [Fig. 10-15]

If focus is at the origin [Fig. 10-16] the equation in polar coordinates is

10.28

$$r = \frac{2a}{1 - \cos \theta}$$

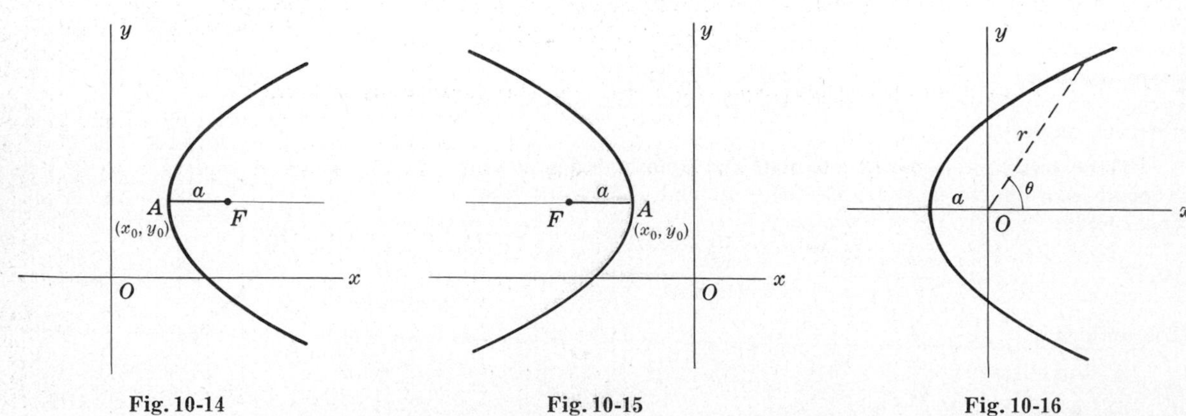

Fig. 10-14 Fig. 10-15 Fig. 10-16

In case the axis is parallel to the y axis, interchange x and y or replace θ by $\frac{1}{2}\pi - \theta$ [or $90° - \theta$].

HYPERBOLA WITH CENTER $C(x_0, y_0)$ AND MAJOR AXIS PARALLEL TO x AXIS

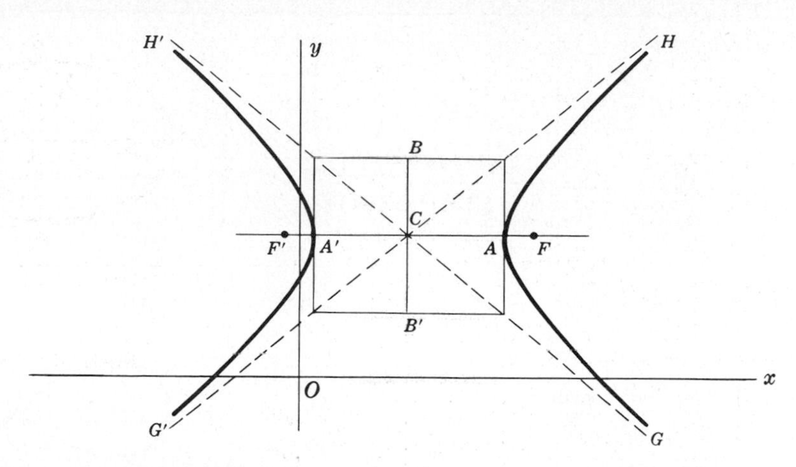

Fig. 10-17

10.29 Length of major axis $A'A = 2a$

10.30 Length of minor axis $B'B = 2b$

10.31 Distance from center C to focus F or $F' = c = \sqrt{a^2 + b^2}$

10.32 Eccentricity $\epsilon = \dfrac{c}{a} = \dfrac{\sqrt{a^2 + b^2}}{a}$

10.33 Equation in rectangular coordinates: $\dfrac{(x - x_0)^2}{a^2} - \dfrac{(y - y_0)^2}{b^2} = 1$

10.34 Slopes of asymptotes $G'H$ and $GH' = \pm \dfrac{b}{a}$

10.35 Equation in polar coordinates if C is at O: $r^2 = \dfrac{a^2b^2}{b^2 \cos^2 \theta - a^2 \sin^2 \theta}$

10.36 Equation in polar coordinates if C is on X axis and F' is at O: $r = \dfrac{a(\epsilon^2 - 1)}{1 - \epsilon \cos \theta}$

10.37 If P is any point on the hyperbola, $PF - PF' = \pm 2a$ [depending on branch]

If the major axis is parallel to the y axis, interchange x and y in the above or replace θ by $\frac{1}{2}\pi - \theta$ [or $90° - \theta$].

11 SPECIAL PLANE CURVES

LEMNISCATE

11.1 Equation in polar coordinates:

$$r^2 = a^2 \cos 2\theta$$

11.2 Equation in rectangular coordinates:

$$(x^2+y^2)^2 = a^2(x^2-y^2)$$

11.3 Angle between AB' or $A'B$ and x axis $= 45°$

11.4 Area of one loop $= \frac{1}{2}a^2$

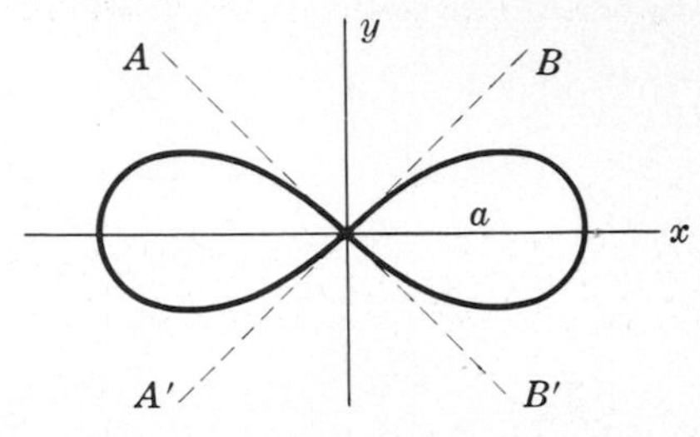

Fig. 11-1

CYCLOID

11.5 Equations in parametric form:

$$\begin{cases} x = a(\phi - \sin\phi) \\ y = a(1-\cos\phi) \end{cases}$$

11.6 Area of one arch $= 3\pi a^2$

11.7 Arc length of one arch $= 8a$

This is a curve described by a point P on a circle of radius a rolling along x axis.

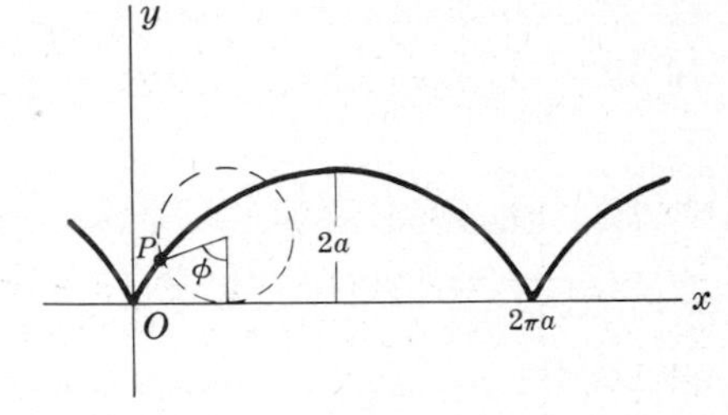

Fig. 11-2

HYPOCYCLOID WITH FOUR CUSPS

11.8 Equation in rectangular coordinates:

$$x^{2/3} + y^{2/3} = a^{2/3}$$

11.9 Equations in parametric form:

$$\begin{cases} x = a\cos^3\theta \\ y = a\sin^3\theta \end{cases}$$

11.10 Area bounded by curve $= \frac{3}{8}\pi a^2$

11.11 Arc length of entire curve $= 6a$

This is a curve described by a point P on a circle of radius $a/4$ as it rolls on the inside of a circle of radius a.

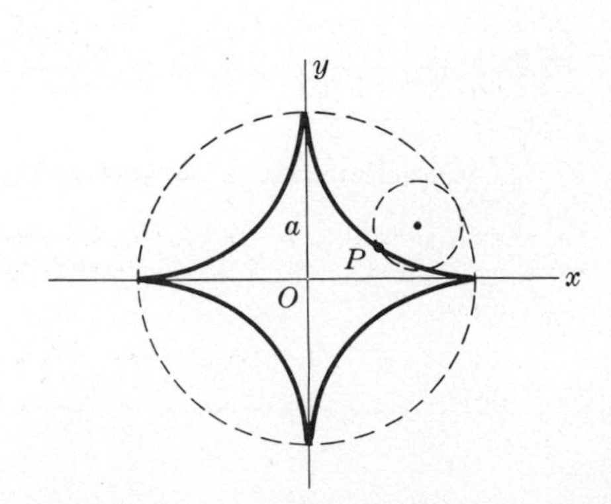

Fig. 11-3

CARDIOID

11.12 Equation: $r = a(1 + \cos\theta)$

11.13 Area bounded by curve $= \frac{3}{2}\pi a^2$

11.14 Arc length of curve $= 8a$

This is the curve described by a point P of a circle of radius a as it rolls on the outside of a fixed circle of radius a. The curve is also a special case of the limacon of Pascal [see 11.32].

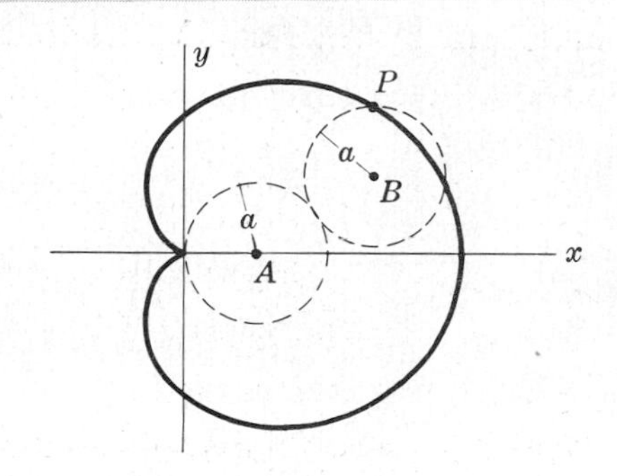

Fig. 11-4

CATENARY

11.15 Equation: $y = \frac{a}{2}(e^{x/a} + e^{-x/a}) = a\cosh\frac{x}{a}$

This is the curve in which a heavy uniform chain would hang if suspended vertically from fixed points A and B.

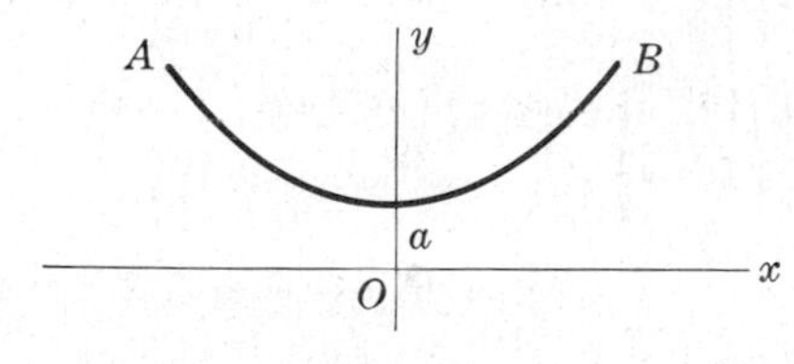

Fig. 11-5

THREE-LEAVED ROSE

11.16 Equation: $r = a\cos 3\theta$

The equation $r = a\sin 3\theta$ is a similar curve obtained by rotating the curve of Fig. 11-6 counterclockwise through 30° or $\pi/6$ radians.

In general $r = a\cos n\theta$ or $r = a\sin n\theta$ has n leaves if n is odd.

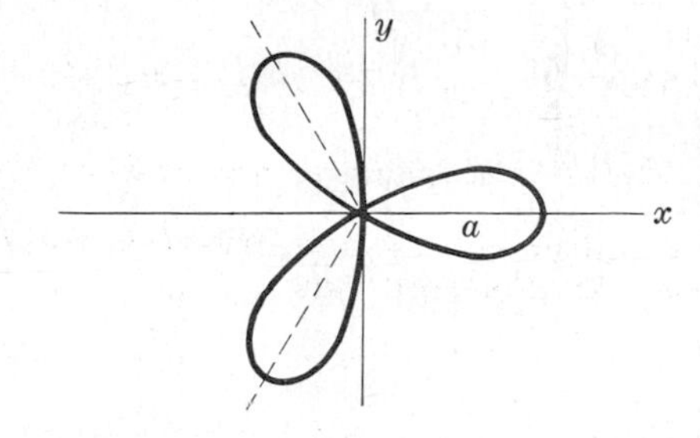

Fig. 11-6

FOUR-LEAVED ROSE

11.17 Equation: $r = a\cos 2\theta$

The equation $r = a\sin 2\theta$ is a similar curve obtained by rotating the curve of Fig. 11-7 counterclockwise through 45° or $\pi/4$ radians.

In general $r = a\cos n\theta$ or $r = a\sin n\theta$ has $2n$ leaves if n is even.

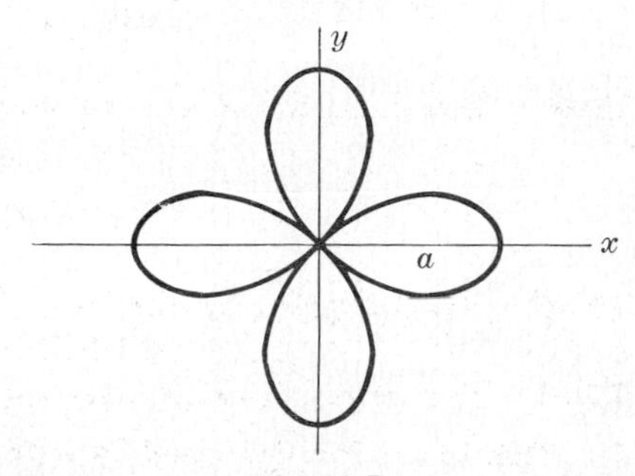

Fig. 11-7

EPICYCLOID

11.18 Parametric equations:

$$\begin{cases} x = (a+b)\cos\theta - b\cos\left(\dfrac{a+b}{b}\right)\theta \\ y = (a+b)\sin\theta - b\sin\left(\dfrac{a+b}{b}\right)\theta \end{cases}$$

This is the curve described by a point P on a circle of radius b as it rolls on the outside of a circle of radius a.

The cardioid [Fig. 11-4] is a special case of an epicycloid.

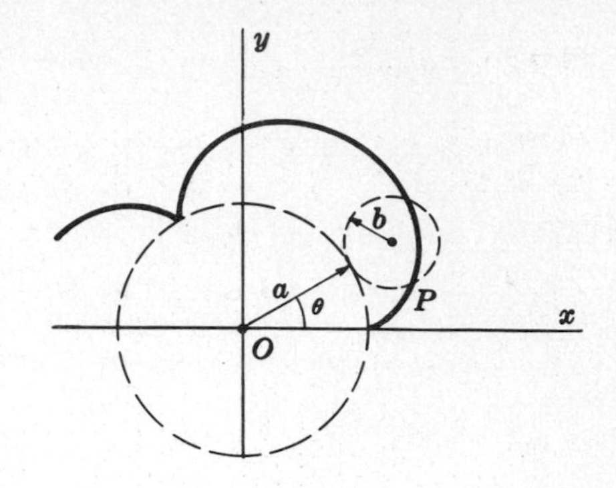

Fig. 11-8

GENERAL HYPOCYCLOID

11.19 Parametric equations:

$$\begin{cases} x = (a-b)\cos\phi + b\cos\left(\dfrac{a-b}{b}\right)\phi \\ y = (a-b)\sin\phi - b\sin\left(\dfrac{a-b}{b}\right)\phi \end{cases}$$

This is the curve described by a point P on a circle of radius b as it rolls on the inside of a circle of radius a.

If $b = a/4$, the curve is that of Fig. 11-3.

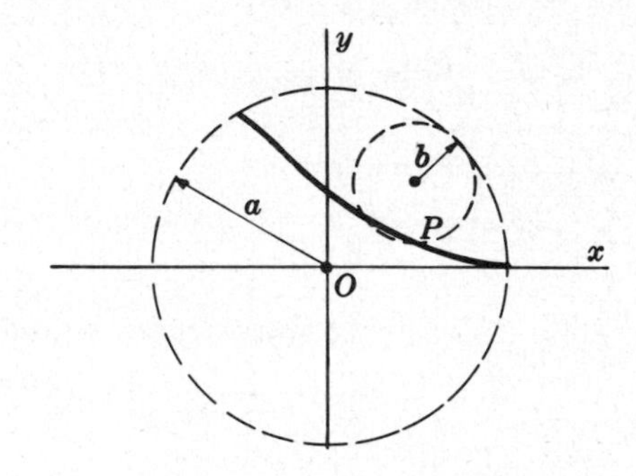

Fig. 11-9

TROCHOID

11.20 Parametric equations: $\begin{cases} x = a\phi - b\sin\phi \\ y = a - b\cos\phi \end{cases}$

This is the curve described by a point P at distance b from the center of a circle of radius a as the circle rolls on the x axis.

If $b < a$, the curve is as shown in Fig. 11-10 and is called a *curtate cycloid.*

If $b > a$, the curve is as shown in Fig. 11-11 and is called a *prolate cycloid.*

If $b = a$, the curve is the cycloid of Fig. 11-2.

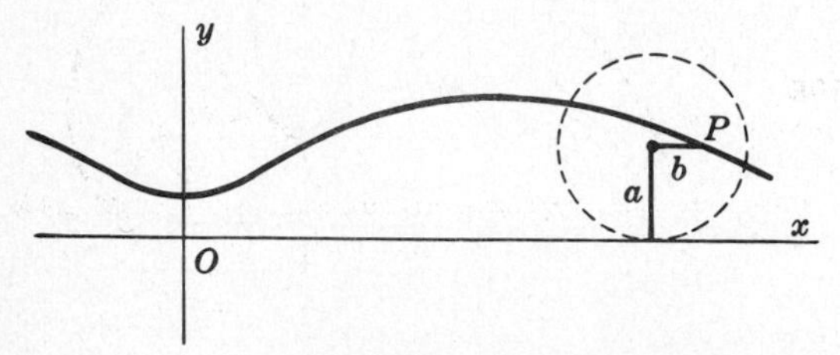

Fig. 11-10

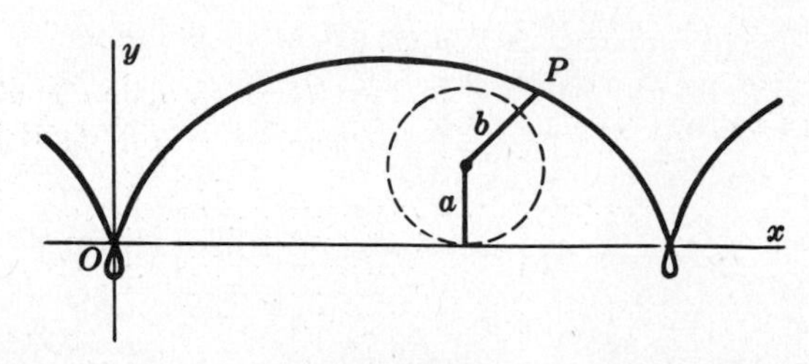

Fig. 11-11

TRACTRIX

11.21 Parametric equations: $\begin{cases} x = a(\ln \cot \frac{1}{2}\phi - \cos \phi) \\ y = a \sin \phi \end{cases}$

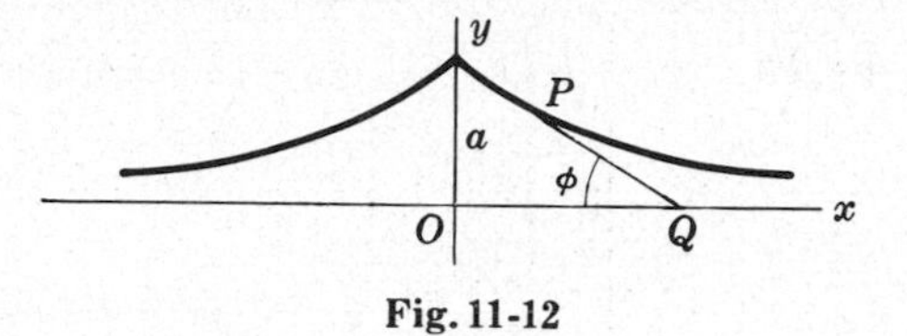

Fig. 11-12

This is the curve described by endpoint P of a taut string PQ of length a as the other end Q is moved along the x axis.

WITCH OF AGNESI

11.22 Equation in rectangular coordinates: $y = \dfrac{8a^3}{x^2 + 4a^2}$

11.23 Parametric equations: $\begin{cases} x = 2a \cot \theta \\ y = a(1 - \cos 2\theta) \end{cases}$

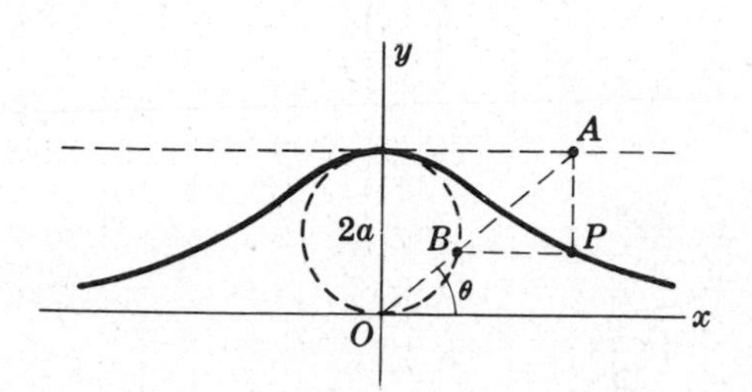

Fig. 11-13

In Fig. 11-13 the variable line OA intersects $y = 2a$ and the circle of radius a with center $(0, a)$ at A and B respectively. Any point P on the "witch" is located by constructing lines parallel to the x and y axes through B and A respectively and determining the point P of intersection.

FOLIUM OF DESCARTES

11.24 Equation in rectangular coordinates:

$$x^3 + y^3 = 3axy$$

11.25 Parametric equations:

$$\begin{cases} x = \dfrac{3at}{1 + t^3} \\ y = \dfrac{3at^2}{1 + t^3} \end{cases}$$

11.26 Area of loop $= \dfrac{3}{2}a^2$

11.27 Equation of asymptote: $x + y + a = 0$

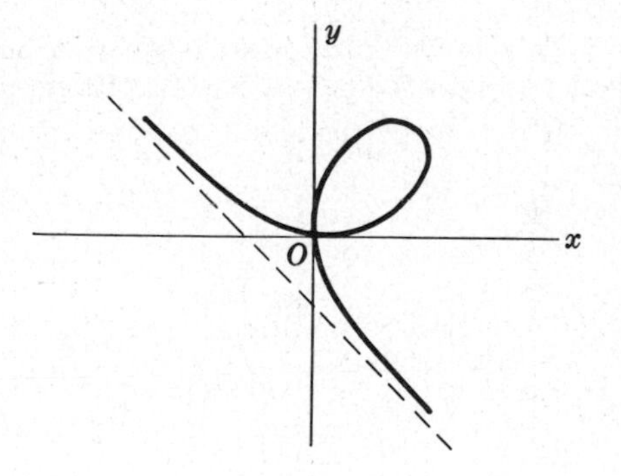

Fig. 11-14

INVOLUTE OF A CIRCLE

11.28 Parametric equations:

$$\begin{cases} x = a(\cos \phi + \phi \sin \phi) \\ y = a(\sin \phi - \phi \cos \phi) \end{cases}$$

This is the curve described by the endpoint P of a string as it unwinds from a circle of radius a while held taut.

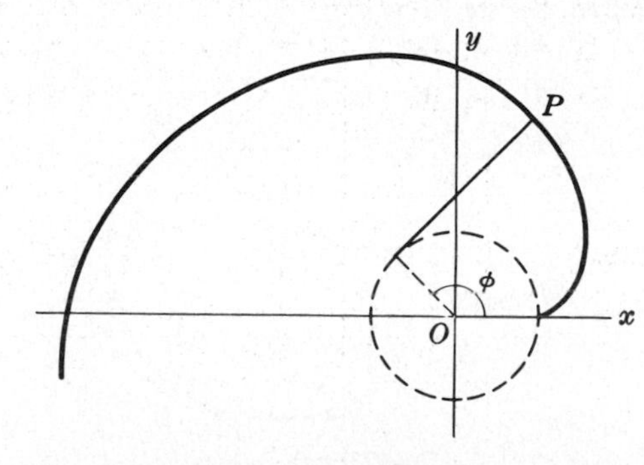

Fig. 11-15

EVOLUTE OF AN ELLIPSE

11.29 Equation in rectangular coordinates:

$$(ax)^{2/3} + (by)^{2/3} = (a^2 - b^2)^{2/3}$$

11.30 Parametric equations:

$$\begin{cases} ax = (a^2 - b^2)\cos^3\theta \\ by = (a^2 - b^2)\sin^3\theta \end{cases}$$

This curve is the envelope of the normals to the ellipse $x^2/a^2 + y^2/b^2 = 1$ shown dashed in Fig. 11-16.

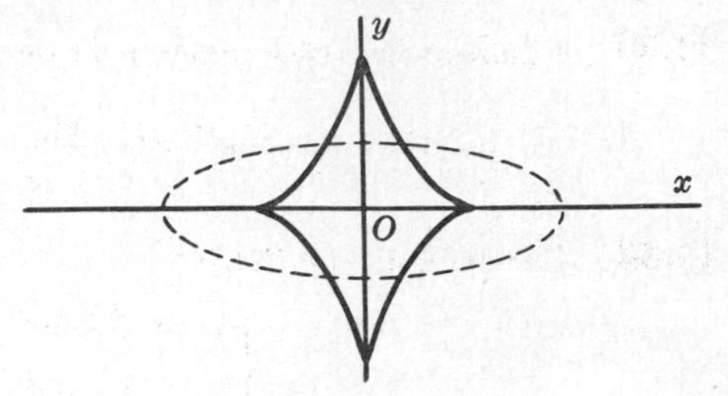

Fig. 11-16

OVALS OF CASSINI

11.31 Polar equation: $r^4 + a^4 - 2a^2r^2\cos 2\theta = b^4$

This is the curve described by a point P such that the product of its distances from two fixed points [distance $2a$ apart] is a constant b^2.

The curve is as in Fig. 11-17 or Fig. 11-18 according as $b < a$ or $b > a$ respectively.

If $b = a$, the curve is a *lemniscate* [Fig. 11-1].

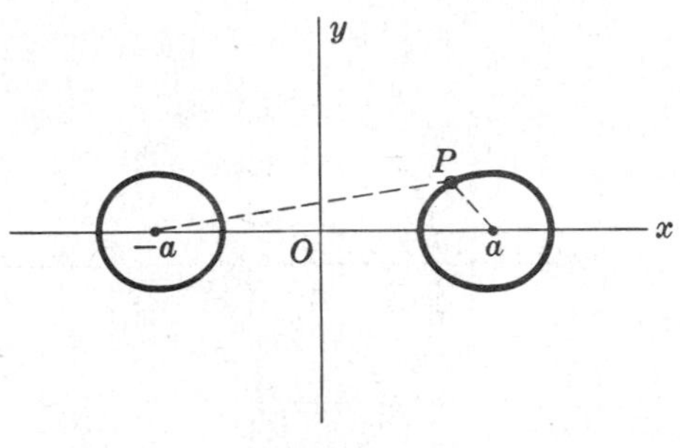

Fig. 11-17

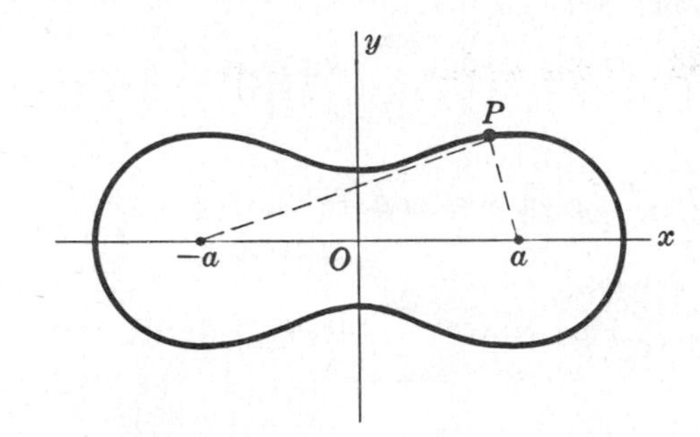

Fig. 11-18

LIMACON OF PASCAL

11.32 Polar equation: $r = b + a\cos\theta$

Let OQ be a line joining origin O to any point Q on a circle of diameter a passing through O. Then the curve is the locus of all points P such that $PQ = b$.

The curve is as in Fig. 11-19 or Fig. 11-20 according as $b > a$ or $b < a$ respectively. If $b = a$, the curve is a *cardioid* [Fig. 11-4].

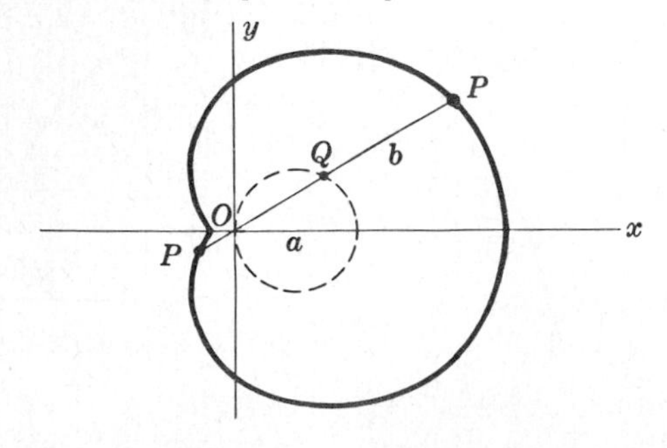

Fig. 11-19

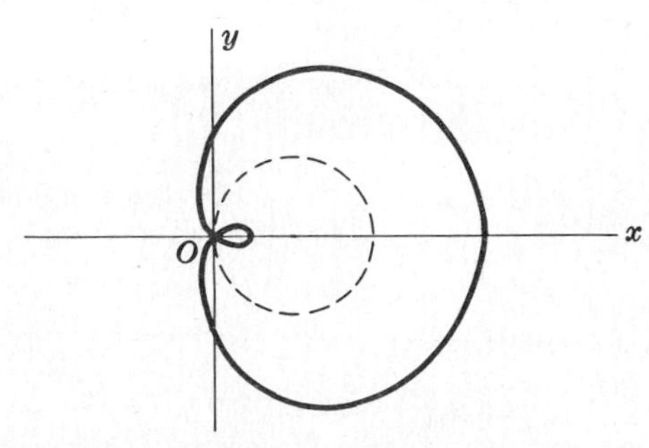

Fig. 11-20

CISSOID OF DIOCLES

11.33 Equation in rectangular coordinates:

$$y^2 = \frac{x^3}{2a - x}$$

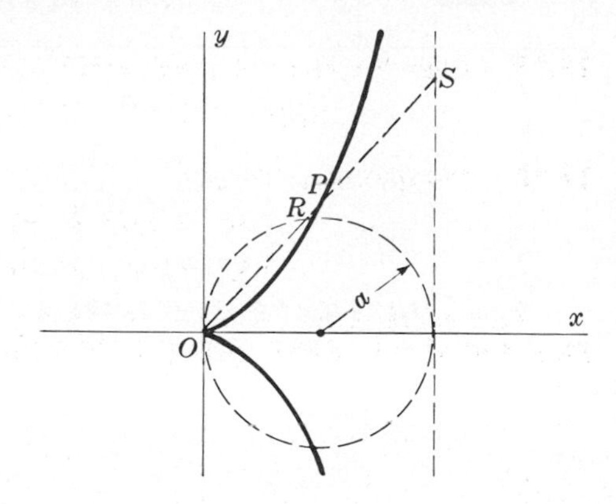

Fig. 11-21

11.34 Parametric equations:

$$\begin{cases} x = 2a \sin^2 \theta \\ y = \dfrac{2a \sin^3 \theta}{\cos \theta} \end{cases}$$

This is the curve described by a point P such that the distance OP = distance RS. It is used in the problem of *duplication of a cube*, i.e. finding the side of a cube which has twice the volume of a given cube.

SPIRAL OF ARCHIMEDES

11.35 Polar equation: $r = a\theta$

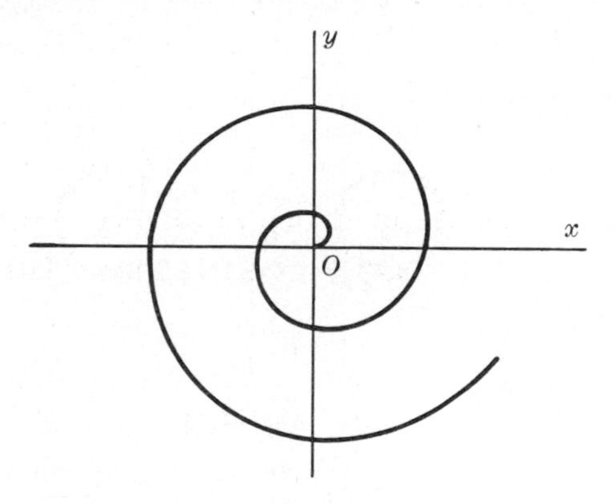

Fig. 11-22

12 FORMULAS from SOLID ANALYTIC GEOMETRY

DISTANCE d BETWEEN TWO POINTS $P_1(x_1, y_1, z_1)$ AND $P_2(x_2, y_2, z_2)$

12.1

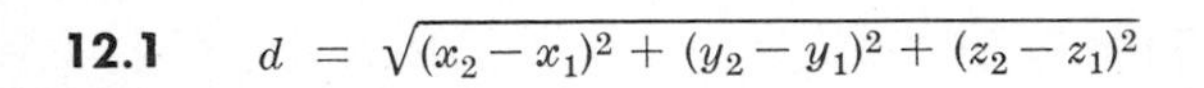

$$d = \sqrt{(x_2 - x_1)^2 + (y_2 - y_1)^2 + (z_2 - z_1)^2}$$

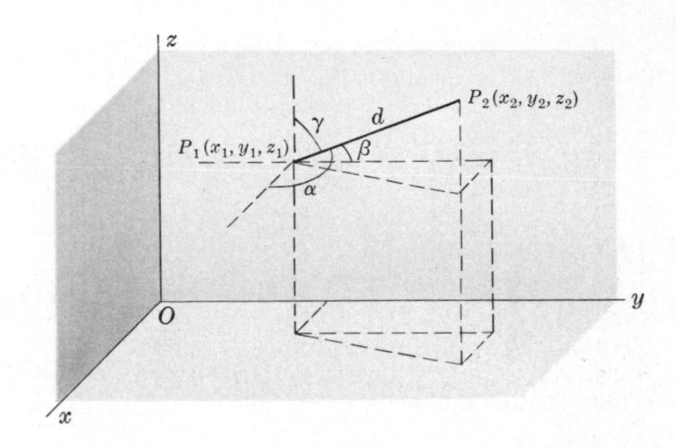

Fig. 12-1

DIRECTION COSINES OF LINE JOINING POINTS $P_1(x_1, y_1, z_1)$ AND $P_2(x_2, y_2, z_2)$

12.2

$$l = \cos\alpha = \frac{x_2 - x_1}{d}, \quad m = \cos\beta = \frac{y_2 - y_1}{d}, \quad n = \cos\gamma = \frac{z_2 - z_1}{d}$$

where α, β, γ are the angles which line P_1P_2 makes with the positive x, y, z axes respectively and d is given by 12.1 [see Fig. 12-1].

RELATIONSHIP BETWEEN DIRECTION COSINES

12.3

$$\cos^2\alpha + \cos^2\beta + \cos^2\gamma = 1 \quad \text{or} \quad l^2 + m^2 + n^2 = 1$$

DIRECTION NUMBERS

Numbers L, M, N which are proportional to the direction cosines l, m, n are called *direction numbers*. The relationship between them is given by

12.4

$$l = \frac{L}{\sqrt{L^2 + M^2 + N^2}}, \quad m = \frac{M}{\sqrt{L^2 + M^2 + N^2}}, \quad n = \frac{N}{\sqrt{L^2 + M^2 + N^2}}$$

EQUATIONS OF LINE JOINING $P_1(x_1, y_1, z_1)$ AND $P_2(x_2, y_2, z_2)$ IN STANDARD FORM

12.5
$$\frac{x - x_1}{x_2 - x_1} = \frac{y - y_1}{y_2 - y_1} = \frac{z - z_1}{z_2 - z_1} \quad \text{or} \quad \frac{x - x_1}{l} = \frac{y - y_1}{m} = \frac{z - z_1}{n}$$

These are also valid if l, m, n are replaced by L, M, N respectively.

EQUATIONS OF LINE JOINING $P_1(x_1, y_1, z_1)$ AND $P_2(x_2, y_2, z_2)$ IN PARAMETRIC FORM

12.6
$$x = x_1 + lt, \quad y = y_1 + mt, \quad z = z_1 + nt$$

These are also valid if l, m, n are replaced by L, M, N respectively.

ANGLE ϕ BETWEEN TWO LINES WITH DIRECTION COSINES l_1, m_1, n_1 AND l_2, m_2, n_2

12.7
$$\cos\phi = l_1 l_2 + m_1 m_2 + n_1 n_2$$

GENERAL EQUATION OF A PLANE

12.8
$$Ax + By + Cz + D = 0 \qquad [A, B, C, D \text{ are constants}]$$

EQUATION OF PLANE PASSING THROUGH POINTS (x_1, y_1, z_1), (x_2, y_2, z_2), (x_3, y_3, z_3)

12.9
$$\begin{vmatrix} x - x_1 & y - y_1 & z - z_1 \\ x_2 - x_1 & y_2 - y_1 & z_2 - z_1 \\ x_3 - x_1 & y_3 - y_1 & z_3 - z_1 \end{vmatrix} = 0$$

or

12.10
$$\begin{vmatrix} y_2 - y_1 & z_2 - z_1 \\ y_3 - y_1 & z_3 - z_1 \end{vmatrix}(x - x_1) + \begin{vmatrix} z_2 - z_1 & x_2 - x_1 \\ z_3 - z_1 & x_3 - x_1 \end{vmatrix}(y - y_1) + \begin{vmatrix} x_2 - x_1 & y_2 - y_1 \\ x_3 - x_1 & y_3 - y_1 \end{vmatrix}(z - z_1) = 0$$

EQUATION OF PLANE IN INTERCEPT FORM

12.11
$$\frac{x}{a} + \frac{y}{b} + \frac{z}{c} = 1$$

where a, b, c are the intercepts on the x, y, z axes respectively.

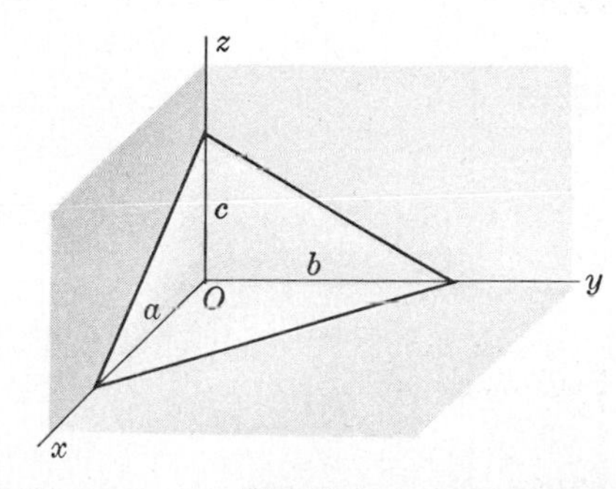

Fig. 12-2

EQUATIONS OF LINE THROUGH (x_0, y_0, z_0) AND PERPENDICULAR TO PLANE $Ax + By + Cz + D = 0$

12.12
$$\frac{x - x_0}{A} = \frac{y - y_0}{B} = \frac{z - z_0}{C} \quad \text{or} \quad x = x_0 + At, \quad y = y_0 + Bt, \quad z = z_0 + Ct$$

Note that the direction numbers for a line perpendicular to the plane $Ax + By + Cz + D = 0$ are A, B, C.

DISTANCE FROM POINT (x_0, y_0, z_0) TO PLANE $Ax + By + Cz + D = 0$

12.13
$$\frac{Ax_0 + By_0 + Cz_0 + D}{\pm\sqrt{A^2 + B^2 + C^2}}$$

where the sign is chosen so that the distance is nonnegative.

NORMAL FORM FOR EQUATION OF PLANE

12.14
$$x \cos\alpha + y \cos\beta + z \cos\gamma = p$$

where p = perpendicular distance from O to plane at P and α, β, γ are angles between OP and positive x, y, z axes.

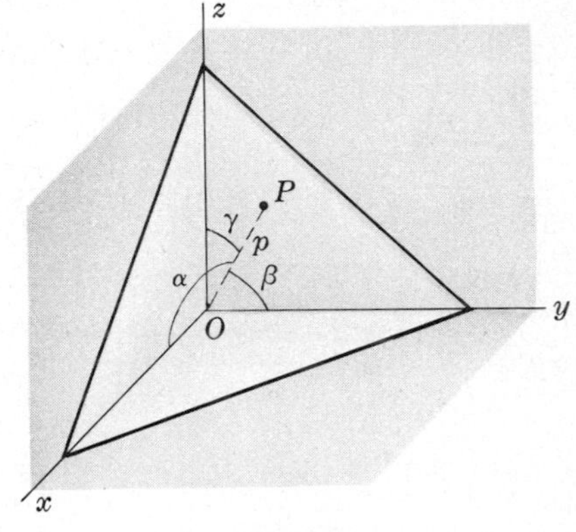

Fig. 12-3

TRANSFORMATION OF COORDINATES INVOLVING PURE TRANSLATION

12.15
$$\begin{cases} x = x' + x_0 \\ y = y' + y_0 \\ z = z' + z_0 \end{cases} \quad \text{or} \quad \begin{cases} x' = x - x_0 \\ y' = y - y_0 \\ z' = z - z_0 \end{cases}$$

where (x, y, z) are old coordinates [i.e. coordinates relative to xyz system], (x', y', z') are new coordinates [relative to $x'y'z'$ system] and (x_0, y_0, z_0) are the coordinates of the new origin O' relative to the old xyz coordinate system.

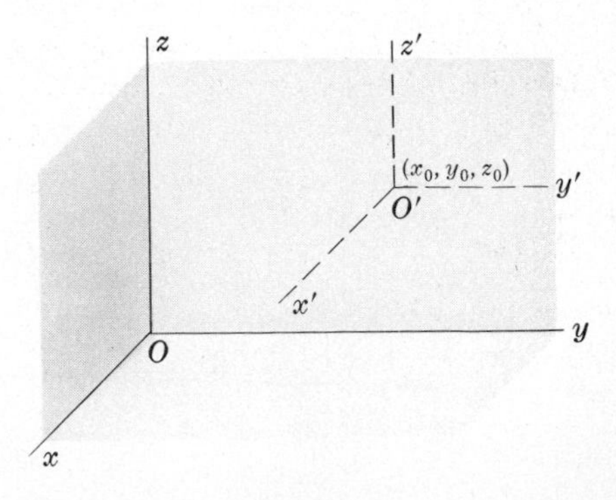

Fig. 12-4

TRANSFORMATION OF COORDINATES INVOLVING PURE ROTATION

12.16
$$\begin{cases} x = l_1x' + l_2y' + l_3z' \\ y = m_1x' + m_2y' + m_3z' \\ z = n_1x' + n_2y' + n_3z' \end{cases}$$

or
$$\begin{cases} x' = l_1x + m_1y + n_1z \\ y' = l_2x + m_2y + n_2z \\ z' = l_3x + m_3y + n_3z \end{cases}$$

where the origins of the xyz and $x'y'z'$ systems are the same and l_1, m_1, n_1; l_2, m_2, n_2; l_3, m_3, n_3 are the direction cosines of the x', y', z' axes relative to the x, y, z axes respectively.

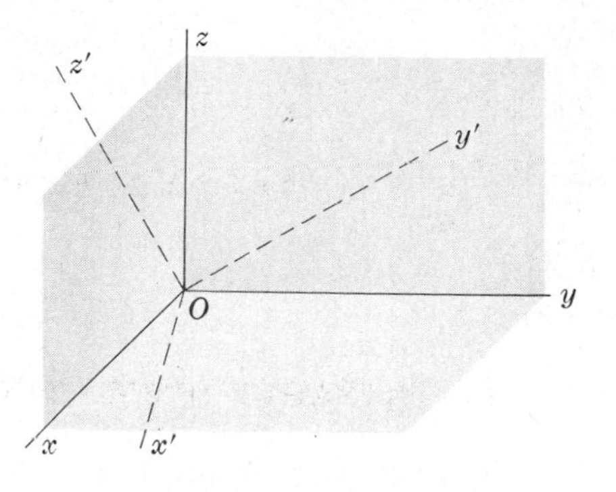

Fig. 12-5

TRANSFORMATION OF COORDINATES INVOLVING TRANSLATION AND ROTATION

12.17
$$\begin{cases} x = l_1x' + l_2y' + l_3z' + x_0 \\ y = m_1x' + m_2y' + m_3z' + y_0 \\ z = n_1x' + n_2y' + n_3z' + z_0 \end{cases}$$

or
$$\begin{cases} x' = l_1(x-x_0) + m_1(y-y_0) + n_1(z-z_0) \\ y' = l_2(x-x_0) + m_2(y-y_0) + n_2(z-z_0) \\ z' = l_3(x-x_0) + m_3(y-y_0) + n_3(z-z_0) \end{cases}$$

where the origin O' of the $x'y'z'$ system has coordinates (x_0, y_0, z_0) relative to the xyz system and l_1, m_1, n_1; l_2, m_2, n_2; l_3, m_3, n_3 are the direction cosines of the x', y', z' axes relative to the x, y, z axes respectively.

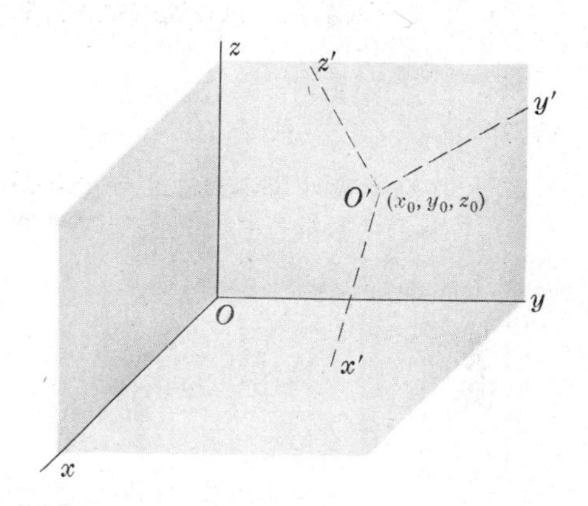

Fig. 12-6

CYLINDRICAL COORDINATES (r, θ, z)

A point P can be located by cylindrical coordinates (r, θ, z) [see Fig. 12-7] as well as rectangular coordinates (x, y, z).

The transformation between these coordinates is

12.18
$$\begin{cases} x = r\cos\theta \\ y = r\sin\theta \\ z = z \end{cases} \quad \text{or} \quad \begin{cases} r = \sqrt{x^2+y^2} \\ \theta = \tan^{-1}(y/x) \\ z = z \end{cases}$$

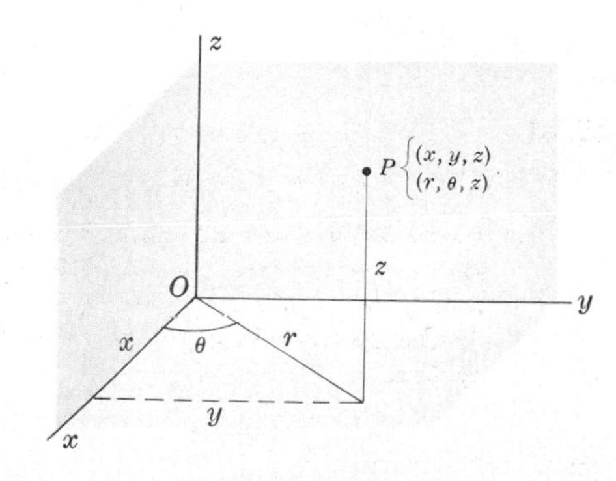

Fig. 12-7

SPHERICAL COORDINATES (r, θ, ϕ)

A point P can be located by spherical coordinates (r, θ, ϕ) [see Fig. 12-8] as well as rectangular coordinates (x, y, z).

The transformation between those coordinates is

12.19
$$\begin{cases} x = r \sin\theta \cos\phi \\ y = r \sin\theta \sin\phi \\ z = r \cos\theta \end{cases}$$

or
$$\begin{cases} r = \sqrt{x^2 + y^2 + z^2} \\ \phi = \tan^{-1}(y/x) \\ \theta = \cos^{-1}(z/\sqrt{x^2 + y^2 + z^2}) \end{cases}$$

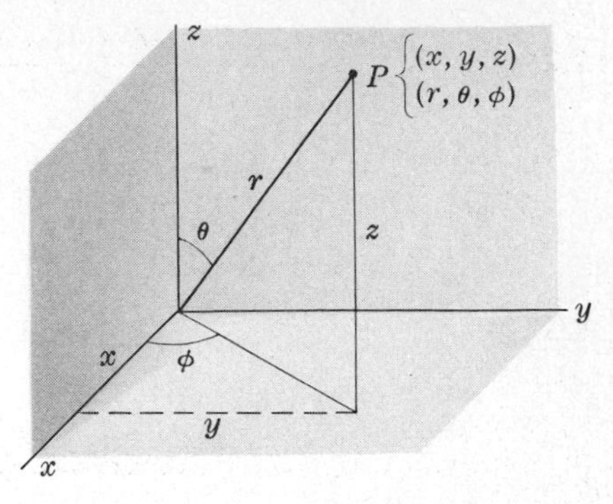

Fig. 12-8

EQUATION OF SPHERE IN RECTANGULAR COORDINATES

12.20
$$(x - x_0)^2 + (y - y_0)^2 + (z - z_0)^2 = R^2$$

where the sphere has center (x_0, y_0, z_0) and radius R.

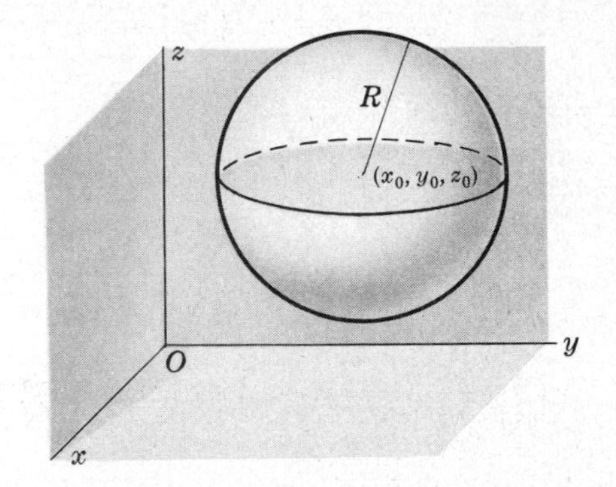

Fig. 12-9

EQUATION OF SPHERE IN CYLINDRICAL COORDINATES

12.21
$$r^2 - 2r_0 r \cos(\theta - \theta_0) + r_0^2 + (z - z_0)^2 = R^2$$

where the sphere has center (r_0, θ_0, z_0) in cylindrical coordinates and radius R.

If the center is at the origin the equation is

12.22
$$r^2 + z^2 = R^2$$

EQUATION OF SPHERE IN SPHERICAL COORDINATES

12.23
$$r^2 + r_0^2 - 2r_0 r \sin\theta \sin\theta_0 \cos(\phi - \phi_0) = R^2$$

where the sphere has center (r_0, θ_0, ϕ_0) in spherical coordinates and radius R.

If the center is at the origin the equation is

12.24
$$r = R$$

EQUATION OF ELLIPSOID WITH CENTER (x_0, y_0, z_0) AND SEMI-AXES a, b, c

12.25 $$\frac{(x-x_0)^2}{a^2} + \frac{(y-y_0)^2}{b^2} + \frac{(z-z_0)^2}{c^2} = 1$$

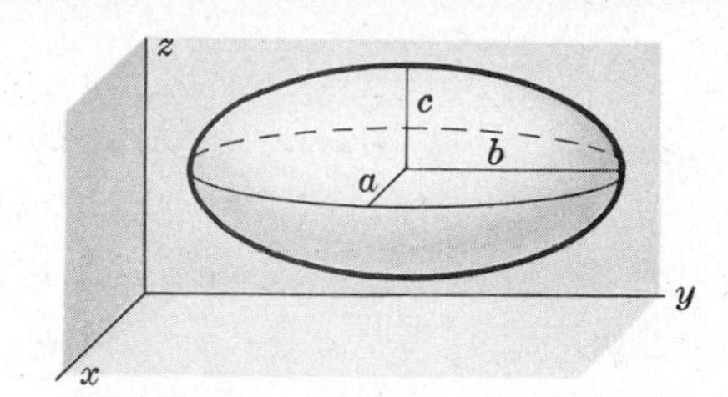

Fig. 12-10

ELLIPTIC CYLINDER WITH AXIS AS z AXIS

12.26 $$\frac{x^2}{a^2} + \frac{y^2}{b^2} = 1$$

where a, b are semi-axes of elliptic cross section.

If $b = a$ it becomes a circular cylinder of radius a.

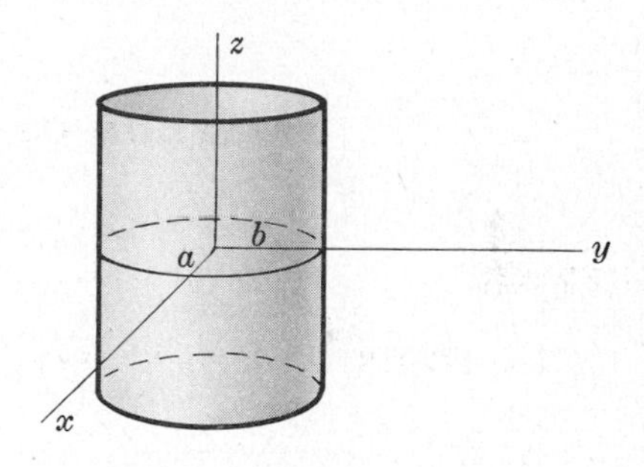

Fig. 12-11

ELLIPTIC CONE WITH AXIS AS z AXIS

12.27 $$\frac{x^2}{a^2} + \frac{y^2}{b^2} = \frac{z^2}{c^2}$$

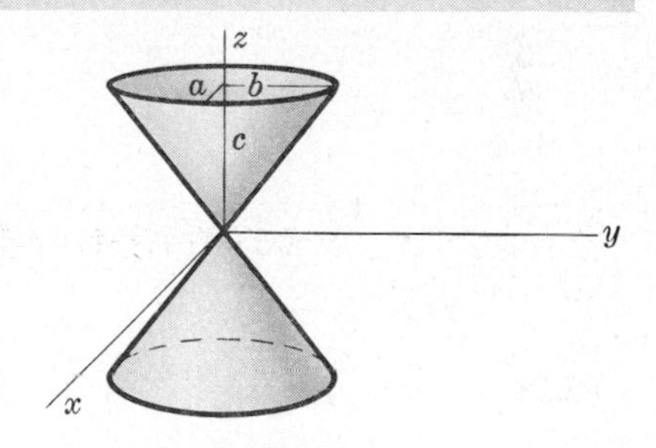

Fig. 12-12

HYPERBOLOID OF ONE SHEET

12.28 $$\frac{x^2}{a^2} + \frac{y^2}{b^2} - \frac{z^2}{c^2} = 1$$

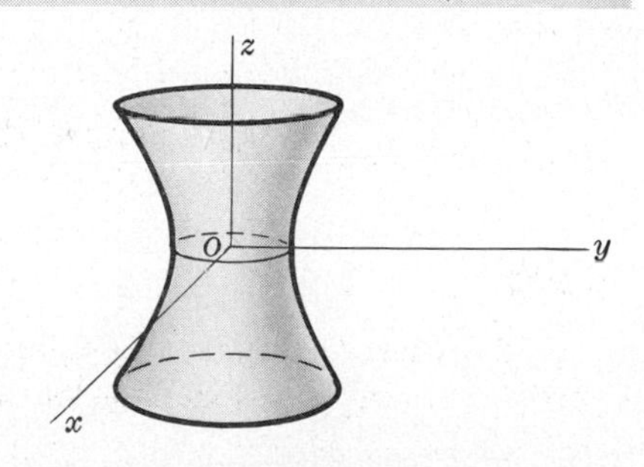

Fig. 12-13

HYPERBOLOID OF TWO SHEETS

12.29 $$\frac{x^2}{a^2} - \frac{y^2}{b^2} - \frac{z^2}{c^2} = 1$$

Note orientation of axes in Fig. 12-14.

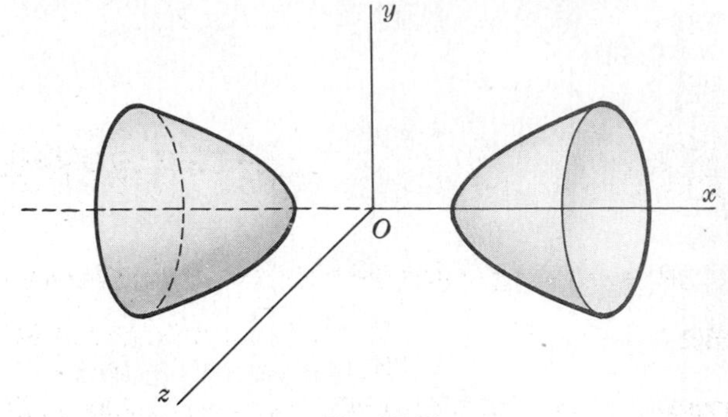

Fig. 12-14

ELLIPTIC PARABOLOID

12.30 $$\frac{x^2}{a^2} + \frac{y^2}{b^2} = \frac{z}{c}$$

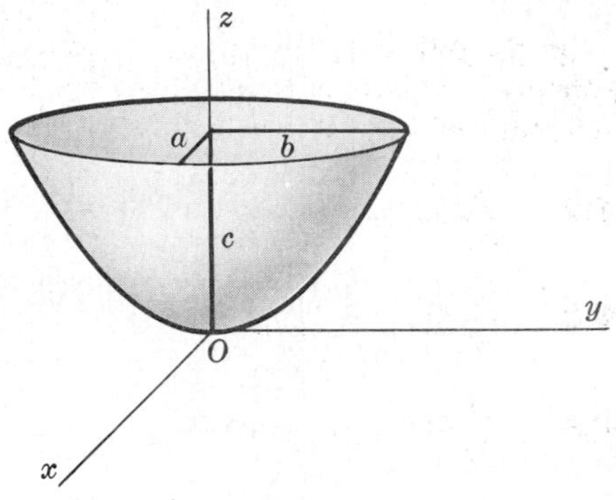

Fig. 12-15

HYPERBOLIC PARABOLOID

12.31 $$\frac{x^2}{a^2} - \frac{y^2}{b^2} = \frac{z}{c}$$

Note orientation of axes in Fig. 12-16.

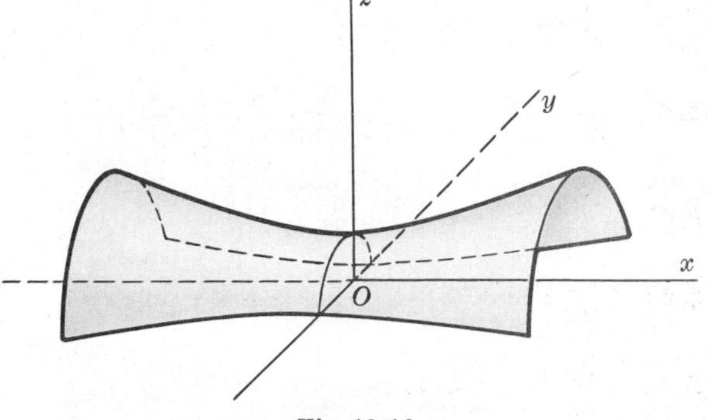

Fig. 12-16

13 DERIVATIVES

DEFINITION OF A DERIVATIVE

If $y = f(x)$, the derivative of y or $f(x)$ with respect to x is defined as

13.1 $$\frac{dy}{dx} = \lim_{h \to 0} \frac{f(x+h) - f(x)}{h} = \lim_{\Delta x \to 0} \frac{f(x+\Delta x) - f(x)}{\Delta x}$$

where $h = \Delta x$. The derivative is also denoted by y', df/dx or $f'(x)$. The process of taking a derivative is called *differentiation*.

GENERAL RULES OF DIFFERENTIATION

In the following, u, v, w are functions of x; a, b, c, n are constants [restricted if indicated]; $e = 2.71828\ldots$ is the natural base of logarithms; $\ln u$ is the natural logarithm of u [i.e. the logarithm to the base e] where it is assumed that $u > 0$ and all angles are in radians.

13.2 $$\frac{d}{dx}(c) = 0$$

13.3 $$\frac{d}{dx}(cx) = c$$

13.4 $$\frac{d}{dx}(cx^n) = ncx^{n-1}$$

13.5 $$\frac{d}{dx}(u \pm v \pm w \pm \cdots) = \frac{du}{dx} \pm \frac{dv}{dx} \pm \frac{dw}{dx} \pm \cdots$$

13.6 $$\frac{d}{dx}(cu) = c\frac{du}{dx}$$

13.7 $$\frac{d}{dx}(uv) = u\frac{dv}{dx} + v\frac{du}{dx}$$

13.8 $$\frac{d}{dx}(uvw) = uv\frac{dw}{dx} + uw\frac{dv}{dx} + vw\frac{du}{dx}$$

13.9 $$\frac{d}{dx}\left(\frac{u}{v}\right) = \frac{v(du/dx) - u(dv/dx)}{v^2}$$

13.10 $$\frac{d}{dx}(u^n) = nu^{n-1}\frac{du}{dx}$$

13.11 $$\frac{dy}{dx} = \frac{dy}{du}\frac{du}{dx} \quad \text{(Chain rule)}$$

13.12 $$\frac{du}{dx} = \frac{1}{dx/du}$$

13.13 $$\frac{dy}{dx} = \frac{dy/du}{dx/du}$$

DERIVATIVES OF TRIGONOMETRIC AND INVERSE TRIGONOMETRIC FUNCTIONS

13.14 $\dfrac{d}{dx}\sin u = \cos u\dfrac{du}{dx}$

13.15 $\dfrac{d}{dx}\cos u = -\sin u\dfrac{du}{dx}$

13.16 $\dfrac{d}{dx}\tan u = \sec^2 u\dfrac{du}{dx}$

13.17 $\dfrac{d}{dx}\cot u = -\csc^2 u\dfrac{du}{dx}$

13.18 $\dfrac{d}{dx}\sec u = \sec u \tan u\dfrac{du}{dx}$

13.19 $\dfrac{d}{dx}\csc u = -\csc u \cot u\dfrac{du}{dx}$

13.20 $\dfrac{d}{dx}\sin^{-1} u = \dfrac{1}{\sqrt{1-u^2}}\dfrac{du}{dx} \qquad \left[-\dfrac{\pi}{2} < \sin^{-1} u < \dfrac{\pi}{2}\right]$

13.21 $\dfrac{d}{dx}\cos^{-1} u = \dfrac{-1}{\sqrt{1-u^2}}\dfrac{du}{dx} \qquad [0 < \cos^{-1} u < \pi]$

13.22 $\dfrac{d}{dx}\tan^{-1} u = \dfrac{1}{1+u^2}\dfrac{du}{dx} \qquad \left[-\dfrac{\pi}{2} < \tan^{-1} u < \dfrac{\pi}{2}\right]$

13.23 $\dfrac{d}{dx}\cot^{-1} u = \dfrac{-1}{1+u^2}\dfrac{du}{dx} \qquad [0 < \cot^{-1} u < \pi]$

13.24 $\dfrac{d}{dx}\sec^{-1} u = \dfrac{1}{|u|\sqrt{u^2-1}}\dfrac{du}{dx} = \dfrac{\pm 1}{u\sqrt{u^2-1}}\dfrac{du}{dx} \qquad \begin{bmatrix} + \text{ if } 0 < \sec^{-1} u < \pi/2 \\ - \text{ if } \pi/2 < \sec^{-1} u < \pi \end{bmatrix}$

13.25 $\dfrac{d}{dx}\csc^{-1} u = \dfrac{-1}{|u|\sqrt{u^2-1}}\dfrac{du}{dx} = \dfrac{\mp 1}{u\sqrt{u^2-1}}\dfrac{du}{dx} \qquad \begin{bmatrix} - \text{ if } 0 < \csc^{-1} u < \pi/2 \\ + \text{ if } -\pi/2 < \csc^{-1} u < 0 \end{bmatrix}$

DERIVATIVES OF EXPONENTIAL AND LOGARITHMIC FUNCTIONS

13.26 $\dfrac{d}{dx}\log_a u = \dfrac{\log_a e}{u}\dfrac{du}{dx} \qquad a \neq 0, 1$

13.27 $\dfrac{d}{dx}\ln u = \dfrac{d}{dx}\log_e u = \dfrac{1}{u}\dfrac{du}{dx}$

13.28 $\dfrac{d}{dx}a^u = a^u \ln a\dfrac{du}{dx}$

13.29 $\dfrac{d}{dx}e^u = e^u\dfrac{du}{dx}$

13.30 $\dfrac{d}{dx}u^v = \dfrac{d}{dx}e^{v \ln u} = e^{v \ln u}\dfrac{d}{dx}[v \ln u] = vu^{v-1}\dfrac{du}{dx} + u^v \ln u\dfrac{dv}{dx}$

DERIVATIVES OF HYPERBOLIC AND INVERSE HYPERBOLIC FUNCTIONS

13.31 $\dfrac{d}{dx}\sinh u = \cosh u\dfrac{du}{dx}$

13.32 $\dfrac{d}{dx}\cosh u = \sinh u\dfrac{du}{dx}$

13.33 $\dfrac{d}{dx}\tanh u = \operatorname{sech}^2 u\dfrac{du}{dx}$

13.34 $\dfrac{d}{dx}\coth u = -\operatorname{csch}^2 u\dfrac{du}{dx}$

13.35 $\dfrac{d}{dx}\operatorname{sech} u = -\operatorname{sech} u \tanh u\dfrac{du}{dx}$

13.36 $\dfrac{d}{dx}\operatorname{csch} u = -\operatorname{csch} u \coth u\dfrac{du}{dx}$

13.37 $$\frac{d}{dx}\sinh^{-1}u = \frac{1}{\sqrt{u^2+1}}\frac{du}{dx}$$

13.38 $$\frac{d}{dx}\cosh^{-1}u = \frac{\pm 1}{\sqrt{u^2-1}}\frac{du}{dx} \qquad \left[\begin{array}{l}+ \text{ if } \cosh^{-1}u > 0,\ u > 1\\ - \text{ if } \cosh^{-1}u < 0,\ u > 1\end{array}\right]$$

13.39 $$\frac{d}{dx}\tanh^{-1}u = \frac{1}{1-u^2}\frac{du}{dx} \qquad [-1 < u < 1]$$

13.40 $$\frac{d}{dx}\coth^{-1}u = \frac{1}{1-u^2}\frac{du}{dx} \qquad [u > 1 \text{ or } u < -1]$$

13.41 $$\frac{d}{dx}\operatorname{sech}^{-1}u = \frac{\mp 1}{u\sqrt{1-u^2}}\frac{du}{dx} \qquad \left[\begin{array}{l}- \text{ if } \operatorname{sech}^{-1}u > 0,\ 0 < u < 1\\ + \text{ if } \operatorname{sech}^{-1}u < 0,\ 0 < u < 1\end{array}\right]$$

13.42 $$\frac{d}{dx}\operatorname{csch}^{-1}u = \frac{-1}{|u|\sqrt{1+u^2}}\frac{du}{dx} = \frac{\mp 1}{u\sqrt{1+u^2}}\frac{du}{dx} \qquad [- \text{ if } u > 0,\ + \text{ if } u < 0]$$

HIGHER DERIVATIVES

The second, third and higher derivatives are defined as follows.

13.43 Second derivative $$= \frac{d}{dx}\left(\frac{dy}{dx}\right) = \frac{d^2y}{dx^2} = f''(x) = y''$$

13.44 Third derivative $$= \frac{d}{dx}\left(\frac{d^2y}{dx^2}\right) = \frac{d^3y}{dx^3} = f'''(x) = y'''$$

13.45 nth derivative $$= \frac{d}{dx}\left(\frac{d^{n-1}y}{dx^{n-1}}\right) = \frac{d^ny}{dx^n} = f^{(n)}(x) = y^{(n)}$$

LEIBNITZ'S RULE FOR HIGHER DERIVATIVES OF PRODUCTS

Let D^p stand for the operator $\frac{d^p}{dx^p}$ so that $D^pu = \frac{d^pu}{dx^p}$ = the pth derivative of u. Then

13.46 $$D^n(uv) = uD^nv + \binom{n}{1}(Du)(D^{n-1}v) + \binom{n}{2}(D^2u)(D^{n-2}v) + \cdots + vD^nu$$

where $\binom{n}{1}, \binom{n}{2}, \ldots$ are the binomial coefficients [page 3].

As special cases we have

13.47 $$\frac{d^2}{dx^2}(uv) = u\frac{d^2v}{dx^2} + 2\frac{du}{dx}\frac{dv}{dx} + v\frac{d^2u}{dx^2}$$

13.48 $$\frac{d^3}{dx^3}(uv) = u\frac{d^3v}{dx^3} + 3\frac{du}{dx}\frac{d^2v}{dx^2} + 3\frac{d^2u}{dx^2}\frac{dv}{dx} + v\frac{d^3u}{dx^3}$$

DIFFERENTIALS

Let $y = f(x)$ and $\Delta y = f(x+\Delta x) - f(x)$. Then

13.49 $$\frac{\Delta y}{\Delta x} = \frac{f(x+\Delta x) - f(x)}{\Delta x} = f'(x) + \epsilon = \frac{dy}{dx} + \epsilon$$

where $\epsilon \to 0$ as $\Delta x \to 0$. Thus

13.50 $$\Delta y = f'(x)\,\Delta x + \epsilon\,\Delta x$$

If we call $\Delta x = dx$ the differential of x, then we define the differential of y to be

13.51 $$dy = f'(x)\,dx$$

RULES FOR DIFFERENTIALS

The rules for differentials are exactly analogous to those for derivatives. As examples we observe that

13.52 $$d(u \pm v \pm w \pm \cdots) = du \pm dv \pm dw \pm \cdots$$

13.53 $$d(uv) = u\,dv + v\,du$$

13.54 $$d\left(\frac{u}{v}\right) = \frac{v\,du - u\,dv}{v^2}$$

13.55 $$d(u^n) = nu^{n-1}\,du$$

13.56 $$d(\sin u) = \cos u\,du$$

13.57 $$d(\cos u) = -\sin u\,du$$

PARTIAL DERIVATIVES

Let $f(x, y)$ be a function of the two variables x and y. Then we define the partial derivative of $f(x, y)$ with respect to x, keeping y constant, to be

13.58 $$\frac{\partial f}{\partial x} = \lim_{\Delta x \to 0} \frac{f(x + \Delta x, y) - f(x, y)}{\Delta x}$$

Similarly the partial derivative of $f(x, y)$ with respect to y, keeping x constant, is defined to be

13.59 $$\frac{\partial f}{\partial y} = \lim_{\Delta y \to 0} \frac{f(x, y + \Delta y) - f(x, y)}{\Delta y}$$

Partial derivatives of higher order can be defined as follows.

13.60 $$\frac{\partial^2 f}{\partial x^2} = \frac{\partial}{\partial x}\left(\frac{\partial f}{\partial x}\right), \qquad \frac{\partial^2 f}{\partial y^2} = \frac{\partial}{\partial y}\left(\frac{\partial f}{\partial y}\right)$$

13.61 $$\frac{\partial^2 f}{\partial x\,\partial y} = \frac{\partial}{\partial x}\left(\frac{\partial f}{\partial y}\right), \qquad \frac{\partial^2 f}{\partial y\,\partial x} = \frac{\partial}{\partial y}\left(\frac{\partial f}{\partial x}\right)$$

The results in 13.61 will be equal if the function and its partial derivatives are continuous, i.e. in such case the order of differentiation makes no difference.

The differential of $f(x, y)$ is defined as

13.62 $$df = \frac{\partial f}{\partial x}dx + \frac{\partial f}{\partial y}dy$$

where $dx = \Delta x$ and $dy = \Delta y$.

Extension to functions of more than two variables are exactly analogous.

14 INDEFINITE INTEGRALS

DEFINITION OF AN INDEFINITE INTEGRAL

If $\dfrac{dy}{dx} = f(x)$, then y is the function whose derivative is $f(x)$ and is called the *anti-derivative* of $f(x)$ or the *indefinite integral* of $f(x)$, denoted by $\int f(x)\,dx$. Similarly if $y = \int f(u)\,du$, then $\dfrac{dy}{du} = f(u)$. Since the derivative of a constant is zero, all indefinite integrals differ by an arbitrary constant.

For the definition of a definite integral, see page 94. The process of finding an integral is called *integration.*

GENERAL RULES OF INTEGRATION

In the following, u, v, w are functions of x; a, b, p, q, n any constants, restricted if indicated; $e = 2.71828\ldots$ is the natural base of logarithms; $\ln u$ denotes the natural logarithm of u where it is assumed that $u > 0$ [in general, to extend formulas to cases where $u < 0$ as well, replace $\ln u$ by $\ln |u|$]; all angles are in radians; all constants of integration are omitted but implied.

14.1 $$\int a\,dx = ax$$

14.2 $$\int af(x)\,dx = a\int f(x)\,dx$$

14.3 $$\int (u \pm v \pm w \pm \cdots)\,dx = \int u\,dx \pm \int v\,dx \pm \int w\,dx \pm \cdots$$

14.4 $$\int u\,dv = uv - \int v\,du \qquad \text{[Integration by parts]}$$

For generalized integration by parts, see 14.48.

14.5 $$\int f(ax)\,dx = \frac{1}{a}\int f(u)\,du$$

14.6 $$\int F\{f(x)\}\,dx = \int F(u)\frac{dx}{du}\,du = \int \frac{F(u)}{f'(x)}\,du \quad \text{where } u = f(x)$$

14.7 $$\int u^n\,du = \frac{u^{n+1}}{n+1}, \quad n \neq -1 \qquad \text{[For } n = -1\text{, see 14.8]}$$

14.8 $$\int \frac{du}{u} = \ln u \quad \text{if } u > 0 \text{ or } \ln(-u) \text{ if } u < 0$$
$$= \ln |u|$$

14.9 $$\int e^u\,du = e^u$$

14.10 $$\int a^u\,du = \int e^{u \ln a}\,du = \frac{e^{u \ln a}}{\ln a} = \frac{a^u}{\ln a}, \quad a > 0,\ a \neq 1$$

14.11 $\int \sin u\, du = -\cos u$

14.12 $\int \cos u\, du = \sin u$

14.13 $\int \tan u\, du = \ln \sec u = -\ln \cos u$

14.14 $\int \cot u\, du = \ln \sin u$

14.15 $\int \sec u\, du = \ln(\sec u + \tan u) = \ln \tan\left(\frac{u}{2} + \frac{\pi}{4}\right)$

14.16 $\int \csc u\, du = \ln(\csc u - \cot u) = \ln \tan \frac{u}{2}$

14.17 $\int \sec^2 u\, du = \tan u$

14.18 $\int \csc^2 u\, du = -\cot u$

14.19 $\int \tan^2 u\, du = \tan u - u$

14.20 $\int \cot^2 u\, du = -\cot u - u$

14.21 $\int \sin^2 u\, du = \frac{u}{2} - \frac{\sin 2u}{4} = \frac{1}{2}(u - \sin u \cos u)$

14.22 $\int \cos^2 u\, du = \frac{u}{2} + \frac{\sin 2u}{4} = \frac{1}{2}(u + \sin u \cos u)$

14.23 $\int \sec u \tan u\, du = \sec u$

14.24 $\int \csc u \cot u\, du = -\csc u$

14.25 $\int \sinh u\, du = \cosh u$

14.26 $\int \cosh u\, du = \sinh u$

14.27 $\int \tanh u\, du = \ln \cosh u$

14.28 $\int \coth u\, du = \ln \sinh u$

14.29 $\int \operatorname{sech} u\, du = \sin^{-1}(\tanh u)$ or $2 \tan^{-1} e^u$

14.30 $\int \operatorname{csch} u\, du = \ln \tanh \frac{u}{2}$ or $-\coth^{-1} e^u$

14.31 $\int \operatorname{sech}^2 u\, du = \tanh u$

14.32 $\int \operatorname{csch}^2 u\, du = -\coth u$

14.33 $\int \tanh^2 u\, du = u - \tanh u$

14.34 $\int \coth^2 u\,du = u - \coth u$

14.35 $\int \sinh^2 u\,du = \frac{\sinh 2u}{4} - \frac{u}{2} = \frac{1}{2}(\sinh u \cosh u - u)$

14.36 $\int \cosh^2 u\,du = \frac{\sinh 2u}{4} + \frac{u}{2} = \frac{1}{2}(\sinh u \cosh u + u)$

14.37 $\int \operatorname{sech} u \tanh u\,du = -\operatorname{sech} u$

14.38 $\int \operatorname{csch} u \coth u\,du = -\operatorname{csch} u$

14.39 $\int \frac{du}{u^2 + a^2} = \frac{1}{a}\tan^{-1}\frac{u}{a}$

14.40 $\int \frac{du}{u^2 - a^2} = \frac{1}{2a}\ln\left(\frac{u-a}{u+a}\right) = -\frac{1}{a}\coth^{-1}\frac{u}{a} \qquad u^2 > a^2$

14.41 $\int \frac{du}{a^2 - u^2} = \frac{1}{2a}\ln\left(\frac{a+u}{a-u}\right) = \frac{1}{a}\tanh^{-1}\frac{u}{a} \qquad u^2 < a^2$

14.42 $\int \frac{du}{\sqrt{a^2 - u^2}} = \sin^{-1}\frac{u}{a}$

14.43 $\int \frac{du}{\sqrt{u^2 + a^2}} = \ln(u + \sqrt{u^2 + a^2}) \quad \text{or} \quad \sinh^{-1}\frac{u}{a}$

14.44 $\int \frac{du}{\sqrt{u^2 - a^2}} = \ln(u + \sqrt{u^2 - a^2})$

14.45 $\int \frac{du}{u\sqrt{u^2 - a^2}} = \frac{1}{a}\sec^{-1}\left|\frac{u}{a}\right|$

14.46 $\int \frac{du}{u\sqrt{u^2 + a^2}} = -\frac{1}{a}\ln\left(\frac{a + \sqrt{u^2 + a^2}}{u}\right)$

14.47 $\int \frac{du}{u\sqrt{a^2 - u^2}} = -\frac{1}{a}\ln\left(\frac{a + \sqrt{a^2 - u^2}}{u}\right)$

14.48 $\int f^{(n)}g\,dx = f^{(n-1)}g - f^{(n-2)}g' + f^{(n-3)}g'' - \cdots (-1)^n \int fg^{(n)}\,dx$

This is called *generalized integration* by parts.

IMPORTANT TRANSFORMATIONS

Often in practice an integral can be simplified by using an appropriate transformation or substitution and formula 14.6, page 57. The following list gives some transformations and their effects.

14.49 $\int F(ax + b)\,dx = \frac{1}{a}\int F(u)\,du \qquad \text{where } u = ax + b$

14.50 $\int F(\sqrt{ax + b})\,dx = \frac{2}{a}\int u\,F(u)\,du \qquad \text{where } u = \sqrt{ax + b}$

14.51 $\int F(\sqrt[n]{ax + b})\,dx = \frac{n}{a}\int u^{n-1}F(u)\,du \qquad \text{where } u = \sqrt[n]{ax + b}$

14.52 $\int F(\sqrt{a^2 - x^2})\,dx = a\int F(a\cos u)\cos u\,du \qquad \text{where } x = a\sin u$

14.53 $\int F(\sqrt{x^2 + a^2})\,dx = a\int F(a\sec u)\sec^2 u\,du \qquad \text{where } x = a\tan u$

14.54 $\int F(\sqrt{x^2-a^2})\,dx = a\int F(a\tan u)\sec u\tan u\,du$ where $x = a\sec u$

14.55 $\int F(e^{ax})\,dx = \frac{1}{a}\int \frac{F(u)}{u}\,du$ where $u = e^{ax}$

14.56 $\int F(\ln x)\,dx = \int F(u)\,e^u\,du$ where $u = \ln x$

14.57 $\int F\left(\sin^{-1}\frac{x}{a}\right)dx = a\int F(u)\cos u\,du$ where $u = \sin^{-1}\frac{x}{a}$

Similar results apply for other inverse trigonometric functions.

14.58 $\int F(\sin x, \cos x)\,dx = 2\int F\left(\frac{2u}{1+u^2}, \frac{1-u^2}{1+u^2}\right)\frac{du}{1+u^2}$ where $u = \tan\frac{x}{2}$

SPECIAL INTEGRALS

Pages 60 through 93 provide a table of integrals classified under special types. The remarks given on page 57 apply here as well. It is assumed in all cases that division by zero is excluded.

INTEGRALS INVOLVING $ax + b$

14.59 $\int \frac{dx}{ax+b} = \frac{1}{a}\ln(ax+b)$

14.60 $\int \frac{x\,dx}{ax+b} = \frac{x}{a} - \frac{b}{a^2}\ln(ax+b)$

14.61 $\int \frac{x^2\,dx}{ax+b} = \frac{(ax+b)^2}{2a^3} - \frac{2b(ax+b)}{a^3} + \frac{b^2}{a^3}\ln(ax+b)$

14.62 $\int \frac{x^3\,dx}{ax+b} = \frac{(ax+b)^3}{3a^4} - \frac{3b(ax+b)^2}{2a^4} + \frac{3b^2(ax+b)}{a^4} - \frac{b^3}{a^4}\ln(ax+b)$

14.63 $\int \frac{dx}{x(ax+b)} = \frac{1}{b}\ln\left(\frac{x}{ax+b}\right)$

14.64 $\int \frac{dx}{x^2(ax+b)} = -\frac{1}{bx} + \frac{a}{b^2}\ln\left(\frac{ax+b}{x}\right)$

14.65 $\int \frac{dx}{x^3(ax+b)} = \frac{2ax-b}{2b^2x^2} + \frac{a^2}{b^3}\ln\left(\frac{x}{ax+b}\right)$

14.66 $\int \frac{dx}{(ax+b)^2} = \frac{-1}{a(ax+b)}$

14.67 $\int \frac{x\,dx}{(ax+b)^2} = \frac{b}{a^2(ax+b)} + \frac{1}{a^2}\ln(ax+b)$

14.68 $\int \frac{x^2\,dx}{(ax+b)^2} = \frac{ax+b}{a^3} - \frac{b^2}{a^3(ax+b)} - \frac{2b}{a^3}\ln(ax+b)$

14.69 $\int \frac{x^3\,dx}{(ax+b)^2} = \frac{(ax+b)^2}{2a^4} - \frac{3b(ax+b)}{a^4} + \frac{b^3}{a^4(ax+b)} + \frac{3b^2}{a^4}\ln(ax+b)$

14.70 $\int \frac{dx}{x(ax+b)^2} = \frac{1}{b(ax+b)} + \frac{1}{b^2}\ln\left(\frac{x}{ax+b}\right)$

14.71 $\int \frac{dx}{x^2(ax+b)^2} = \frac{-a}{b^2(ax+b)} - \frac{1}{b^2x} + \frac{2a}{b^3}\ln\left(\frac{ax+b}{x}\right)$

14.72 $$\int \frac{dx}{x^3(ax+b)^2} = -\frac{(ax+b)^2}{2b^4x^2} + \frac{3a(ax+b)}{b^4x} - \frac{a^3x}{b^4(ax+b)} - \frac{3a^2}{b^4}\ln\left(\frac{ax+b}{x}\right)$$

14.73 $$\int \frac{dx}{(ax+b)^3} = \frac{-1}{2(ax+b)^2}$$

14.74 $$\int \frac{x\,dx}{(ax+b)^3} = \frac{-1}{a^2(ax+b)} + \frac{b}{2a^2(ax+b)^2}$$

14.75 $$\int \frac{x^2\,dx}{(ax+b)^3} = \frac{2b}{a^3(ax+b)} - \frac{b^2}{2a^3(ax+b)^2} + \frac{1}{a^3}\ln(ax+b)$$

14.76 $$\int \frac{x^3\,dx}{(ax+b)^3} = \frac{x}{a^3} - \frac{3b^2}{a^4(ax+b)} + \frac{b^3}{2a^4(ax+b)^2} - \frac{3b}{a^4}\ln(ax+b)$$

14.77 $$\int \frac{dx}{x(ax+b)^3} = \frac{a^2x^2}{2b^3(ax+b)^2} - \frac{2ax}{b^3(ax+b)} - \frac{1}{b^3}\ln\left(\frac{ax+b}{x}\right)$$

14.78 $$\int \frac{dx}{x^2(ax+b)^3} = \frac{-a}{2b^2(ax+b)^2} - \frac{2a}{b^3(ax+b)} - \frac{1}{b^3x} + \frac{3a}{b^4}\ln\left(\frac{ax+b}{x}\right)$$

14.79 $$\int \frac{dx}{x^3(ax+b)^3} = \frac{a^4x^2}{2b^5(ax+b)^2} - \frac{4a^3x}{b^5(ax+b)} - \frac{(ax+b)^2}{2b^5x^2} - \frac{6a^2}{b^5}\ln\left(\frac{ax+b}{x}\right)$$

14.80 $$\int (ax+b)^n\,dx = \frac{(ax+b)^{n+1}}{(n+1)a}.$$ If $n=-1$, see 14.59.

14.81 $$\int x(ax+b)^n\,dx = \frac{(ax+b)^{n+2}}{(n+2)a^2} - \frac{b(ax+b)^{n+1}}{(n+1)a^2}, \quad n \neq -1,-2$$

If $n=-1,-2$, see 14.60, 14.67.

14.82 $$\int x^2(ax+b)^n\,dx = \frac{(ax+b)^{n+3}}{(n+3)a^3} - \frac{2b(ax+b)^{n+2}}{(n+2)a^3} + \frac{b^2(ax+b)^{n+1}}{(n+1)a^3}$$

If $n=-1,-2,-3$, see 14.61, 14.68, 14.75.

14.83 $$\int x^m(ax+b)^n\,dx = \begin{cases} \dfrac{x^{m+1}(ax+b)^n}{m+n+1} + \dfrac{nb}{m+n+1}\displaystyle\int x^m(ax+b)^{n-1}\,dx \\ \dfrac{x^m(ax+b)^{n+1}}{(m+n+1)a} - \dfrac{mb}{(m+n+1)a}\displaystyle\int x^{m-1}(ax+b)^n\,dx \\ \dfrac{-x^{m+1}(ax+b)^{n+1}}{(n+1)b} + \dfrac{m+n+2}{(n+1)b}\displaystyle\int x^m(ax+b)^{n+1}\,dx \end{cases}$$

INTEGRALS INVOLVING $\sqrt{ax+b}$

14.84 $$\int \frac{dx}{\sqrt{ax+b}} = \frac{2\sqrt{ax+b}}{a}$$

14.85 $$\int \frac{x\,dx}{\sqrt{ax+b}} = \frac{2(ax-2b)}{3a^2}\sqrt{ax+b}$$

14.86 $$\int \frac{x^2\,dx}{\sqrt{ax+b}} = \frac{2(3a^2x^2-4abx+8b^2)}{15a^3}\sqrt{ax+b}$$

14.87 $$\int \frac{dx}{x\sqrt{ax+b}} = \begin{cases} \dfrac{1}{\sqrt{b}}\ln\left(\dfrac{\sqrt{ax+b}-\sqrt{b}}{\sqrt{ax+b}+\sqrt{b}}\right) \\ \dfrac{2}{\sqrt{-b}}\tan^{-1}\sqrt{\dfrac{ax+b}{-b}} \end{cases}$$

14.88 $$\int \frac{dx}{x^2\sqrt{ax+b}} = -\frac{\sqrt{ax+b}}{bx} - \frac{a}{2b}\int \frac{dx}{x\sqrt{ax+b}}$$ [See 14.87]

14.89 $$\int \sqrt{ax+b}\,dx = \frac{2\sqrt{(ax+b)^3}}{3a}$$

14.90 $$\int x\sqrt{ax+b}\,dx = \frac{2(3ax-2b)}{15a^2}\sqrt{(ax+b)^3}$$

14.91 $$\int x^2\sqrt{ax+b}\,dx = \frac{2(15a^2x^2-12abx+8b^2)}{105a^3}\sqrt{(ax+b)^3}$$

14.92 $$\int \frac{\sqrt{ax+b}}{x}\,dx = 2\sqrt{ax+b} + b\int \frac{dx}{x\sqrt{ax+b}}$$ [See 14.87]

14.93 $$\int \frac{\sqrt{ax+b}}{x^2}\,dx = -\frac{\sqrt{ax+b}}{x} + \frac{a}{2}\int \frac{dx}{x\sqrt{ax+b}}$$ [See 14.87]

14.94 $$\int \frac{x^m}{\sqrt{ax+b}}\,dx = \frac{2x^m\sqrt{ax+b}}{(2m+1)a} - \frac{2mb}{(2m+1)a}\int \frac{x^{m-1}}{\sqrt{ax+b}}\,dx$$

14.95 $$\int \frac{dx}{x^m\sqrt{ax+b}} = -\frac{\sqrt{ax+b}}{(m-1)bx^{m-1}} - \frac{(2m-3)a}{(2m-2)b}\int \frac{dx}{x^{m-1}\sqrt{ax+b}}$$

14.96 $$\int x^m\sqrt{ax+b}\,dx = \frac{2x^m}{(2m+3)a}(ax+b)^{3/2} - \frac{2mb}{(2m+3)a}\int x^{m-1}\sqrt{ax+b}\,dx$$

14.97 $$\int \frac{\sqrt{ax+b}}{x^m}\,dx = -\frac{\sqrt{ax+b}}{(m-1)x^{m-1}} + \frac{a}{2(m-1)}\int \frac{dx}{x^{m-1}\sqrt{ax+b}}$$

14.98 $$\int \frac{\sqrt{ax+b}}{x^m}\,dx = \frac{-(ax+b)^{3/2}}{(m-1)bx^{m-1}} - \frac{(2m-5)a}{(2m-2)b}\int \frac{\sqrt{ax+b}}{x^{m-1}}\,dx$$

14.99 $$\int (ax+b)^{m/2}\,dx = \frac{2(ax+b)^{(m+2)/2}}{a(m+2)}$$

14.100 $$\int x(ax+b)^{m/2}\,dx = \frac{2(ax+b)^{(m+4)/2}}{a^2(m+4)} - \frac{2b(ax+b)^{(m+2)/2}}{a^2(m+2)}$$

14.101 $$\int x^2(ax+b)^{m/2}\,dx = \frac{2(ax+b)^{(m+6)/2}}{a^3(m+6)} - \frac{4b(ax+b)^{(m+4)/2}}{a^3(m+4)} + \frac{2b^2(ax+b)^{(m+2)/2}}{a^3(m+2)}$$

14.102 $$\int \frac{(ax+b)^{m/2}}{x}\,dx = \frac{2(ax+b)^{m/2}}{m} + b\int \frac{(ax+b)^{(m-2)/2}}{x}\,dx$$

14.103 $$\int \frac{(ax+b)^{m/2}}{x^2}\,dx = -\frac{(ax+b)^{(m+2)/2}}{bx} + \frac{ma}{2b}\int \frac{(ax+b)^{m/2}}{x}\,dx$$

14.104 $$\int \frac{dx}{x(ax+b)^{m/2}} = \frac{2}{(m-2)b(ax+b)^{(m-2)/2}} + \frac{1}{b}\int \frac{dx}{x(ax+b)^{(m-2)/2}}$$

INTEGRALS INVOLVING $ax+b$ AND $px+q$

14.105 $$\int \frac{dx}{(ax+b)(px+q)} = \frac{1}{bp-aq}\ln\left(\frac{px+q}{ax+b}\right)$$

14.106 $$\int \frac{x\,dx}{(ax+b)(px+q)} = \frac{1}{bp-aq}\left\{\frac{b}{a}\ln(ax+b) - \frac{q}{p}\ln(px+q)\right\}$$

14.107 $$\int \frac{dx}{(ax+b)^2(px+q)} = \frac{1}{bp-aq}\left\{\frac{1}{ax+b} + \frac{p}{bp-aq}\ln\left(\frac{px+q}{ax+b}\right)\right\}$$

14.108 $$\int \frac{x\,dx}{(ax+b)^2(px+q)} = \frac{1}{bp-aq}\left\{\frac{q}{bp-aq}\ln\left(\frac{ax+b}{px+q}\right) - \frac{b}{a(ax+b)}\right\}$$

14.109 $$\int \frac{x^2\,dx}{(ax+b)^2(px+q)} = \frac{b^2}{(bp-aq)a^2(ax+b)} + \frac{1}{(bp-aq)^2}\left\{\frac{q^2}{p}\ln(px+q) + \frac{b(bp-2aq)}{a^2}\ln(ax+b)\right\}$$

14.110 $$\int \frac{dx}{(ax+b)^m(px+q)^n} = \frac{-1}{(n-1)(bp-aq)}\left\{\frac{1}{(ax+b)^{m-1}(px+q)^{n-1}} + a(m+n-2)\int \frac{dx}{(ax+b)^m(px+q)^{n-1}}\right\}$$

14.111 $$\int \frac{ax+b}{px+q}\,dx = \frac{ax}{p} + \frac{bp-aq}{p^2}\ln(px+q)$$

14.112 $$\int \frac{(ax+b)^m}{(px+q)^n}\,dx = \begin{cases} \dfrac{-1}{(n-1)(bp-aq)}\left\{\dfrac{(ax+b)^{m+1}}{(px+q)^{n-1}} + (n-m-2)a\displaystyle\int \frac{(ax+b)^m}{(px+q)^{n-1}}\,dx\right\} \\ \dfrac{-1}{(n-m-1)p}\left\{\dfrac{(ax+b)^m}{(px+q)^{n-1}} + m(bp-aq)\displaystyle\int \frac{(ax+b)^{m-1}}{(px+q)^n}\,dx\right\} \\ \dfrac{-1}{(n-1)p}\left\{\dfrac{(ax+b)^m}{(px+q)^{n-1}} - ma\displaystyle\int \frac{(ax+b)^{m-1}}{(px+q)^{n-1}}\,dx\right\} \end{cases}$$

INTEGRALS INVOLVING $\sqrt{ax+b}$ AND $px+q$

14.113 $$\int \frac{px+q}{\sqrt{ax+b}}\,dx = \frac{2(apx+3aq-2bp)}{3a^2}\sqrt{ax+b}$$

14.114 $$\int \frac{dx}{(px+q)\sqrt{ax+b}} = \begin{cases} \dfrac{1}{\sqrt{bp-aq}\sqrt{p}}\ln\left(\dfrac{\sqrt{p(ax+b)}-\sqrt{bp-aq}}{\sqrt{p(ax+b)}+\sqrt{bp-aq}}\right) \\ \dfrac{2}{\sqrt{aq-bp}\sqrt{p}}\tan^{-1}\sqrt{\dfrac{p(ax+b)}{aq-bp}} \end{cases}$$

14.115 $$\int \frac{\sqrt{ax+b}}{px+q}\,dx = \begin{cases} \dfrac{2\sqrt{ax+b}}{p} + \dfrac{\sqrt{bp-aq}}{p\sqrt{p}}\ln\left(\dfrac{\sqrt{p(ax+b)}-\sqrt{bp-aq}}{\sqrt{p(ax+b)}+\sqrt{bp-aq}}\right) \\ \dfrac{2\sqrt{ax+b}}{p} - \dfrac{2\sqrt{aq-bp}}{p\sqrt{p}}\tan^{-1}\sqrt{\dfrac{p(ax+b)}{aq-bp}} \end{cases}$$

14.116 $$\int (px+q)^n\sqrt{ax+b}\,dx = \frac{2(px+q)^{n+1}\sqrt{ax+b}}{(2n+3)p} + \frac{bp-aq}{(2n+3)p}\int \frac{(px+q)^n}{\sqrt{ax+b}}\,dx$$

14.117 $$\int \frac{dx}{(px+q)^n\sqrt{ax+b}} = \frac{\sqrt{ax+b}}{(n-1)(aq-bp)(px+q)^{n-1}} + \frac{(2n-3)a}{2(n-1)(aq-bp)}\int \frac{dx}{(px+q)^{n-1}\sqrt{ax+b}}$$

14.118 $$\int \frac{(px+q)^n}{\sqrt{ax+b}}\,dx = \frac{2(px+q)^n\sqrt{ax+b}}{(2n+1)a} + \frac{2n(aq-bp)}{(2n+1)a}\int \frac{(px+q)^{n-1}\,dx}{\sqrt{ax+b}}$$

14.119 $$\int \frac{\sqrt{ax+b}}{(px+q)^n}\,dx = \frac{-\sqrt{ax+b}}{(n-1)p(px+q)^{n-1}} + \frac{a}{2(n-1)p}\int \frac{dx}{(px+q)^{n-1}\sqrt{ax+b}}$$

INTEGRALS INVOLVING $\sqrt{ax+b}$ AND $\sqrt{px+q}$

14.120 $$\int \frac{dx}{\sqrt{(ax+b)(px+q)}} = \begin{cases} \dfrac{2}{\sqrt{ap}}\ln\left(\sqrt{a(px+q)}+\sqrt{p(ax+b)}\right) \\ \dfrac{2}{\sqrt{-ap}}\tan^{-1}\sqrt{\dfrac{-p(ax+b)}{a(px+q)}} \end{cases}$$

14.121 $$\int \frac{x\,dx}{\sqrt{(ax+b)(px+q)}} = \frac{\sqrt{(ax+b)(px+q)}}{ap} - \frac{bp+aq}{2ap}\int \frac{dx}{\sqrt{(ax+b)(px+q)}}$$

14.122 $$\int \sqrt{(ax+b)(px+q)}\,dx = \frac{2apx+bp+aq}{4ap}\sqrt{(ax+b)(px+q)} - \frac{(bp-aq)^2}{8ap}\int \frac{dx}{\sqrt{(ax+b)(px+q)}}$$

14.123 $$\int \sqrt{\frac{px+q}{ax+b}}\,dx = \frac{\sqrt{(ax+b)(px+q)}}{a} + \frac{aq-bp}{2a}\int \frac{dx}{\sqrt{(ax+b)(px+q)}}$$

14.124 $$\int \frac{dx}{(px+q)\sqrt{(ax+b)(px+q)}} = \frac{2\sqrt{ax+b}}{(aq-bp)\sqrt{px+q}}$$

INTEGRALS INVOLVING $x^2 + a^2$

14.125 $$\int \frac{dx}{x^2+a^2} = \frac{1}{a}\tan^{-1}\frac{x}{a}$$

14.126 $$\int \frac{x\,dx}{x^2+a^2} = \frac{1}{2}\ln(x^2+a^2)$$

14.127 $$\int \frac{x^2\,dx}{x^2+a^2} = x - a\tan^{-1}\frac{x}{a}$$

14.128 $$\int \frac{x^3\,dx}{x^2+a^2} = \frac{x^2}{2} - \frac{a^2}{2}\ln(x^2+a^2)$$

14.129 $$\int \frac{dx}{x(x^2+a^2)} = \frac{1}{2a^2}\ln\left(\frac{x^2}{x^2+a^2}\right)$$

14.130 $$\int \frac{dx}{x^2(x^2+a^2)} = -\frac{1}{a^2x} - \frac{1}{a^3}\tan^{-1}\frac{x}{a}$$

14.131 $$\int \frac{dx}{x^3(x^2+a^2)} = -\frac{1}{2a^2x^2} - \frac{1}{2a^4}\ln\left(\frac{x^2}{x^2+a^2}\right)$$

14.132 $$\int \frac{dx}{(x^2+a^2)^2} = \frac{x}{2a^2(x^2+a^2)} + \frac{1}{2a^3}\tan^{-1}\frac{x}{a}$$

14.133 $$\int \frac{x\,dx}{(x^2+a^2)^2} = \frac{-1}{2(x^2+a^2)}$$

14.134 $$\int \frac{x^2\,dx}{(x^2+a^2)^2} = \frac{-x}{2(x^2+a^2)} + \frac{1}{2a}\tan^{-1}\frac{x}{a}$$

14.135 $$\int \frac{x^3\,dx}{(x^2+a^2)^2} = \frac{a^2}{2(x^2+a^2)} + \frac{1}{2}\ln(x^2+a^2)$$

14.136 $$\int \frac{dx}{x(x^2+a^2)^2} = \frac{1}{2a^2(x^2+a^2)} + \frac{1}{2a^4}\ln\left(\frac{x^2}{x^2+a^2}\right)$$

14.137 $$\int \frac{dx}{x^2(x^2+a^2)^2} = -\frac{1}{a^4x} - \frac{x}{2a^4(x^2+a^2)} - \frac{3}{2a^5}\tan^{-1}\frac{x}{a}$$

14.138 $$\int \frac{dx}{x^3(x^2+a^2)^2} = -\frac{1}{2a^4x^2} - \frac{1}{2a^4(x^2+a^2)} - \frac{1}{a^6}\ln\left(\frac{x^2}{x^2+a^2}\right)$$

14.139 $$\int \frac{dx}{(x^2+a^2)^n} = \frac{x}{2(n-1)a^2(x^2+a^2)^{n-1}} + \frac{2n-3}{(2n-2)a^2}\int \frac{dx}{(x^2+a^2)^{n-1}}$$

14.140 $$\int \frac{x\,dx}{(x^2+a^2)^n} = \frac{-1}{2(n-1)(x^2+a^2)^{n-1}}$$

14.141 $$\int \frac{dx}{x(x^2+a^2)^n} = \frac{1}{2(n-1)a^2(x^2+a^2)^{n-1}} + \frac{1}{a^2}\int \frac{dx}{x(x^2+a^2)^{n-1}}$$

14.142 $$\int \frac{x^m\,dx}{(x^2+a^2)^n} = \int \frac{x^{m-2}\,dx}{(x^2+a^2)^{n-1}} - a^2\int \frac{x^{m-2}\,dx}{(x^2+a^2)^n}$$

14.143 $$\int \frac{dx}{x^m(x^2+a^2)^n} = \frac{1}{a^2}\int \frac{dx}{x^m(x^2+a^2)^{n-1}} - \frac{1}{a^2}\int \frac{dx}{x^{m-2}(x^2+a^2)^n}$$

INTEGRALS INVOLVING $x^2 - a^2$, $x^2 > a^2$

14.144 $$\int \frac{dx}{x^2 - a^2} = \frac{1}{2a} \ln\left(\frac{x-a}{x+a}\right) \quad \text{or} \quad -\frac{1}{a}\coth^{-1}\frac{x}{a}$$

14.145 $$\int \frac{x\,dx}{x^2 - a^2} = \frac{1}{2}\ln(x^2 - a^2)$$

14.146 $$\int \frac{x^2\,dx}{x^2 - a^2} = x + \frac{a}{2}\ln\left(\frac{x-a}{x+a}\right)$$

14.147 $$\int \frac{x^3\,dx}{x^2 - a^2} = \frac{x^2}{2} + \frac{a^2}{2}\ln(x^2 - a^2)$$

14.148 $$\int \frac{dx}{x(x^2 - a^2)} = \frac{1}{2a^2}\ln\left(\frac{x^2 - a^2}{x^2}\right)$$

14.149 $$\int \frac{dx}{x^2(x^2 - a^2)} = \frac{1}{a^2 x} + \frac{1}{2a^3}\ln\left(\frac{x-a}{x+a}\right)$$

14.150 $$\int \frac{dx}{x^3(x^2 - a^2)} = \frac{1}{2a^2x^2} - \frac{1}{2a^4}\ln\left(\frac{x^2}{x^2 - a^2}\right)$$

14.151 $$\int \frac{dx}{(x^2 - a^2)^2} = \frac{-x}{2a^2(x^2 - a^2)} - \frac{1}{4a^3}\ln\left(\frac{x-a}{x+a}\right)$$

14.152 $$\int \frac{x\,dx}{(x^2 - a^2)^2} = \frac{-1}{2(x^2 - a^2)}$$

14.153 $$\int \frac{x^2\,dx}{(x^2 - a^2)^2} = \frac{-x}{2(x^2 - a^2)} + \frac{1}{4a}\ln\left(\frac{x-a}{x+a}\right)$$

14.154 $$\int \frac{x^3\,dx}{(x^2 - a^2)^2} = \frac{-a^2}{2(x^2 - a^2)} + \frac{1}{2}\ln(x^2 - a^2)$$

14.155 $$\int \frac{dx}{x(x^2 - a^2)^2} = \frac{-1}{2a^2(x^2 - a^2)} + \frac{1}{2a^4}\ln\left(\frac{x^2}{x^2 - a^2}\right)$$

14.156 $$\int \frac{dx}{x^2(x^2 - a^2)^2} = -\frac{1}{a^4 x} - \frac{x}{2a^4(x^2 - a^2)} - \frac{3}{4a^5}\ln\left(\frac{x-a}{x+a}\right)$$

14.157 $$\int \frac{dx}{x^3(x^2 - a^2)^2} = -\frac{1}{2a^4x^2} - \frac{1}{2a^4(x^2 - a^2)} + \frac{1}{a^6}\ln\left(\frac{x^2}{x^2 - a^2}\right)$$

14.158 $$\int \frac{dx}{(x^2 - a^2)^n} = \frac{-x}{2(n-1)a^2(x^2 - a^2)^{n-1}} - \frac{2n-3}{(2n-2)a^2}\int \frac{dx}{(x^2 - a^2)^{n-1}}$$

14.159 $$\int \frac{x\,dx}{(x^2 - a^2)^n} = \frac{-1}{2(n-1)(x^2 - a^2)^{n-1}}$$

14.160 $$\int \frac{dx}{x(x^2 - a^2)^n} = \frac{-1}{2(n-1)a^2(x^2 - a^2)^{n-1}} - \frac{1}{a^2}\int \frac{dx}{x(x^2 - a^2)^{n-1}}$$

14.161 $$\int \frac{x^m\,dx}{(x^2 - a^2)^n} = \int \frac{x^{m-2}\,dx}{(x^2 - a^2)^{n-1}} + a^2\int \frac{x^{m-2}\,dx}{(x^2 - a^2)^n}$$

14.162 $$\int \frac{dx}{x^m(x^2 - a^2)^n} = \frac{1}{a^2}\int \frac{dx}{x^{m-2}(x^2 - a^2)^n} - \frac{1}{a^2}\int \frac{dx}{x^m(x^2 - a^2)^{n-1}}$$

INTEGRALS INVOLVING $a^2 - x^2, \quad x^2 < a^2$

14.163 $$\int \frac{dx}{a^2 - x^2} = \frac{1}{2a} \ln\left(\frac{a+x}{a-x}\right) \quad \text{or} \quad \frac{1}{a}\tanh^{-1}\frac{x}{a}$$

14.164 $$\int \frac{x\,dx}{a^2 - x^2} = -\frac{1}{2}\ln(a^2 - x^2)$$

14.165 $$\int \frac{x^2\,dx}{a^2 - x^2} = -x + \frac{a}{2}\ln\left(\frac{a+x}{a-x}\right)$$

14.166 $$\int \frac{x^3\,dx}{a^2 - x^2} = -\frac{x^2}{2} - \frac{a^2}{2}\ln(a^2 - x^2)$$

14.167 $$\int \frac{dx}{x(a^2 - x^2)} = \frac{1}{2a^2}\ln\left(\frac{x^2}{a^2 - x^2}\right)$$

14.168 $$\int \frac{dx}{x^2(a^2 - x^2)} = -\frac{1}{a^2 x} + \frac{1}{2a^3}\ln\left(\frac{a+x}{a-x}\right)$$

14.169 $$\int \frac{dx}{x^3(a^2 - x^2)} = -\frac{1}{2a^2x^2} + \frac{1}{2a^4}\ln\left(\frac{x^2}{a^2 - x^2}\right)$$

14.170 $$\int \frac{dx}{(a^2 - x^2)^2} = \frac{x}{2a^2(a^2 - x^2)} + \frac{1}{4a^3}\ln\left(\frac{a+x}{a-x}\right)$$

14.171 $$\int \frac{x\,dx}{(a^2 - x^2)^2} = \frac{1}{2(a^2 - x^2)}$$

14.172 $$\int \frac{x^2\,dx}{(a^2 - x^2)^2} = \frac{x}{2(a^2 - x^2)} - \frac{1}{4a}\ln\left(\frac{a+x}{a-x}\right)$$

14.173 $$\int \frac{x^3\,dx}{(a^2 - x^2)^2} = \frac{a^2}{2(a^2 - x^2)} + \frac{1}{2}\ln(a^2 - x^2)$$

14.174 $$\int \frac{dx}{x(a^2 - x^2)^2} = \frac{1}{2a^2(a^2 - x^2)} + \frac{1}{2a^4}\ln\left(\frac{x^2}{a^2 - x^2}\right)$$

14.175 $$\int \frac{dx}{x^2(a^2 - x^2)^2} = \frac{-1}{a^4 x} + \frac{x}{2a^4(a^2 - x^2)} + \frac{3}{4a^5}\ln\left(\frac{a+x}{a-x}\right)$$

14.176 $$\int \frac{dx}{x^3(a^2 - x^2)^2} = \frac{-1}{2a^4x^2} + \frac{1}{2a^4(a^2 - x^2)} + \frac{1}{a^6}\ln\left(\frac{x^2}{a^2 - x^2}\right)$$

14.177 $$\int \frac{dx}{(a^2 - x^2)^n} = \frac{x}{2(n-1)a^2(a^2 - x^2)^{n-1}} + \frac{2n-3}{(2n-2)a^2}\int \frac{dx}{(a^2 - x^2)^{n-1}}$$

14.178 $$\int \frac{x\,dx}{(a^2 - x^2)^n} = \frac{1}{2(n-1)(a^2 - x^2)^{n-1}}$$

14.179 $$\int \frac{dx}{x(a^2 - x^2)^n} = \frac{1}{2(n-1)a^2(a^2 - x^2)^{n-1}} + \frac{1}{a^2}\int \frac{dx}{x(a^2 - x^2)^{n-1}}$$

14.180 $$\int \frac{x^m\,dx}{(a^2 - x^2)^n} = a^2\int \frac{x^{m-2}dx}{(a^2 - x^2)^n} - \int \frac{x^{m-2}dx}{(a^2 - x^2)^{n-1}}$$

14.181 $$\int \frac{dx}{x^m(a^2 - x^2)^n} = \frac{1}{a^2}\int \frac{dx}{x^m(a^2 - x^2)^{n-1}} + \frac{1}{a^2}\int \frac{dx}{x^{m-2}(a^2 - x^2)^n}$$

INTEGRALS INVOLVING $\sqrt{x^2+a^2}$

14.182 $$\int \frac{dx}{\sqrt{x^2+a^2}} = \ln\left(x+\sqrt{x^2+a^2}\right) \quad \text{or} \quad \sinh^{-1}\frac{x}{a}$$

14.183 $$\int \frac{x\,dx}{\sqrt{x^2+a^2}} = \sqrt{x^2+a^2}$$

14.184 $$\int \frac{x^2\,dx}{\sqrt{x^2+a^2}} = \frac{x\sqrt{x^2+a^2}}{2} - \frac{a^2}{2}\ln\left(x+\sqrt{x^2+a^2}\right)$$

14.185 $$\int \frac{x^3\,dx}{\sqrt{x^2+a^2}} = \frac{(x^2+a^2)^{3/2}}{3} - a^2\sqrt{x^2+a^2}$$

14.186 $$\int \frac{dx}{x\sqrt{x^2+a^2}} = -\frac{1}{a}\ln\left(\frac{a+\sqrt{x^2+a^2}}{x}\right)$$

14.187 $$\int \frac{dx}{x^2\sqrt{x^2+a^2}} = -\frac{\sqrt{x^2+a^2}}{a^2x}$$

14.188 $$\int \frac{dx}{x^3\sqrt{x^2+a^2}} = -\frac{\sqrt{x^2+a^2}}{2a^2x^2} + \frac{1}{2a^3}\ln\left(\frac{a+\sqrt{x^2+a^2}}{x}\right)$$

14.189 $$\int \sqrt{x^2+a^2}\,dx = \frac{x\sqrt{x^2+a^2}}{2} + \frac{a^2}{2}\ln\left(x+\sqrt{x^2+a^2}\right)$$

14.190 $$\int x\sqrt{x^2+a^2}\,dx = \frac{(x^2+a^2)^{3/2}}{3}$$

14.191 $$\int x^2\sqrt{x^2+a^2}\,dx = \frac{x(x^2+a^2)^{3/2}}{4} - \frac{a^2x\sqrt{x^2+a^2}}{8} - \frac{a^4}{8}\ln\left(x+\sqrt{x^2+a^2}\right)$$

14.192 $$\int x^3\sqrt{x^2+a^2}\,dx = \frac{(x^2+a^2)^{5/2}}{5} - \frac{a^2(x^2+a^2)^{3/2}}{3}$$

14.193 $$\int \frac{\sqrt{x^2+a^2}}{x}\,dx = \sqrt{x^2+a^2} - a\ln\left(\frac{a+\sqrt{x^2+a^2}}{x}\right)$$

14.194 $$\int \frac{\sqrt{x^2+a^2}}{x^2}\,dx = -\frac{\sqrt{x^2+a^2}}{x} + \ln\left(x+\sqrt{x^2+a^2}\right)$$

14.195 $$\int \frac{\sqrt{x^2+a^2}}{x^3}\,dx = -\frac{\sqrt{x^2+a^2}}{2x^2} - \frac{1}{2a}\ln\left(\frac{a+\sqrt{x^2+a^2}}{x}\right)$$

14.196 $$\int \frac{dx}{(x^2+a^2)^{3/2}} = \frac{x}{a^2\sqrt{x^2+a^2}}$$

14.197 $$\int \frac{x\,dx}{(x^2+a^2)^{3/2}} = \frac{-1}{\sqrt{x^2+a^2}}$$

14.198 $$\int \frac{x^2\,dx}{(x^2+a^2)^{3/2}} = \frac{-x}{\sqrt{x^2+a^2}} + \ln\left(x+\sqrt{x^2+a^2}\right)$$

14.199 $$\int \frac{x^3\,dx}{(x^2+a^2)^{3/2}} = \sqrt{x^2+a^2} + \frac{a^2}{\sqrt{x^2+a^2}}$$

14.200 $$\int \frac{dx}{x(x^2+a^2)^{3/2}} = \frac{1}{a^2\sqrt{x^2+a^2}} - \frac{1}{a^3}\ln\left(\frac{a+\sqrt{x^2+a^2}}{x}\right)$$

14.201 $$\int \frac{dx}{x^2(x^2+a^2)^{3/2}} = -\frac{\sqrt{x^2+a^2}}{a^4x} - \frac{x}{a^4\sqrt{x^2+a^2}}$$

14.202 $$\int \frac{dx}{x^3(x^2+a^2)^{3/2}} = \frac{-1}{2a^2x^2\sqrt{x^2+a^2}} - \frac{3}{2a^4\sqrt{x^2+a^2}} + \frac{3}{2a^5}\ln\left(\frac{a+\sqrt{x^2+a^2}}{x}\right)$$

14.203 $$\int (x^2+a^2)^{3/2}\,dx = \frac{x(x^2+a^2)^{3/2}}{4} + \frac{3a^2x\sqrt{x^2+a^2}}{8} + \frac{3}{8}a^4 \ln(x+\sqrt{x^2+a^2})$$

14.204 $$\int x(x^2+a^2)^{3/2}\,dx = \frac{(x^2+a^2)^{5/2}}{5}$$

14.205 $$\int x^2(x^2+a^2)^{3/2}\,dx = \frac{x(x^2+a^2)^{5/2}}{6} - \frac{a^2x(x^2+a^2)^{3/2}}{24} - \frac{a^4x\sqrt{x^2+a^2}}{16} - \frac{a^6}{16}\ln(x+\sqrt{x^2+a^2})$$

14.206 $$\int x^3(x^2+a^2)^{3/2}\,dx = \frac{(x^2+a^2)^{7/2}}{7} - \frac{a^2(x^2+a^2)^{5/2}}{5}$$

14.207 $$\int \frac{(x^2+a^2)^{3/2}}{x}\,dx = \frac{(x^2+a^2)^{3/2}}{3} + a^2\sqrt{x^2+a^2} - a^3\ln\left(\frac{a+\sqrt{x^2+a^2}}{x}\right)$$

14.208 $$\int \frac{(x^2+a^2)^{3/2}}{x^2}\,dx = -\frac{(x^2+a^2)^{3/2}}{x} + \frac{3x\sqrt{x^2+a^2}}{2} + \frac{3}{2}a^2\ln(x+\sqrt{x^2+a^2})$$

14.209 $$\int \frac{(x^2+a^2)^{3/2}}{x^3}\,dx = -\frac{(x^2+a^2)^{3/2}}{2x^2} + \frac{3}{2}\sqrt{x^2+a^2} - \frac{3}{2}a\ln\left(\frac{a+\sqrt{x^2+a^2}}{x}\right)$$

INTEGRALS INVOLVING $\sqrt{x^2-a^2}$

14.210 $$\int \frac{dx}{\sqrt{x^2-a^2}} = \ln(x+\sqrt{x^2-a^2}), \qquad \int \frac{x\,dx}{\sqrt{x^2-a^2}} = \sqrt{x^2-a^2}$$

14.211 $$\int \frac{x^2\,dx}{\sqrt{x^2-a^2}} = \frac{x\sqrt{x^2-a^2}}{2} + \frac{a^2}{2}\ln(x+\sqrt{x^2-a^2})$$

14.212 $$\int \frac{x^3\,dx}{\sqrt{x^2-a^2}} = \frac{(x^2-a^2)^{3/2}}{3} + a^2\sqrt{x^2-a^2}$$

14.213 $$\int \frac{dx}{x\sqrt{x^2-a^2}} = \frac{1}{a}\sec^{-1}\left|\frac{x}{a}\right|$$

14.214 $$\int \frac{dx}{x^2\sqrt{x^2-a^2}} = \frac{\sqrt{x^2-a^2}}{a^2x}$$

14.215 $$\int \frac{dx}{x^3\sqrt{x^2-a^2}} = \frac{\sqrt{x^2-a^2}}{2a^2x^2} + \frac{1}{2a^3}\sec^{-1}\left|\frac{x}{a}\right|$$

14.216 $$\int \sqrt{x^2-a^2}\,dx = \frac{x\sqrt{x^2-a^2}}{2} - \frac{a^2}{2}\ln(x+\sqrt{x^2-a^2})$$

14.217 $$\int x\sqrt{x^2-a^2}\,dx = \frac{(x^2-a^2)^{3/2}}{3}$$

14.218 $$\int x^2\sqrt{x^2-a^2}\,dx = \frac{x(x^2-a^2)^{3/2}}{4} + \frac{a^2x\sqrt{x^2-a^2}}{8} - \frac{a^4}{8}\ln(x+\sqrt{x^2-a^2})$$

14.219 $$\int x^3\sqrt{x^2-a^2}\,dx = \frac{(x^2-a^2)^{5/2}}{5} + \frac{a^2(x^2-a^2)^{3/2}}{3}$$

14.220 $$\int \frac{\sqrt{x^2-a^2}}{x}\,dx = \sqrt{x^2-a^2} - a\sec^{-1}\left|\frac{x}{a}\right|$$

14.221 $$\int \frac{\sqrt{x^2-a^2}}{x^2}\,dx = -\frac{\sqrt{x^2-a^2}}{x} + \ln(x+\sqrt{x^2-a^2})$$

14.222 $$\int \frac{\sqrt{x^2-a^2}}{x^3}\,dx = -\frac{\sqrt{x^2-a^2}}{2x^2} + \frac{1}{2a}\sec^{-1}\left|\frac{x}{a}\right|$$

14.223 $$\int \frac{dx}{(x^2-a^2)^{3/2}} = -\frac{x}{a^2\sqrt{x^2-a^2}}$$

14.224 $$\int \frac{x\,dx}{(x^2-a^2)^{3/2}} = \frac{-1}{\sqrt{x^2-a^2}}$$

14.225 $$\int \frac{x^2\,dx}{(x^2-a^2)^{3/2}} = -\frac{x}{\sqrt{x^2-a^2}} + \ln\left(x+\sqrt{x^2-a^2}\right)$$

14.226 $$\int \frac{x^3\,dx}{(x^2-a^2)^{3/2}} = \sqrt{x^2-a^2} - \frac{a^2}{\sqrt{x^2-a^2}}$$

14.227 $$\int \frac{dx}{x(x^2-a^2)^{3/2}} = \frac{-1}{a^2\sqrt{x^2-a^2}} - \frac{1}{a^3}\sec^{-1}\left|\frac{x}{a}\right|$$

14.228 $$\int \frac{dx}{x^2(x^2-a^2)^{3/2}} = -\frac{\sqrt{x^2-a^2}}{a^4x} - \frac{x}{a^4\sqrt{x^2-a^2}}$$

14.229 $$\int \frac{dx}{x^3(x^2-a^2)^{3/2}} = \frac{1}{2a^2x^2\sqrt{x^2-a^2}} - \frac{3}{2a^4\sqrt{x^2-a^2}} - \frac{3}{2a^5}\sec^{-1}\left|\frac{x}{a}\right|$$

14.230 $$\int (x^2-a^2)^{3/2}\,dx = \frac{x(x^2-a^2)^{3/2}}{4} - \frac{3a^2x\sqrt{x^2-a^2}}{8} + \frac{3}{8}a^4\ln\left(x+\sqrt{x^2-a^2}\right)$$

14.231 $$\int x(x^2-a^2)^{3/2}\,dx = \frac{(x^2-a^2)^{5/2}}{5}$$

14.232 $$\int x^2(x^2-a^2)^{3/2}\,dx = \frac{x(x^2-a^2)^{5/2}}{6} + \frac{a^2x(x^2-a^2)^{3/2}}{24} - \frac{a^4x\sqrt{x^2-a^2}}{16} + \frac{a^6}{16}\ln\left(x+\sqrt{x^2-a^2}\right)$$

14.233 $$\int x^3(x^2-a^2)^{3/2}\,dx = \frac{(x^2-a^2)^{7/2}}{7} + \frac{a^2(x^2-a^2)^{5/2}}{5}$$

14.234 $$\int \frac{(x^2-a^2)^{3/2}}{x}\,dx = \frac{(x^2-a^2)^{3/2}}{3} - a^2\sqrt{x^2-a^2} + a^3\sec^{-1}\left|\frac{x}{a}\right|$$

14.235 $$\int \frac{(x^2-a^2)^{3/2}}{x^2}\,dx = -\frac{(x^2-a^2)^{3/2}}{x} + \frac{3x\sqrt{x^2-a^2}}{2} - \frac{3}{2}a^2\ln\left(x+\sqrt{x^2-a^2}\right)$$

14.236 $$\int \frac{(x^2-a^2)^{3/2}}{x^3}\,dx = -\frac{(x^2-a^2)^{3/2}}{2x^2} + \frac{3\sqrt{x^2-a^2}}{2} - \frac{3}{2}a\sec^{-1}\left|\frac{x}{a}\right|$$

INTEGRALS INVOLVING $\sqrt{a^2-x^2}$

14.237 $$\int \frac{dx}{\sqrt{a^2-x^2}} = \sin^{-1}\frac{x}{a}$$

14.238 $$\int \frac{x\,dx}{\sqrt{a^2-x^2}} = -\sqrt{a^2-x^2}$$

14.239 $$\int \frac{x^2\,dx}{\sqrt{a^2-x^2}} = -\frac{x\sqrt{a^2-x^2}}{2} + \frac{a^2}{2}\sin^{-1}\frac{x}{a}$$

14.240 $$\int \frac{x^3\,dx}{\sqrt{a^2-x^2}} = \frac{(a^2-x^2)^{3/2}}{3} - a^2\sqrt{a^2-x^2}$$

14.241 $$\int \frac{dx}{x\sqrt{a^2-x^2}} = -\frac{1}{a}\ln\left(\frac{a+\sqrt{a^2-x^2}}{x}\right)$$

14.242 $$\int \frac{dx}{x^2\sqrt{a^2-x^2}} = -\frac{\sqrt{a^2-x^2}}{a^2x}$$

14.243 $$\int \frac{dx}{x^3\sqrt{a^2-x^2}} = -\frac{\sqrt{a^2-x^2}}{2a^2x^2} - \frac{1}{2a^3}\ln\left(\frac{a+\sqrt{a^2-x^2}}{x}\right)$$

14.244 $$\int \sqrt{a^2-x^2}\,dx = \frac{x\sqrt{a^2-x^2}}{2} + \frac{a^2}{2}\sin^{-1}\frac{x}{a}$$

14.245 $$\int x\sqrt{a^2-x^2}\,dx = -\frac{(a^2-x^2)^{3/2}}{3}$$

14.246 $$\int x^2\sqrt{a^2-x^2}\,dx = -\frac{x(a^2-x^2)^{3/2}}{4} + \frac{a^2x\sqrt{a^2-x^2}}{8} + \frac{a^4}{8}\sin^{-1}\frac{x}{a}$$

14.247 $$\int x^3\sqrt{a^2-x^2}\,dx = \frac{(a^2-x^2)^{5/2}}{5} - \frac{a^2(a^2-x^2)^{3/2}}{3}$$

14.248 $$\int \frac{\sqrt{a^2-x^2}}{x}\,dx = \sqrt{a^2-x^2} - a\ln\left(\frac{a+\sqrt{a^2-x^2}}{x}\right)$$

14.249 $$\int \frac{\sqrt{a^2-x^2}}{x^2}\,dx = -\frac{\sqrt{a^2-x^2}}{x} - \sin^{-1}\frac{x}{a}$$

14.250 $$\int \frac{\sqrt{a^2-x^2}}{x^3}\,dx = -\frac{\sqrt{a^2-x^2}}{2x^2} + \frac{1}{2a}\ln\left(\frac{a+\sqrt{a^2-x^2}}{x}\right)$$

14.251 $$\int \frac{dx}{(a^2-x^2)^{3/2}} = \frac{x}{a^2\sqrt{a^2-x^2}}$$

14.252 $$\int \frac{x\,dx}{(a^2-x^2)^{3/2}} = \frac{1}{\sqrt{a^2-x^2}}$$

14.253 $$\int \frac{x^2\,dx}{(a^2-x^2)^{3/2}} = \frac{x}{\sqrt{a^2-x^2}} - \sin^{-1}\frac{x}{a}$$

14.254 $$\int \frac{x^3\,dx}{(a^2-x^2)^{3/2}} = \sqrt{a^2-x^2} + \frac{a^2}{\sqrt{a^2-x^2}}$$

14.255 $$\int \frac{dx}{x(a^2-x^2)^{3/2}} = \frac{1}{a^2\sqrt{a^2-x^2}} - \frac{1}{a^3}\ln\left(\frac{a+\sqrt{a^2-x^2}}{x}\right)$$

14.256 $$\int \frac{dx}{x^2(a^2-x^2)^{3/2}} = -\frac{\sqrt{a^2-x^2}}{a^4x} + \frac{x}{a^4\sqrt{a^2-x^2}}$$

14.257 $$\int \frac{dx}{x^3(a^2-x^2)^{3/2}} = \frac{-1}{2a^2x^2\sqrt{a^2-x^2}} + \frac{3}{2a^4\sqrt{a^2-x^2}} - \frac{3}{2a^5}\ln\left(\frac{a+\sqrt{a^2-x^2}}{x}\right)$$

14.258 $$\int (a^2-x^2)^{3/2}\,dx = \frac{x(a^2-x^2)^{3/2}}{4} + \frac{3a^2x\sqrt{a^2-x^2}}{8} + \frac{3}{8}a^4\sin^{-1}\frac{x}{a}$$

14.259 $$\int x(a^2-x^2)^{3/2}\,dx = -\frac{(a^2-x^2)^{5/2}}{5}$$

14.260 $$\int x^2(a^2-x^2)^{3/2}\,dx = -\frac{x(a^2-x^2)^{5/2}}{6} + \frac{a^2x(a^2-x^2)^{3/2}}{24} + \frac{a^4x\sqrt{a^2-x^2}}{16} + \frac{a^6}{16}\sin^{-1}\frac{x}{a}$$

14.261 $$\int x^3(a^2-x^2)^{3/2}\,dx = \frac{(a^2-x^2)^{7/2}}{7} - \frac{a^2(a^2-x^2)^{5/2}}{5}$$

14.262 $$\int \frac{(a^2-x^2)^{3/2}}{x}\,dx = \frac{(a^2-x^2)^{3/2}}{3} + a^2\sqrt{a^2-x^2} - a^3\ln\left(\frac{a+\sqrt{a^2-x^2}}{x}\right)$$

14.263 $$\int \frac{(a^2-x^2)^{3/2}}{x^2}\,dx = -\frac{(a^2-x^2)^{3/2}}{x} - \frac{3x\sqrt{a^2-x^2}}{2} - \frac{3}{2}a^2\sin^{-1}\frac{x}{a}$$

14.264 $$\int \frac{(a^2-x^2)^{3/2}}{x^3}\,dx = -\frac{(a^2-x^2)^{3/2}}{2x^2} - \frac{3\sqrt{a^2-x^2}}{2} + \frac{3}{2}a\ln\left(\frac{a+\sqrt{a^2-x^2}}{x}\right)$$

INTEGRALS INVOLVING $ax^2 + bx + c$

14.265 $$\int \frac{dx}{ax^2+bx+c} = \begin{cases} \dfrac{2}{\sqrt{4ac-b^2}} \tan^{-1} \dfrac{2ax+b}{\sqrt{4ac-b^2}} \\[2ex] \dfrac{1}{\sqrt{b^2-4ac}} \ln \left(\dfrac{2ax+b-\sqrt{b^2-4ac}}{2ax+b+\sqrt{b^2-4ac}} \right) \end{cases}$$

If $b^2 = 4ac$, $ax^2+bx+c = a(x+b/2a)^2$ and the results on pages 60-61 can be used. If $b = 0$ use results on page 64. If a or $c = 0$ use results on pages 60-61.

14.266 $$\int \frac{x\,dx}{ax^2+bx+c} = \frac{1}{2a} \ln (ax^2+bx+c) - \frac{b}{2a} \int \frac{dx}{ax^2+bx+c}$$

14.267 $$\int \frac{x^2\,dx}{ax^2+bx+c} = \frac{x}{a} - \frac{b}{2a^2} \ln (ax^2+bx+c) + \frac{b^2-2ac}{2a^2} \int \frac{dx}{ax^2+bx+c}$$

14.268 $$\int \frac{x^m\,dx}{ax^2+bx+c} = \frac{x^{m-1}}{(m-1)a} - \frac{c}{a} \int \frac{x^{m-2}\,dx}{ax^2+bx+c} - \frac{b}{a} \int \frac{x^{m-1}\,dx}{ax^2+bx+c}$$

14.269 $$\int \frac{dx}{x(ax^2+bx+c)} = \frac{1}{2c} \ln \left(\frac{x^2}{ax^2+bx+c} \right) - \frac{b}{2c} \int \frac{dx}{ax^2+bx+c}$$

14.270 $$\int \frac{dx}{x^2(ax^2+bx+c)} = \frac{b}{2c^2} \ln \left(\frac{ax^2+bx+c}{x^2} \right) - \frac{1}{cx} + \frac{b^2-2ac}{2c^2} \int \frac{dx}{ax^2+bx+c}$$

14.271 $$\int \frac{dx}{x^n(ax^2+bx+c)} = -\frac{1}{(n-1)cx^{n-1}} - \frac{b}{c} \int \frac{dx}{x^{n-1}(ax^2+bx+c)} - \frac{a}{c} \int \frac{dx}{x^{n-2}(ax^2+bx+c)}$$

14.272 $$\int \frac{dx}{(ax^2+bx+c)^2} = \frac{2ax+b}{(4ac-b^2)(ax^2+bx+c)} + \frac{2a}{4ac-b^2} \int \frac{dx}{ax^2+bx+c}$$

14.273 $$\int \frac{x\,dx}{(ax^2+bx+c)^2} = -\frac{bx+2c}{(4ac-b^2)(ax^2+bx+c)} - \frac{b}{4ac-b^2} \int \frac{dx}{ax^2+bx+c}$$

14.274 $$\int \frac{x^2\,dx}{(ax^2+bx+c)^2} = \frac{(b^2-2ac)x+bc}{a(4ac-b^2)(ax^2+bx+c)} + \frac{2c}{4ac-b^2} \int \frac{dx}{ax^2+bx+c}$$

14.275 $$\int \frac{x^m\,dx}{(ax^2+bx+c)^n} = -\frac{x^{m-1}}{(2n-m-1)a(ax^2+bx+c)^{n-1}} + \frac{(m-1)c}{(2n-m-1)a} \int \frac{x^{m-2}\,dx}{(ax^2+bx+c)^n}$$
$$- \frac{(n-m)b}{(2n-m-1)a} \int \frac{x^{m-1}\,dx}{(ax^2+bx+c)^n}$$

14.276 $$\int \frac{x^{2n-1}\,dx}{(ax^2+bx+c)^n} = \frac{1}{a} \int \frac{x^{2n-3}\,dx}{(ax^2+bx+c)^{n-1}} - \frac{c}{a} \int \frac{x^{2n-3}\,dx}{(ax^2+bx+c)^n} - \frac{b}{a} \int \frac{x^{2n-2}\,dx}{(ax^2+bx+c)^n}$$

14.277 $$\int \frac{dx}{x(ax^2+bx+c)^2} = \frac{1}{2c(ax^2+bx+c)} - \frac{b}{2c} \int \frac{dx}{(ax^2+bx+c)^2} + \frac{1}{c} \int \frac{dx}{x(ax^2+bx+c)}$$

14.278 $$\int \frac{dx}{x^2(ax^2+bx+c)^2} = -\frac{1}{cx(ax^2+bx+c)} - \frac{3a}{c} \int \frac{dx}{(ax^2+bx+c)^2} - \frac{2b}{c} \int \frac{dx}{x(ax^2+bx+c)^2}$$

14.279 $$\int \frac{dx}{x^m(ax^2+bx+c)^n} = -\frac{1}{(m-1)cx^{m-1}(ax^2+bx+c)^{n-1}} - \frac{(m+2n-3)a}{(m-1)c} \int \frac{dx}{x^{m-2}(ax^2+bx+c)^n}$$
$$- \frac{(m+n-2)b}{(m-1)c} \int \frac{dx}{x^{m-1}(ax^2+bx+c)^n}$$

INTEGRALS INVOLVING $\sqrt{ax^2+bx+c}$

In the following results if $b^2 = 4ac$, $\sqrt{ax^2+bx+c} = \sqrt{a}\,(x+b/2a)$ and the results on pages 60-61 can be used. If $b = 0$ use the results on pages 67-70. If $a = 0$ or $c = 0$ use the results on pages 61-62.

14.280 $$\int \frac{dx}{\sqrt{ax^2+bx+c}} = \begin{cases} \dfrac{1}{\sqrt{a}} \ln\left(2\sqrt{a}\sqrt{ax^2+bx+c} + 2ax + b\right) \\ -\dfrac{1}{\sqrt{-a}} \sin^{-1}\left(\dfrac{2ax+b}{\sqrt{b^2-4ac}}\right) \quad \text{or} \quad \dfrac{1}{\sqrt{a}} \sinh^{-1}\left(\dfrac{2ax+b}{\sqrt{4ac-b^2}}\right) \end{cases}$$

14.281 $$\int \frac{x\,dx}{\sqrt{ax^2+bx+c}} = \frac{\sqrt{ax^2+bx+c}}{a} - \frac{b}{2a}\int \frac{dx}{\sqrt{ax^2+bx+c}}$$

14.282 $$\int \frac{x^2\,dx}{\sqrt{ax^2+bx+c}} = \frac{2ax-3b}{4a^2}\sqrt{ax^2+bx+c} + \frac{3b^2-4ac}{8a^2}\int \frac{dx}{\sqrt{ax^2+bx+c}}$$

14.283 $$\int \frac{dx}{x\sqrt{ax^2+bx+c}} = \begin{cases} -\dfrac{1}{\sqrt{c}} \ln\left(\dfrac{2\sqrt{c}\sqrt{ax^2+bx+c} + bx + 2c}{x}\right) \\ \dfrac{1}{\sqrt{-c}} \sin^{-1}\left(\dfrac{bx+2c}{|x|\sqrt{b^2-4ac}}\right) \quad \text{or} \quad -\dfrac{1}{\sqrt{c}} \sinh^{-1}\left(\dfrac{bx+2c}{|x|\sqrt{4ac-b^2}}\right) \end{cases}$$

14.284 $$\int \frac{dx}{x^2\sqrt{ax^2+bx+c}} = -\frac{\sqrt{ax^2+bx+c}}{cx} - \frac{b}{2c}\int \frac{dx}{x\sqrt{ax^2+bx+c}}$$

14.285 $$\int \sqrt{ax^2+bx+c}\,dx = \frac{(2ax+b)\sqrt{ax^2+bx+c}}{4a} + \frac{4ac-b^2}{8a}\int \frac{dx}{\sqrt{ax^2+bx+c}}$$

14.286 $$\int x\sqrt{ax^2+bx+c}\,dx = \frac{(ax^2+bx+c)^{3/2}}{3a} - \frac{b(2ax+b)}{8a^2}\sqrt{ax^2+bx+c} - \frac{b(4ac-b^2)}{16a^2}\int \frac{dx}{\sqrt{ax^2+bx+c}}$$

14.287 $$\int x^2\sqrt{ax^2+bx+c}\,dx = \frac{6ax-5b}{24a^2}(ax^2+bx+c)^{3/2} + \frac{5b^2-4ac}{16a^2}\int \sqrt{ax^2+bx+c}\,dx$$

14.288 $$\int \frac{\sqrt{ax^2+bx+c}}{x}\,dx = \sqrt{ax^2+bx+c} + \frac{b}{2}\int \frac{dx}{\sqrt{ax^2+bx+c}} + c\int \frac{dx}{x\sqrt{ax^2+bx+c}}$$

14.289 $$\int \frac{\sqrt{ax^2+bx+c}}{x^2}\,dx = -\frac{\sqrt{ax^2+bx+c}}{x} + a\int \frac{dx}{\sqrt{ax^2+bx+c}} + \frac{b}{2}\int \frac{dx}{x\sqrt{ax^2+bx+c}}$$

14.290 $$\int \frac{dx}{(ax^2+bx+c)^{3/2}} = \frac{2(2ax+b)}{(4ac-b^2)\sqrt{ax^2+bx+c}}$$

14.291 $$\int \frac{x\,dx}{(ax^2+bx+c)^{3/2}} = \frac{2(bx+2c)}{(b^2-4ac)\sqrt{ax^2+bx+c}}$$

14.292 $$\int \frac{x^2\,dx}{(ax^2+bx+c)^{3/2}} = \frac{(2b^2-4ac)x+2bc}{a(4ac-b^2)\sqrt{ax^2+bx+c}} + \frac{1}{a}\int \frac{dx}{\sqrt{ax^2+bx+c}}$$

14.293 $$\int \frac{dx}{x(ax^2+bx+c)^{3/2}} = \frac{1}{c\sqrt{ax^2+bx+c}} + \frac{1}{c}\int \frac{dx}{x\sqrt{ax^2+bx+c}} - \frac{b}{2c}\int \frac{dx}{(ax^2+bx+c)^{3/2}}$$

14.294 $$\int \frac{dx}{x^2(ax^2+bx+c)^{3/2}} = -\frac{ax^2+2bx+c}{c^2x\sqrt{ax^2+bx+c}} + \frac{b^2-2ac}{2c^2}\int \frac{dx}{(ax^2+bx+c)^{3/2}} - \frac{3b}{2c^2}\int \frac{dx}{x\sqrt{ax^2+bx+c}}$$

14.295 $$\int (ax^2+bx+c)^{n+1/2}\,dx = \frac{(2ax+b)(ax^2+bx+c)^{n+1/2}}{4a(n+1)} + \frac{(2n+1)(4ac-b^2)}{8a(n+1)}\int (ax^2+bx+c)^{n-1/2}\,dx$$

14.296 $$\int x(ax^2+bx+c)^{n+1/2}\,dx = \frac{(ax^2+bx+c)^{n+3/2}}{a(2n+3)} - \frac{b}{2a}\int (ax^2+bx+c)^{n+1/2}\,dx$$

14.297 $$\int \frac{dx}{(ax^2+bx+c)^{n+1/2}} = \frac{2(2ax+b)}{(2n-1)(4ac-b^2)(ax^2+bx+c)^{n-1/2}} + \frac{8a(n-1)}{(2n-1)(4ac-b^2)}\int \frac{dx}{(ax^2+bx+c)^{n-1/2}}$$

14.298 $$\int \frac{dx}{x(ax^2+bx+c)^{n+1/2}} = \frac{1}{(2n-1)c(ax^2+bx+c)^{n-1/2}} + \frac{1}{c}\int \frac{dx}{x(ax^2+bx+c)^{n-1/2}} - \frac{b}{2c}\int \frac{dx}{(ax^2+bx+c)^{n+1/2}}$$

INTEGRALS INVOLVING $x^3 + a^3$

Note that for formulas involving $x^3 - a^3$ replace a by $-a$.

14.299 $$\int \frac{dx}{x^3+a^3} = \frac{1}{6a^2}\ln\frac{(x+a)^2}{x^2-ax+a^2} + \frac{1}{a^2\sqrt{3}}\tan^{-1}\frac{2x-a}{a\sqrt{3}}$$

14.300 $$\int \frac{x\,dx}{x^3+a^3} = \frac{1}{6a}\ln\frac{x^2-ax+a^2}{(x+a)^2} + \frac{1}{a\sqrt{3}}\tan^{-1}\frac{2x-a}{a\sqrt{3}}$$

14.301 $$\int \frac{x^2\,dx}{x^3+a^3} = \frac{1}{3}\ln(x^3+a^3)$$

14.302 $$\int \frac{dx}{x(x^3+a^3)} = \frac{1}{3a^3}\ln\left(\frac{x^3}{x^3+a^3}\right)$$

14.303 $$\int \frac{dx}{x^2(x^3+a^3)} = -\frac{1}{a^3x} - \frac{1}{6a^4}\ln\frac{x^2-ax+a^2}{(x+a)^2} - \frac{1}{a^4\sqrt{3}}\tan^{-1}\frac{2x-a}{a\sqrt{3}}$$

14.304 $$\int \frac{dx}{(x^3+a^3)^2} = \frac{x}{3a^3(x^3+a^3)} + \frac{1}{9a^5}\ln\frac{(x+a)^2}{x^2-ax+a^2} + \frac{2}{3a^5\sqrt{3}}\tan^{-1}\frac{2x-a}{a\sqrt{3}}$$

14.305 $$\int \frac{x\,dx}{(x^3+a^3)^2} = \frac{x^2}{3a^3(x^3+a^3)} + \frac{1}{18a^4}\ln\frac{x^2-ax+a^2}{(x+a)^2} + \frac{1}{3a^4\sqrt{3}}\tan^{-1}\frac{2x-a}{a\sqrt{3}}$$

14.306 $$\int \frac{x^2\,dx}{(x^3+a^3)^2} = -\frac{1}{3(x^3+a^3)}$$

14.307 $$\int \frac{dx}{x(x^3+a^3)^2} = \frac{1}{3a^3(x^3+a^3)} + \frac{1}{3a^6}\ln\left(\frac{x^3}{x^3+a^3}\right)$$

14.308 $$\int \frac{dx}{x^2(x^3+a^3)^2} = -\frac{1}{a^6x} - \frac{x^2}{3a^6(x^3+a^3)} - \frac{4}{3a^6}\int \frac{x\,dx}{x^3+a^3}$$ [See 14.300]

14.309 $$\int \frac{x^m\,dx}{x^3+a^3} = \frac{x^{m-2}}{m-2} - a^3\int \frac{x^{m-3}\,dx}{x^3+a^3}$$

14.310 $$\int \frac{dx}{x^n(x^3+a^3)} = \frac{-1}{a^3(n-1)x^{n-1}} - \frac{1}{a^3}\int \frac{dx}{x^{n-3}(x^3+a^3)}$$

INTEGRALS INVOLVING $x^4 \pm a^4$

14.311 $$\int \frac{dx}{x^4+a^4} = \frac{1}{4a^3\sqrt{2}}\ln\left(\frac{x^2+ax\sqrt{2}+a^2}{x^2-ax\sqrt{2}+a^2}\right) - \frac{1}{2a^3\sqrt{2}}\tan^{-1}\frac{ax\sqrt{2}}{x^2-a^2}$$

14.312 $$\int \frac{x\,dx}{x^4+a^4} = \frac{1}{2a^2}\tan^{-1}\frac{x^2}{a^2}$$

14.313 $$\int \frac{x^2\,dx}{x^4+a^4} = \frac{1}{4a\sqrt{2}}\ln\left(\frac{x^2-ax\sqrt{2}+a^2}{x^2+ax\sqrt{2}+a^2}\right) - \frac{1}{2a\sqrt{2}}\tan^{-1}\frac{ax\sqrt{2}}{x^2-a^2}$$

14.314 $$\int \frac{x^3\,dx}{x^4+a^4} = \frac{1}{4}\ln(x^4+a^4)$$

14.315 $\displaystyle\int \frac{dx}{x(x^4+a^4)} = \frac{1}{4a^4}\ln\left(\frac{x^4}{x^4+a^4}\right)$

14.316 $\displaystyle\int \frac{dx}{x^2(x^4+a^4)} = -\frac{1}{a^4x} - \frac{1}{4a^5\sqrt{2}}\ln\left(\frac{x^2-ax\sqrt{2}+a^2}{x^2+ax\sqrt{2}+a^2}\right) + \frac{1}{2a^5\sqrt{2}}\tan^{-1}\frac{ax\sqrt{2}}{x^2-a^2}$

14.317 $\displaystyle\int \frac{dx}{x^3(x^4+a^4)} = -\frac{1}{2a^4x^2} - \frac{1}{2a^6}\tan^{-1}\frac{x^2}{a^2}$

14.318 $\displaystyle\int \frac{dx}{x^4-a^4} = \frac{1}{4a^3}\ln\left(\frac{x-a}{x+a}\right) - \frac{1}{2a^3}\tan^{-1}\frac{x}{a}$

14.319 $\displaystyle\int \frac{x\,dx}{x^4-a^4} = \frac{1}{4a^2}\ln\left(\frac{x^2-a^2}{x^2+a^2}\right)$

14.320 $\displaystyle\int \frac{x^2\,dx}{x^4-a^4} = \frac{1}{4a}\ln\left(\frac{x-a}{x+a}\right) + \frac{1}{2a}\tan^{-1}\frac{x}{a}$

14.321 $\displaystyle\int \frac{x^3\,dx}{x^4-a^4} = \frac{1}{4}\ln(x^4-a^4)$

14.322 $\displaystyle\int \frac{dx}{x(x^4-a^4)} = \frac{1}{4a^4}\ln\left(\frac{x^4-a^4}{x^4}\right)$

14.323 $\displaystyle\int \frac{dx}{x^2(x^4-a^4)} = \frac{1}{a^4x} + \frac{1}{4a^5}\ln\left(\frac{x-a}{x+a}\right) + \frac{1}{2a^5}\tan^{-1}\frac{x}{a}$

14.324 $\displaystyle\int \frac{dx}{x^3(x^4-a^4)} = \frac{1}{2a^4x^2} + \frac{1}{4a^6}\ln\left(\frac{x^2-a^2}{x^2+a^2}\right)$

INTEGRALS INVOLVING $x^n \pm a^n$

14.325 $\displaystyle\int \frac{dx}{x(x^n+a^n)} = \frac{1}{na^n}\ln\frac{x^n}{x^n+a^n}$

14.326 $\displaystyle\int \frac{x^{n-1}\,dx}{x^n+a^n} = \frac{1}{n}\ln(x^n+a^n)$

14.327 $\displaystyle\int \frac{x^m\,dx}{(x^n+a^n)^r} = \int \frac{x^{m-n}\,dx}{(x^n+a^n)^{r-1}} - a^n\int \frac{x^{m-n}\,dx}{(x^n+a^n)^r}$

14.328 $\displaystyle\int \frac{dx}{x^m(x^n+a^n)^r} = \frac{1}{a^n}\int \frac{dx}{x^m(x^n+a^n)^{r-1}} - \frac{1}{a^n}\int \frac{dx}{x^{m-n}(x^n+a^n)^r}$

14.329 $\displaystyle\int \frac{dx}{x\sqrt{x^n+a^n}} = \frac{1}{n\sqrt{a^n}}\ln\left(\frac{\sqrt{x^n+a^n}-\sqrt{a^n}}{\sqrt{x^n+a^n}+\sqrt{a^n}}\right)$

14.330 $\displaystyle\int \frac{dx}{x(x^n-a^n)} = \frac{1}{na^n}\ln\left(\frac{x^n-a^n}{x^n}\right)$

14.331 $\displaystyle\int \frac{x^{n-1}\,dx}{x^n-a^n} = \frac{1}{n}\ln(x^n-a^n)$

14.332 $\displaystyle\int \frac{x^m\,dx}{(x^n-a^n)^r} = a^n\int \frac{x^{m-n}\,dx}{(x^n-a^n)^r} + \int \frac{x^{m-n}\,dx}{(x^n-a^n)^{r-1}}$

14.333 $\displaystyle\int \frac{dx}{x^m(x^n-a^n)^r} = \frac{1}{a^n}\int \frac{dx}{x^{m-n}(x^n-a^n)^r} - \frac{1}{a^n}\int \frac{dx}{x^m(x^n-a^n)^{r-1}}$

14.334 $\displaystyle\int \frac{dx}{x\sqrt{x^n-a^n}} = \frac{2}{n\sqrt{a^n}}\cos^{-1}\sqrt{\frac{a^n}{x^n}}$

14.335 $$\int \frac{x^{p-1}\,dx}{x^{2m}+a^{2m}} = \frac{1}{ma^{2m-p}}\sum_{k=1}^{m} \sin\frac{(2k-1)p\pi}{2m}\tan^{-1}\left(\frac{x+a\cos[(2k-1)\pi/2m]}{a\sin[(2k-1)\pi/2m]}\right) - \frac{1}{2ma^{2m-p}}\sum_{k=1}^{m}\cos\frac{(2k-1)p\pi}{2m}\ln\left(x^2+2ax\cos\frac{(2k-1)\pi}{2m}+a^2\right)$$

where $0 < p \leqq 2m$.

14.336 $$\int \frac{x^{p-1}\,dx}{x^{2m}-a^{2m}} = \frac{1}{2ma^{2m-p}}\sum_{k=1}^{m-1}\cos\frac{kp\pi}{m}\ln\left(x^2-2ax\cos\frac{k\pi}{m}+a^2\right) - \frac{1}{ma^{2m-p}}\sum_{k=1}^{m-1}\sin\frac{kp\pi}{m}\tan^{-1}\left(\frac{x-a\cos(k\pi/m)}{a\sin(k\pi/m)}\right) + \frac{1}{2ma^{2m-p}}\{\ln(x-a)+(-1)^p\ln(x+a)\}$$

where $0 < p \leqq 2m$.

14.337 $$\int \frac{x^{p-1}\,dx}{x^{2m+1}+a^{2m+1}} = \frac{2(-1)^{p-1}}{(2m+1)a^{2m-p+1}}\sum_{k=1}^{m}\sin\frac{2kp\pi}{2m+1}\tan^{-1}\left(\frac{x+a\cos[2k\pi/(2m+1)]}{a\sin[2k\pi/(2m+1)]}\right) - \frac{(-1)^{p-1}}{(2m+1)a^{2m-p+1}}\sum_{k=1}^{m}\cos\frac{2kp\pi}{2m+1}\ln\left(x^2+2ax\cos\frac{2k\pi}{2m+1}+a^2\right) + \frac{(-1)^{p-1}\ln(x+a)}{(2m+1)a^{2m-p+1}}$$

where $0 < p \leqq 2m+1$.

14.338 $$\int \frac{x^{p-1}\,dx}{x^{2m+1}-a^{2m+1}} = \frac{-2}{(2m+1)a^{2m-p+1}}\sum_{k=1}^{m}\sin\frac{2kp\pi}{2m+1}\tan^{-1}\left(\frac{x-a\cos[2k\pi/(2m+1)]}{a\sin[2k\pi/(2m+1)]}\right) + \frac{1}{(2m+1)a^{2m-p+1}}\sum_{k=1}^{m}\cos\frac{2kp\pi}{2m+1}\ln\left(x^2-2ax\cos\frac{2k\pi}{2m+1}+a^2\right) + \frac{\ln(x-a)}{(2m+1)a^{2m-p+1}}$$

where $0 < p \leqq 2m+1$.

INTEGRALS INVOLVING $\sin ax$

14.339 $$\int \sin ax\,dx = -\frac{\cos ax}{a}$$

14.340 $$\int x\sin ax\,dx = \frac{\sin ax}{a^2} - \frac{x\cos ax}{a}$$

14.341 $$\int x^2\sin ax\,dx = \frac{2x}{a^2}\sin ax + \left(\frac{2}{a^3}-\frac{x^2}{a}\right)\cos ax$$

14.342 $$\int x^3\sin ax\,dx = \left(\frac{3x^2}{a^2}-\frac{6}{a^4}\right)\sin ax + \left(\frac{6x}{a^3}-\frac{x^3}{a}\right)\cos ax$$

14.343 $$\int \frac{\sin ax}{x}\,dx = ax - \frac{(ax)^3}{3\cdot 3!} + \frac{(ax)^5}{5\cdot 5!} - \cdots$$

14.344 $$\int \frac{\sin ax}{x^2}\,dx = -\frac{\sin ax}{x} + a\int\frac{\cos ax}{x}\,dx$$ [see 14.373]

14.345 $$\int \frac{dx}{\sin ax} = \frac{1}{a}\ln(\csc ax - \cot ax) = \frac{1}{a}\ln\tan\frac{ax}{2}$$

14.346 $$\int \frac{x\,dx}{\sin ax} = \frac{1}{a^2}\left\{ax + \frac{(ax)^3}{18} + \frac{7(ax)^5}{1800} + \cdots + \frac{2(2^{2n-1}-1)B_n(ax)^{2n+1}}{(2n+1)!} + \cdots\right\}$$

14.347 $$\int \sin^2 ax\,dx = \frac{x}{2} - \frac{\sin 2ax}{4a}$$

14.348 $$\int x \sin^2 ax\,dx = \frac{x^2}{4} - \frac{x \sin 2ax}{4a} - \frac{\cos 2ax}{8a^2}$$

14.349 $$\int \sin^3 ax\,dx = -\frac{\cos ax}{a} + \frac{\cos^3 ax}{3a}$$

14.350 $$\int \sin^4 ax\,dx = \frac{3x}{8} - \frac{\sin 2ax}{4a} + \frac{\sin 4ax}{32a}$$

14.351 $$\int \frac{dx}{\sin^2 ax} = -\frac{1}{a}\cot ax$$

14.352 $$\int \frac{dx}{\sin^3 ax} = -\frac{\cos ax}{2a \sin^2 ax} + \frac{1}{2a}\ln \tan \frac{ax}{2}$$

14.353 $$\int \sin px \sin qx\,dx = \frac{\sin (p-q)x}{2(p-q)} - \frac{\sin (p+q)x}{2(p+q)}$$ [If $p = \pm q$, see 14.368.]

14.354 $$\int \frac{dx}{1-\sin ax} = \frac{1}{a}\tan\left(\frac{\pi}{4}+\frac{ax}{2}\right)$$

14.355 $$\int \frac{x\,dx}{1-\sin ax} = \frac{x}{a}\tan\left(\frac{\pi}{4}+\frac{ax}{2}\right) + \frac{2}{a^2}\ln \sin\left(\frac{\pi}{4}-\frac{ax}{2}\right)$$

14.356 $$\int \frac{dx}{1+\sin ax} = -\frac{1}{a}\tan\left(\frac{\pi}{4}-\frac{ax}{2}\right)$$

14.357 $$\int \frac{x\,dx}{1+\sin ax} = -\frac{x}{a}\tan\left(\frac{\pi}{4}-\frac{ax}{2}\right) + \frac{2}{a^2}\ln \sin\left(\frac{\pi}{4}+\frac{ax}{2}\right)$$

14.358 $$\int \frac{dx}{(1-\sin ax)^2} = \frac{1}{2a}\tan\left(\frac{\pi}{4}+\frac{ax}{2}\right) + \frac{1}{6a}\tan^3\left(\frac{\pi}{4}+\frac{ax}{2}\right)$$

14.359 $$\int \frac{dx}{(1+\sin ax)^2} = -\frac{1}{2a}\tan\left(\frac{\pi}{4}-\frac{ax}{2}\right) - \frac{1}{6a}\tan^3\left(\frac{\pi}{4}-\frac{ax}{2}\right)$$

14.360 $$\int \frac{dx}{p+q\sin ax} = \begin{cases} \dfrac{2}{a\sqrt{p^2-q^2}}\tan^{-1}\dfrac{p\tan\frac{1}{2}ax+q}{\sqrt{p^2-q^2}} \\[2ex] \dfrac{1}{a\sqrt{q^2-p^2}}\ln\left(\dfrac{p\tan\frac{1}{2}ax+q-\sqrt{q^2-p^2}}{p\tan\frac{1}{2}ax+q+\sqrt{q^2-p^2}}\right) \end{cases}$$

If $p = \pm q$ see 14.354 and 14.356.

14.361 $$\int \frac{dx}{(p+q\sin ax)^2} = \frac{q\cos ax}{a(p^2-q^2)(p+q\sin ax)} + \frac{p}{p^2-q^2}\int \frac{dx}{p+q\sin ax}$$

If $p = \pm q$ see 14.358 and 14.359.

14.362 $$\int \frac{dx}{p^2+q^2\sin^2 ax} = \frac{1}{ap\sqrt{p^2+q^2}}\tan^{-1}\frac{\sqrt{p^2+q^2}\tan ax}{p}$$

14.363 $$\int \frac{dx}{p^2-q^2\sin^2 ax} = \begin{cases} \dfrac{1}{ap\sqrt{p^2-q^2}}\tan^{-1}\dfrac{\sqrt{p^2-q^2}\tan ax}{p} \\[2ex] \dfrac{1}{2ap\sqrt{q^2-p^2}}\ln\left(\dfrac{\sqrt{q^2-p^2}\tan ax+p}{\sqrt{q^2-p^2}\tan ax-p}\right) \end{cases}$$

14.364 $$\int x^m \sin ax\,dx = -\frac{x^m\cos ax}{a} + \frac{mx^{m-1}\sin ax}{a^2} - \frac{m(m-1)}{a^2}\int x^{m-2}\sin ax\,dx$$

14.365 $$\int \frac{\sin ax}{x^n}\,dx = -\frac{\sin ax}{(n-1)x^{n-1}} + \frac{a}{n-1}\int \frac{\cos ax}{x^{n-1}}\,dx$$ [see 14.395]

14.366 $$\int \sin^n ax\,dx = -\frac{\sin^{n-1} ax\cos ax}{an} + \frac{n-1}{n}\int \sin^{n-2} ax\,dx$$

14.367 $$\int \frac{dx}{\sin^n ax} = \frac{-\cos ax}{a(n-1)\sin^{n-1} ax} + \frac{n-2}{n-1}\int \frac{dx}{\sin^{n-2} ax}$$

14.368 $$\int \frac{x\,dx}{\sin^n ax} = \frac{-x\cos ax}{a(n-1)\sin^{n-1} ax} - \frac{1}{a^2(n-1)(n-2)\sin^{n-2} ax} + \frac{n-2}{n-1}\int \frac{x\,dx}{\sin^{n-2} ax}$$

INTEGRALS INVOLVING $\cos ax$

14.369 $$\int \cos ax\,dx = \frac{\sin ax}{a}$$

14.370 $$\int x\cos ax\,dx = \frac{\cos ax}{a^2} + \frac{x\sin ax}{a}$$

14.371 $$\int x^2\cos ax\,dx = \frac{2x}{a^2}\cos ax + \left(\frac{x^2}{a} - \frac{2}{a^3}\right)\sin ax$$

14.372 $$\int x^3\cos ax\,dx = \left(\frac{3x^2}{a^2} - \frac{6}{a^4}\right)\cos ax + \left(\frac{x^3}{a} - \frac{6x}{a^3}\right)\sin ax$$

14.373 $$\int \frac{\cos ax}{x}\,dx = \ln x - \frac{(ax)^2}{2\cdot 2!} + \frac{(ax)^4}{4\cdot 4!} - \frac{(ax)^6}{6\cdot 6!} + \cdots$$

14.374 $$\int \frac{\cos ax}{x^2}\,dx = -\frac{\cos ax}{x} - a\int \frac{\sin ax}{x}\,dx$$ [See 14.343]

14.375 $$\int \frac{dx}{\cos ax} = \frac{1}{a}\ln(\sec ax + \tan ax) = \frac{1}{a}\ln\tan\left(\frac{\pi}{4} + \frac{ax}{2}\right)$$

14.376 $$\int \frac{x\,dx}{\cos ax} = \frac{1}{a^2}\left\{\frac{(ax)^2}{2} + \frac{(ax)^4}{8} + \frac{5(ax)^6}{144} + \cdots + \frac{E_n(ax)^{2n+2}}{(2n+2)(2n)!} + \cdots\right\}$$

14.377 $$\int \cos^2 ax\,dx = \frac{x}{2} + \frac{\sin 2ax}{4a}$$

14.378 $$\int x\cos^2 ax\,dx = \frac{x^2}{4} + \frac{x\sin 2ax}{4a} + \frac{\cos 2ax}{8a^2}$$

14.379 $$\int \cos^3 ax\,dx = \frac{\sin ax}{a} - \frac{\sin^3 ax}{3a}$$

14.380 $$\int \cos^4 ax\,dx = \frac{3x}{8} + \frac{\sin 2ax}{4a} + \frac{\sin 4ax}{32a}$$

14.381 $$\int \frac{dx}{\cos^2 ax} = \frac{\tan ax}{a}$$

14.382 $$\int \frac{dx}{\cos^3 ax} = \frac{\sin ax}{2a\cos^2 ax} + \frac{1}{2a}\ln\tan\left(\frac{\pi}{4} + \frac{ax}{2}\right)$$

14.383 $$\int \cos ax\cos px\,dx = \frac{\sin(a-p)x}{2(a-p)} + \frac{\sin(a+p)x}{2(a+p)}$$ [If $a = \pm p$, see 14.377.]

14.384 $$\int \frac{dx}{1-\cos ax} = -\frac{1}{a}\cot\frac{ax}{2}$$

14.385 $$\int \frac{x\,dx}{1-\cos ax} = -\frac{x}{a}\cot\frac{ax}{2} + \frac{2}{a^2}\ln\sin\frac{ax}{2}$$

14.386 $$\int \frac{dx}{1+\cos ax} = \frac{1}{a}\tan\frac{ax}{2}$$

14.387 $$\int \frac{x\,dx}{1+\cos ax} = \frac{x}{a}\tan\frac{ax}{2} + \frac{2}{a^2}\ln\cos\frac{ax}{2}$$

14.388 $$\int \frac{dx}{(1-\cos ax)^2} = -\frac{1}{2a}\cot\frac{ax}{2} - \frac{1}{6a}\cot^3\frac{ax}{2}$$

14.389 $$\int \frac{dx}{(1+\cos ax)^2} = \frac{1}{2a}\tan\frac{ax}{2} + \frac{1}{6a}\tan^3\frac{ax}{2}$$

14.390 $$\int \frac{dx}{p + q\cos ax} = \begin{cases} \dfrac{2}{a\sqrt{p^2 - q^2}} \tan^{-1} \sqrt{(p-q)/(p+q)} \tan \tfrac{1}{2}ax \\ \dfrac{1}{a\sqrt{q^2 - p^2}} \ln\left(\dfrac{\tan \frac{1}{2}ax + \sqrt{(q+p)/(q-p)}}{\tan \frac{1}{2}ax - \sqrt{(q+p)/(q-p)}}\right) \end{cases}$$ [If $p = \pm q$ see 14.384 and 14.386.]

14.391 $$\int \frac{dx}{(p + q\cos ax)^2} = \frac{q \sin ax}{a(q^2 - p^2)(p + q\cos ax)} - \frac{p}{q^2 - p^2}\int \frac{dx}{p + q\cos ax}$$ [If $p = \pm q$ see 14.388 and 14.389.]

14.392 $$\int \frac{dx}{p^2 + q^2 \cos^2 ax} = \frac{1}{ap\sqrt{p^2 + q^2}} \tan^{-1} \frac{p \tan ax}{\sqrt{p^2 + q^2}}$$

14.393 $$\int \frac{dx}{p^2 - q^2 \cos^2 ax} = \begin{cases} \dfrac{1}{ap\sqrt{p^2 - q^2}} \tan^{-1} \dfrac{p \tan ax}{\sqrt{p^2 - q^2}} \\ \dfrac{1}{2ap\sqrt{q^2 - p^2}} \ln\left(\dfrac{p \tan ax - \sqrt{q^2 - p^2}}{p \tan ax + \sqrt{q^2 - p^2}}\right) \end{cases}$$

14.394 $$\int x^m \cos ax\, dx = \frac{x^m \sin ax}{a} + \frac{m x^{m-1}}{a^2} \cos ax - \frac{m(m-1)}{a^2}\int x^{m-2} \cos ax\, dx$$

14.395 $$\int \frac{\cos ax}{x^n}\, dx = -\frac{\cos ax}{(n-1)x^{n-1}} - \frac{a}{n-1}\int \frac{\sin ax}{x^{n-1}}\, dx$$ [See 14.365]

14.396 $$\int \cos^n ax\, dx = \frac{\sin ax \cos^{n-1} ax}{an} + \frac{n-1}{n}\int \cos^{n-2} ax\, dx$$

14.397 $$\int \frac{dx}{\cos^n ax} = \frac{\sin ax}{a(n-1)\cos^{n-1} ax} + \frac{n-2}{n-1}\int \frac{dx}{\cos^{n-2} ax}$$

14.398 $$\int \frac{x\, dx}{\cos^n ax} = \frac{x \sin ax}{a(n-1)\cos^{n-1} ax} - \frac{1}{a^2(n-1)(n-2)\cos^{n-2} ax} + \frac{n-2}{n-1}\int \frac{x\, dx}{\cos^{n-2} ax}$$

INTEGRALS INVOLVING $\sin ax$ AND $\cos ax$

14.399 $$\int \sin ax \cos ax\, dx = \frac{\sin^2 ax}{2a}$$

14.400 $$\int \sin px \cos qx\, dx = -\frac{\cos(p-q)x}{2(p-q)} - \frac{\cos(p+q)x}{2(p+q)}$$

14.401 $$\int \sin^n ax \cos ax\, dx = \frac{\sin^{n+1} ax}{(n+1)a}$$ [If $n = -1$, see 14.440.]

14.402 $$\int \cos^n ax \sin ax\, dx = -\frac{\cos^{n+1} ax}{(n+1)a}$$ [If $n = -1$, see 14.429.]

14.403 $$\int \sin^2 ax \cos^2 ax\, dx = \frac{x}{8} - \frac{\sin 4ax}{32a}$$

14.404 $$\int \frac{dx}{\sin ax \cos ax} = \frac{1}{a} \ln \tan ax$$

14.405 $$\int \frac{dx}{\sin^2 ax \cos ax} = \frac{1}{a} \ln \tan\left(\frac{\pi}{4} + \frac{ax}{2}\right) - \frac{1}{a \sin ax}$$

14.406 $$\int \frac{dx}{\sin ax \cos^2 ax} = \frac{1}{a} \ln \tan \frac{ax}{2} + \frac{1}{a \cos ax}$$

14.407 $$\int \frac{dx}{\sin^2 ax \cos^2 ax} = -\frac{2 \cot 2ax}{a}$$

14.408 $$\int \frac{\sin^2 ax}{\cos ax}\,dx = -\frac{\sin ax}{a} + \frac{1}{a}\ln\tan\left(\frac{ax}{2}+\frac{\pi}{4}\right)$$

14.409 $$\int \frac{\cos^2 ax}{\sin ax}\,dx = \frac{\cos ax}{a} + \frac{1}{a}\ln\tan\frac{ax}{2}$$

14.410 $$\int \frac{dx}{\cos ax(1 \pm \sin ax)} = \mp\frac{1}{2a(1 \pm \sin ax)} + \frac{1}{2a}\ln\tan\left(\frac{ax}{2}+\frac{\pi}{4}\right)$$

14.411 $$\int \frac{dx}{\sin ax(1 \pm \cos ax)} = \pm\frac{1}{2a(1 \pm \cos ax)} + \frac{1}{2a}\ln\tan\frac{ax}{2}$$

14.412 $$\int \frac{dx}{\sin ax \pm \cos ax} = \frac{1}{a\sqrt{2}}\ln\tan\left(\frac{ax}{2}\pm\frac{\pi}{8}\right)$$

14.413 $$\int \frac{\sin ax\,dx}{\sin ax \pm \cos ax} = \frac{x}{2} \mp \frac{1}{2a}\ln(\sin ax \pm \cos ax)$$

14.414 $$\int \frac{\cos ax\,dx}{\sin ax \pm \cos ax} = \pm\frac{x}{2} + \frac{1}{2a}\ln(\sin ax \pm \cos ax)$$

14.415 $$\int \frac{\sin ax\,dx}{p + q\cos ax} = -\frac{1}{aq}\ln(p + q\cos ax)$$

14.416 $$\int \frac{\cos ax\,dx}{p + q\sin ax} = \frac{1}{aq}\ln(p + q\sin ax)$$

14.417 $$\int \frac{\sin ax\,dx}{(p + q\cos ax)^n} = \frac{1}{aq(n-1)(p + q\cos ax)^{n-1}}$$

14.418 $$\int \frac{\cos ax\,dx}{(p + q\sin ax)^n} = \frac{-1}{aq(n-1)(p + q\sin ax)^{n-1}}$$

14.419 $$\int \frac{dx}{p\sin ax + q\cos ax} = \frac{1}{a\sqrt{p^2+q^2}}\ln\tan\left(\frac{ax + \tan^{-1}(q/p)}{2}\right)$$

14.420 $$\int \frac{dx}{p\sin ax + q\cos ax + r} = \begin{cases} \dfrac{2}{a\sqrt{r^2-p^2-q^2}}\tan^{-1}\left(\dfrac{p + (r-q)\tan(ax/2)}{\sqrt{r^2-p^2-q^2}}\right) \\[2ex] \dfrac{1}{a\sqrt{p^2+q^2-r^2}}\ln\left(\dfrac{p - \sqrt{p^2+q^2-r^2} + (r-q)\tan(ax/2)}{p + \sqrt{p^2+q^2-r^2} + (r-q)\tan(ax/2)}\right) \end{cases}$$

If $r = q$ see 14.421. If $r^2 = p^2 + q^2$ see 14.422.

14.421 $$\int \frac{dx}{p\sin ax + q(1 + \cos ax)} = \frac{1}{ap}\ln\left(q + p\tan\frac{ax}{2}\right)$$

14.422 $$\int \frac{dx}{p\sin ax + q\cos ax \pm \sqrt{p^2+q^2}} = \frac{-1}{a\sqrt{p^2+q^2}}\tan\left(\frac{\pi}{4} \mp \frac{ax + \tan^{-1}(q/p)}{2}\right)$$

14.423 $$\int \frac{dx}{p^2\sin^2 ax + q^2\cos^2 ax} = \frac{1}{apq}\tan^{-1}\left(\frac{p\tan ax}{q}\right)$$

14.424 $$\int \frac{dx}{p^2\sin^2 ax - q^2\cos^2 ax} = \frac{1}{2apq}\ln\left(\frac{p\tan ax - q}{p\tan ax + q}\right)$$

14.425 $$\int \sin^m ax\cos^n ax\,dx = \begin{cases} -\dfrac{\sin^{m-1} ax\cos^{n+1} ax}{a(m+n)} + \dfrac{m-1}{m+n}\displaystyle\int \sin^{m-2} ax\cos^n ax\,dx \\[2ex] \dfrac{\sin^{m+1} ax\cos^{n-1} ax}{a(m+n)} + \dfrac{n-1}{m+n}\displaystyle\int \sin^m ax\cos^{n-2} ax\,dx \end{cases}$$

14.426 $$\int \frac{\sin^m ax}{\cos^n ax}\,dx = \begin{cases} \dfrac{\sin^{m-1} ax}{a(n-1)\cos^{n-1} ax} - \dfrac{m-1}{n-1}\displaystyle\int \frac{\sin^{m-2} ax}{\cos^{n-2} ax}\,dx \\ \dfrac{\sin^{m+1} ax}{a(n-1)\cos^{n-1} ax} - \dfrac{m-n+2}{n-1}\displaystyle\int \frac{\sin^m ax}{\cos^{n-2} ax}\,dx \\ \dfrac{-\sin^{m-1} ax}{a(m-n)\cos^{n-1} ax} + \dfrac{m-1}{m-n}\displaystyle\int \frac{\sin^{m-2} ax}{\cos^n ax}\,dx \end{cases}$$

14.427 $$\int \frac{\cos^m ax}{\sin^n ax}\,dx = \begin{cases} \dfrac{-\cos^{m-1} ax}{a(n-1)\sin^{n-1} ax} - \dfrac{m-1}{n-1}\displaystyle\int \frac{\cos^{m-2} ax}{\sin^{n-2} ax}\,dx \\ \dfrac{-\cos^{m+1} ax}{a(n-1)\sin^{n-1} ax} - \dfrac{m-n+2}{n-1}\displaystyle\int \frac{\cos^m ax}{\sin^{n-2} ax}\,dx \\ \dfrac{\cos^{m-1} ax}{a(m-n)\sin^{n-1} ax} + \dfrac{m-1}{m-n}\displaystyle\int \frac{\cos^{m-2} ax}{\sin^n ax}\,dx \end{cases}$$

14.428 $$\int \frac{dx}{\sin^m ax \cos^n ax} = \begin{cases} \dfrac{1}{a(n-1)\sin^{m-1} ax \cos^{n-1} ax} + \dfrac{m+n-2}{n-1}\displaystyle\int \frac{dx}{\sin^m ax \cos^{n-2} ax} \\ \dfrac{-1}{a(m-1)\sin^{m-1} ax \cos^{n-1} ax} + \dfrac{m+n-2}{m-1}\displaystyle\int \frac{dx}{\sin^{m-2} ax \cos^n ax} \end{cases}$$

INTEGRALS INVOLVING tan ax

14.429 $$\int \tan ax\,dx = -\frac{1}{a}\ln\cos ax = \frac{1}{a}\ln\sec ax$$

14.430 $$\int \tan^2 ax\,dx = \frac{\tan ax}{a} - x$$

14.431 $$\int \tan^3 ax\,dx = \frac{\tan^2 ax}{2a} + \frac{1}{a}\ln\cos ax$$

14.432 $$\int \tan^n ax \sec^2 ax\,dx = \frac{\tan^{n+1} ax}{(n+1)a}$$

14.433 $$\int \frac{\sec^2 ax}{\tan ax}\,dx = \frac{1}{a}\ln\tan ax$$

14.434 $$\int \frac{dx}{\tan ax} = \frac{1}{a}\ln\sin ax$$

14.435 $$\int x\tan ax\,dx = \frac{1}{a^2}\left\{\frac{(ax)^3}{3} + \frac{(ax)^5}{15} + \frac{2(ax)^7}{105} + \cdots + \frac{2^{2n}(2^{2n}-1)B_n(ax)^{2n+1}}{(2n+1)!} + \cdots\right\}$$

14.436 $$\int \frac{\tan ax}{x}\,dx = ax + \frac{(ax)^3}{9} + \frac{2(ax)^5}{75} + \cdots + \frac{2^{2n}(2^{2n}-1)B_n(ax)^{2n-1}}{(2n-1)(2n)!} + \cdots$$

14.437 $$\int x\tan^2 ax\,dx = \frac{x\tan ax}{a} + \frac{1}{a^2}\ln\cos ax - \frac{x^2}{2}$$

14.438 $$\int \frac{dx}{p+q\tan ax} = \frac{px}{p^2+q^2} + \frac{q}{a(p^2+q^2)}\ln(q\sin ax + p\cos ax)$$

14.439 $$\int \tan^n ax\,dx = \frac{\tan^{n-1} ax}{(n-1)a} - \int \tan^{n-2} ax\,dx$$

INTEGRALS INVOLVING $\cot ax$

14.440 $$\int \cot ax\,dx = \frac{1}{a}\ln \sin ax$$

14.441 $$\int \cot^2 ax\,dx = -\frac{\cot ax}{a} - x$$

14.442 $$\int \cot^3 ax\,dx = -\frac{\cot^2 ax}{2a} - \frac{1}{a}\ln \sin ax$$

14.443 $$\int \cot^n ax \csc^2 ax\,dx = -\frac{\cot^{n+1} ax}{(n+1)a}$$

14.444 $$\int \frac{\csc^2 ax}{\cot ax}\,dx = -\frac{1}{a}\ln \cot ax$$

14.445 $$\int \frac{dx}{\cot ax} = -\frac{1}{a}\ln \cos ax$$

14.446 $$\int x \cot ax\,dx = \frac{1}{a^2}\left\{ax - \frac{(ax)^3}{9} - \frac{(ax)^5}{225} - \cdots - \frac{2^{2n}B_n(ax)^{2n+1}}{(2n+1)!} - \cdots\right\}$$

14.447 $$\int \frac{\cot ax}{x}\,dx = -\frac{1}{ax} - \frac{ax}{3} - \frac{(ax)^3}{135} - \cdots - \frac{2^{2n}B_n(ax)^{2n-1}}{(2n-1)(2n)!} - \cdots$$

14.448 $$\int x \cot^2 ax\,dx = -\frac{x \cot ax}{a} + \frac{1}{a^2}\ln \sin ax - \frac{x^2}{2}$$

14.449 $$\int \frac{dx}{p + q \cot ax} = \frac{px}{p^2+q^2} - \frac{q}{a(p^2+q^2)}\ln(p \sin ax + q \cos ax)$$

14.450 $$\int \cot^n ax\,dx = -\frac{\cot^{n-1} ax}{(n-1)a} - \int \cot^{n-2} ax\,dx$$

INTEGRALS INVOLVING $\sec ax$

14.451 $$\int \sec ax\,dx = \frac{1}{a}\ln(\sec ax + \tan ax) = \frac{1}{a}\ln \tan\left(\frac{ax}{2} + \frac{\pi}{4}\right)$$

14.452 $$\int \sec^2 ax\,dx = \frac{\tan ax}{a}$$

14.453 $$\int \sec^3 ax\,dx = \frac{\sec ax \tan ax}{2a} + \frac{1}{2a}\ln(\sec ax + \tan ax)$$

14.454 $$\int \sec^n ax \tan ax\,dx = \frac{\sec^n ax}{na}$$

14.455 $$\int \frac{dx}{\sec ax} = \frac{\sin ax}{a}$$

14.456 $$\int x \sec ax\,dx = \frac{1}{a^2}\left\{\frac{(ax)^2}{2} + \frac{(ax)^4}{8} + \frac{5(ax)^6}{144} + \cdots + \frac{E_n(ax)^{2n+2}}{(2n+2)(2n)!} + \cdots\right\}$$

14.457 $$\int \frac{\sec ax}{x}\,dx = \ln x + \frac{(ax)^2}{4} + \frac{5(ax)^4}{96} + \frac{61(ax)^6}{4320} + \cdots + \frac{E_n(ax)^{2n}}{2n(2n)!} + \cdots$$

14.458 $$\int x \sec^2 ax\,dx = \frac{x}{a}\tan ax + \frac{1}{a^2}\ln \cos ax$$

14.459 $$\int \frac{dx}{q + p \sec ax} = \frac{x}{q} - \frac{p}{q}\int \frac{dx}{p + q \cos ax}$$

14.460 $$\int \sec^n ax\, dx = \frac{\sec^{n-2} ax \tan ax}{a(n-1)} + \frac{n-2}{n-1}\int \sec^{n-2} ax\, dx$$

INTEGRALS INVOLVING csc ax

14.461 $$\int \csc ax\, dx = \frac{1}{a}\ln(\csc ax - \cot ax) = \frac{1}{a}\ln\tan\frac{ax}{2}$$

14.462 $$\int \csc^2 ax\, dx = -\frac{\cot ax}{a}$$

14.463 $$\int \csc^3 ax\, dx = -\frac{\csc ax \cot ax}{2a} + \frac{1}{2a}\ln\tan\frac{ax}{2}$$

14.464 $$\int \csc^n ax \cot ax\, dx = -\frac{\csc^n ax}{na}$$

14.465 $$\int \frac{dx}{\csc ax} = -\frac{\cos ax}{a}$$

14.466 $$\int x \csc ax\, dx = \frac{1}{a^2}\left\{ax + \frac{(ax)^3}{18} + \frac{7(ax)^5}{1800} + \cdots + \frac{2(2^{2n-1}-1)B_n(ax)^{2n+1}}{(2n+1)!} + \cdots\right\}$$

14.467 $$\int \frac{\csc ax}{x}\, dx = -\frac{1}{ax} + \frac{ax}{6} + \frac{7(ax)^3}{1080} + \cdots + \frac{2(2^{2n-1}-1)B_n(ax)^{2n-1}}{(2n-1)(2n)!} + \cdots$$

14.468 $$\int x \csc^2 ax\, dx = -\frac{x \cot ax}{a} + \frac{1}{a^2}\ln\sin ax$$

14.469 $$\int \frac{dx}{q + p \csc ax} = \frac{x}{q} - \frac{p}{q}\int \frac{dx}{p + q \sin ax}$$ [See 14.360]

14.470 $$\int \csc^n ax\, dx = -\frac{\csc^{n-2} ax \cot ax}{a(n-1)} + \frac{n-2}{n-1}\int \csc^{n-2} ax\, dx$$

INTEGRALS INVOLVING INVERSE TRIGONOMETRIC FUNCTIONS

14.471 $$\int \sin^{-1}\frac{x}{a}\, dx = x\sin^{-1}\frac{x}{a} + \sqrt{a^2 - x^2}$$

14.472 $$\int x \sin^{-1}\frac{x}{a}\, dx = \left(\frac{x^2}{2} - \frac{a^2}{4}\right)\sin^{-1}\frac{x}{a} + \frac{x\sqrt{a^2 - x^2}}{4}$$

14.473 $$\int x^2 \sin^{-1}\frac{x}{a}\, dx = \frac{x^3}{3}\sin^{-1}\frac{x}{a} + \frac{(x^2 + 2a^2)\sqrt{a^2 - x^2}}{9}$$

14.474 $$\int \frac{\sin^{-1}(x/a)}{x}\, dx = \frac{x}{a} + \frac{(x/a)^3}{2\cdot 3\cdot 3} + \frac{1\cdot 3(x/a)^5}{2\cdot 4\cdot 5\cdot 5} + \frac{1\cdot 3\cdot 5(x/a)^7}{2\cdot 4\cdot 6\cdot 7\cdot 7} + \cdots$$

14.475 $$\int \frac{\sin^{-1}(x/a)}{x^2}\, dx = -\frac{\sin^{-1}(x/a)}{x} - \frac{1}{a}\ln\left(\frac{a + \sqrt{a^2 - x^2}}{x}\right)$$

14.476 $$\int \left(\sin^{-1}\frac{x}{a}\right)^2 dx = x\left(\sin^{-1}\frac{x}{a}\right)^2 - 2x + 2\sqrt{a^2 - x^2}\sin^{-1}\frac{x}{a}$$

14.477 $\int \cos^{-1}\frac{x}{a}\,dx = x\cos^{-1}\frac{x}{a} - \sqrt{a^2-x^2}$

14.478 $\int x\cos^{-1}\frac{x}{a}\,dx = \left(\frac{x^2}{2}-\frac{a^2}{4}\right)\cos^{-1}\frac{x}{a} - \frac{x\sqrt{a^2-x^2}}{4}$

14.479 $\int x^2\cos^{-1}\frac{x}{a}\,dx = \frac{x^3}{3}\cos^{-1}\frac{x}{a} - \frac{(x^2+2a^2)\sqrt{a^2-x^2}}{9}$

14.480 $\int \frac{\cos^{-1}(x/a)}{x}\,dx = \frac{\pi}{2}\ln x - \int \frac{\sin^{-1}(x/a)}{x}\,dx$ [See 14.474]

14.481 $\int \frac{\cos^{-1}(x/a)}{x^2}\,dx = -\frac{\cos^{-1}(x/a)}{x} + \frac{1}{a}\ln\left(\frac{a+\sqrt{a^2-x^2}}{x}\right)$

14.482 $\int \left(\cos^{-1}\frac{x}{a}\right)^2 dx = x\left(\cos^{-1}\frac{x}{a}\right)^2 - 2x - 2\sqrt{a^2-x^2}\cos^{-1}\frac{x}{a}$

14.483 $\int \tan^{-1}\frac{x}{a}\,dx = x\tan^{-1}\frac{x}{a} - \frac{a}{2}\ln(x^2+a^2)$

14.484 $\int x\tan^{-1}\frac{x}{a}\,dx = \frac{1}{2}(x^2+a^2)\tan^{-1}\frac{x}{a} - \frac{ax}{2}$

14.485 $\int x^2\tan^{-1}\frac{x}{a}\,dx = \frac{x^3}{3}\tan^{-1}\frac{x}{a} - \frac{ax^2}{6} + \frac{a^3}{6}\ln(x^2+a^2)$

14.486 $\int \frac{\tan^{-1}(x/a)}{x}\,dx = \frac{x}{a} - \frac{(x/a)^3}{3^2} + \frac{(x/a)^5}{5^2} - \frac{(x/a)^7}{7^2} + \cdots$

14.487 $\int \frac{\tan^{-1}(x/a)}{x^2}\,dx = -\frac{1}{x}\tan^{-1}\frac{x}{a} - \frac{1}{2a}\ln\left(\frac{x^2+a^2}{x^2}\right)$

14.488 $\int \cot^{-1}\frac{x}{a}\,dx = x\cot^{-1}\frac{x}{a} + \frac{a}{2}\ln(x^2+a^2)$

14.489 $\int x\cot^{-1}\frac{x}{a}\,dx = \frac{1}{2}(x^2+a^2)\cot^{-1}\frac{x}{a} + \frac{ax}{2}$

14.490 $\int x^2\cot^{-1}\frac{x}{a}\,dx = \frac{x^3}{3}\cot^{-1}\frac{x}{a} + \frac{ax^2}{6} - \frac{a^3}{6}\ln(x^2+a^2)$

14.491 $\int \frac{\cot^{-1}(x/a)}{x}\,dx = \frac{\pi}{2}\ln x - \int \frac{\tan^{-1}(x/a)}{x}\,dx$ [See 14.486]

14.492 $\int \frac{\cot^{-1}(x/a)}{x^2}\,dx = -\frac{\cot^{-1}(x/a)}{x} + \frac{1}{2a}\ln\left(\frac{x^2+a^2}{x^2}\right)$

14.493 $$\int \sec^{-1}\frac{x}{a}\,dx = \begin{cases} x\sec^{-1}\frac{x}{a} - a\ln(x+\sqrt{x^2-a^2}) & 0 < \sec^{-1}\frac{x}{a} < \frac{\pi}{2} \\ x\sec^{-1}\frac{x}{a} + a\ln(x+\sqrt{x^2-a^2}) & \frac{\pi}{2} < \sec^{-1}\frac{x}{a} < \pi \end{cases}$$

14.494 $$\int x\sec^{-1}\frac{x}{a}\,dx = \begin{cases} \frac{x^2}{2}\sec^{-1}\frac{x}{a} - \frac{a\sqrt{x^2-a^2}}{2} & 0 < \sec^{-1}\frac{x}{a} < \frac{\pi}{2} \\ \frac{x^2}{2}\sec^{-1}\frac{x}{a} + \frac{a\sqrt{x^2-a^2}}{2} & \frac{\pi}{2} < \sec^{-1}\frac{x}{a} < \pi \end{cases}$$

14.495 $$\int x^2\sec^{-1}\frac{x}{a}\,dx = \begin{cases} \frac{x^3}{3}\sec^{-1}\frac{x}{a} - \frac{ax\sqrt{x^2-a^2}}{6} - \frac{a^3}{6}\ln(x+\sqrt{x^2-a^2}) & 0 < \sec^{-1}\frac{x}{a} < \frac{\pi}{2} \\ \frac{x^3}{3}\sec^{-1}\frac{x}{a} + \frac{ax\sqrt{x^2-a^2}}{6} + \frac{a^3}{6}\ln(x+\sqrt{x^2-a^2}) & \frac{\pi}{2} < \sec^{-1}\frac{x}{a} < \pi \end{cases}$$

14.496 $$\int \frac{\sec^{-1}(x/a)}{x}\,dx = \frac{\pi}{2}\ln x + \frac{a}{x} + \frac{(a/x)^3}{2\cdot 3\cdot 3} + \frac{1\cdot 3(a/x)^5}{2\cdot 4\cdot 5\cdot 5} + \frac{1\cdot 3\cdot 5(a/x)^7}{2\cdot 4\cdot 6\cdot 7\cdot 7} + \cdots$$

14.497 $$\int \frac{\sec^{-1}(x/a)}{x^2}\,dx = \begin{cases} -\dfrac{\sec^{-1}(x/a)}{x} + \dfrac{\sqrt{x^2-a^2}}{ax} & 0 < \sec^{-1}\dfrac{x}{a} < \dfrac{\pi}{2} \\ -\dfrac{\sec^{-1}(x/a)}{x} - \dfrac{\sqrt{x^2-a^2}}{ax} & \dfrac{\pi}{2} < \sec^{-1}\dfrac{x}{a} < \pi \end{cases}$$

14.498 $$\int \csc^{-1}\frac{x}{a}\,dx = \begin{cases} x\csc^{-1}\dfrac{x}{a} + a\ln(x+\sqrt{x^2-a^2}) & 0 < \csc^{-1}\dfrac{x}{a} < \dfrac{\pi}{2} \\ x\csc^{-1}\dfrac{x}{a} - a\ln(x+\sqrt{x^2-a^2}) & -\dfrac{\pi}{2} < \csc^{-1}\dfrac{x}{a} < 0 \end{cases}$$

14.499 $$\int x\csc^{-1}\frac{x}{a}\,dx = \begin{cases} \dfrac{x^2}{2}\csc^{-1}\dfrac{x}{a} + \dfrac{a\sqrt{x^2-a^2}}{2} & 0 < \csc^{-1}\dfrac{x}{a} < \dfrac{\pi}{2} \\ \dfrac{x^2}{2}\csc^{-1}\dfrac{x}{a} - \dfrac{a\sqrt{x^2-a^2}}{2} & -\dfrac{\pi}{2} < \csc^{-1}\dfrac{x}{a} < 0 \end{cases}$$

14.500 $$\int x^2\csc^{-1}\frac{x}{a}\,dx = \begin{cases} \dfrac{x^3}{3}\csc^{-1}\dfrac{x}{a} + \dfrac{ax\sqrt{x^2-a^2}}{6} + \dfrac{a^3}{6}\ln(x+\sqrt{x^2-a^2}) & 0 < \csc^{-1}\dfrac{x}{a} < \dfrac{\pi}{2} \\ \dfrac{x^3}{3}\csc^{-1}\dfrac{x}{a} - \dfrac{ax\sqrt{x^2-a^2}}{6} - \dfrac{a^3}{6}\ln(x+\sqrt{x^2-a^2}) & -\dfrac{\pi}{2} < \csc^{-1}\dfrac{x}{a} < 0 \end{cases}$$

14.501 $$\int \frac{\csc^{-1}(x/a)}{x}\,dx = -\left(\frac{a}{x} + \frac{(a/x)^3}{2\cdot 3\cdot 3} + \frac{1\cdot 3(a/x)^5}{2\cdot 4\cdot 5\cdot 5} + \frac{1\cdot 3\cdot 5(a/x)^7}{2\cdot 4\cdot 6\cdot 7\cdot 7} + \cdots\right)$$

14.502 $$\int \frac{\csc^{-1}(x/a)}{x^2}\,dx = \begin{cases} -\dfrac{\csc^{-1}(x/a)}{x} - \dfrac{\sqrt{x^2-a^2}}{ax} & 0 < \csc^{-1}\dfrac{x}{a} < \dfrac{\pi}{2} \\ -\dfrac{\csc^{-1}(x/a)}{x} + \dfrac{\sqrt{x^2-a^2}}{ax} & -\dfrac{\pi}{2} < \csc^{-1}\dfrac{x}{a} < 0 \end{cases}$$

14.503 $$\int x^m\sin^{-1}\frac{x}{a}\,dx = \frac{x^{m+1}}{m+1}\sin^{-1}\frac{x}{a} - \frac{1}{m+1}\int\frac{x^{m+1}}{\sqrt{a^2-x^2}}\,dx$$

14.504 $$\int x^m\cos^{-1}\frac{x}{a}\,dx = \frac{x^{m+1}}{m+1}\cos^{-1}\frac{x}{a} + \frac{1}{m+1}\int\frac{x^{m+1}}{\sqrt{a^2-x^2}}\,dx$$

14.505 $$\int x^m\tan^{-1}\frac{x}{a}\,dx = \frac{x^{m+1}}{m+1}\tan^{-1}\frac{x}{a} - \frac{a}{m+1}\int\frac{x^{m+1}}{x^2+a^2}\,dx$$

14.506 $$\int x^m\cot^{-1}\frac{x}{a}\,dx = \frac{x^{m+1}}{m+1}\cot^{-1}\frac{x}{a} + \frac{a}{m+1}\int\frac{x^{m+1}}{x^2+a^2}\,dx$$

14.507 $$\int x^m\sec^{-1}\frac{x}{a}\,dx = \begin{cases} \dfrac{x^{m+1}\sec^{-1}(x/a)}{m+1} - \dfrac{a}{m+1}\displaystyle\int\frac{x^m\,dx}{\sqrt{x^2-a^2}} & 0 < \sec^{-1}\dfrac{x}{a} < \dfrac{\pi}{2} \\ \dfrac{x^{m+1}\sec^{-1}(x/a)}{m+1} + \dfrac{a}{m+1}\displaystyle\int\frac{x^m\,dx}{\sqrt{x^2-a^2}} & \dfrac{\pi}{2} < \sec^{-1}\dfrac{x}{a} < \pi \end{cases}$$

14.508 $$\int x^m\csc^{-1}\frac{x}{a}\,dx = \begin{cases} \dfrac{x^{m+1}\csc^{-1}(x/a)}{m+1} + \dfrac{a}{m+1}\displaystyle\int\frac{x^m\,dx}{\sqrt{x^2-a^2}} & 0 < \csc^{-1}\dfrac{x}{a} < \dfrac{\pi}{2} \\ \dfrac{x^{m+1}\csc^{-1}(x/a)}{m+1} - \dfrac{a}{m+1}\displaystyle\int\frac{x^m\,dx}{\sqrt{x^2-a^2}} & -\dfrac{\pi}{2} < \csc^{-1}\dfrac{x}{a} < 0 \end{cases}$$

INTEGRALS INVOLVING e^{ax}

14.509 $$\int e^{ax}\,dx = \frac{e^{ax}}{a}$$

14.510 $$\int xe^{ax}\,dx = \frac{e^{ax}}{a}\left(x - \frac{1}{a}\right)$$

14.511 $$\int x^2e^{ax}\,dx = \frac{e^{ax}}{a}\left(x^2 - \frac{2x}{a} + \frac{2}{a^2}\right)$$

14.512 $$\int x^ne^{ax}\,dx = \frac{x^ne^{ax}}{a} - \frac{n}{a}\int x^{n-1}e^{ax}\,dx$$
$$= \frac{e^{ax}}{a}\left(x^n - \frac{nx^{n-1}}{a} + \frac{n(n-1)x^{n-2}}{a^2} - \cdots \frac{(-1)^n n!}{a^n}\right) \quad \text{if } n = \text{positive integer}$$

14.513 $$\int \frac{e^{ax}}{x}\,dx = \ln x + \frac{ax}{1\cdot 1!} + \frac{(ax)^2}{2\cdot 2!} + \frac{(ax)^3}{3\cdot 3!} + \cdots$$

14.514 $$\int \frac{e^{ax}}{x^n}\,dx = \frac{-e^{ax}}{(n-1)x^{n-1}} + \frac{a}{n-1}\int \frac{e^{ax}}{x^{n-1}}\,dx$$

14.515 $$\int \frac{dx}{p + qe^{ax}} = \frac{x}{p} - \frac{1}{ap}\ln(p + qe^{ax})$$

14.516 $$\int \frac{dx}{(p + qe^{ax})^2} = \frac{x}{p^2} + \frac{1}{ap(p + qe^{ax})} - \frac{1}{ap^2}\ln(p + qe^{ax})$$

14.517 $$\int \frac{dx}{pe^{ax} + qe^{-ax}} = \begin{cases} \dfrac{1}{a\sqrt{pq}}\tan^{-1}\left(\sqrt{\dfrac{p}{q}}\,e^{ax}\right) \\ \dfrac{1}{2a\sqrt{-pq}}\ln\left(\dfrac{e^{ax} - \sqrt{-q/p}}{e^{ax} + \sqrt{-q/p}}\right) \end{cases}$$

14.518 $$\int e^{ax}\sin bx\,dx = \frac{e^{ax}(a\sin bx - b\cos bx)}{a^2 + b^2}$$

14.519 $$\int e^{ax}\cos bx\,dx = \frac{e^{ax}(a\cos bx + b\sin bx)}{a^2 + b^2}$$

14.520 $$\int xe^{ax}\sin bx\,dx = \frac{xe^{ax}(a\sin bx - b\cos bx)}{a^2 + b^2} - \frac{e^{ax}\{(a^2 - b^2)\sin bx - 2ab\cos bx\}}{(a^2 + b^2)^2}$$

14.521 $$\int xe^{ax}\cos bx\,dx = \frac{xe^{ax}(a\cos bx + b\sin bx)}{a^2 + b^2} - \frac{e^{ax}\{(a^2 - b^2)\cos bx + 2ab\sin bx\}}{(a^2 + b^2)^2}$$

14.522 $$\int e^{ax}\ln x\,dx = \frac{e^{ax}\ln x}{a} - \frac{1}{a}\int \frac{e^{ax}}{x}\,dx$$

14.523 $$\int e^{ax}\sin^n bx\,dx = \frac{e^{ax}\sin^{n-1} bx}{a^2 + n^2b^2}(a\sin bx - nb\cos bx) + \frac{n(n-1)b^2}{a^2 + n^2b^2}\int e^{ax}\sin^{n-2} bx\,dx$$

14.524 $$\int e^{ax}\cos^n bx\,dx = \frac{e^{ax}\cos^{n-1} bx}{a^2 + n^2b^2}(a\cos bx + nb\sin bx) + \frac{n(n-1)b^2}{a^2 + n^2b^2}\int e^{ax}\cos^{n-2} bx\,dx$$

INTEGRALS INVOLVING $\ln x$

14.525 $$\int \ln x\,dx = x\ln x - x$$

14.526 $$\int x\ln x\,dx = \frac{x^2}{2}(\ln x - \tfrac{1}{2})$$

14.527 $$\int x^m \ln x\,dx = \frac{x^{m+1}}{m+1}\left(\ln x - \frac{1}{m+1}\right)$$ [If $m=-1$ see 14.528.]

14.528 $$\int \frac{\ln x}{x}\,dx = \frac{1}{2}\ln^2 x$$

14.529 $$\int \frac{\ln x}{x^2}\,dx = -\frac{\ln x}{x} - \frac{1}{x}$$

14.530 $$\int \ln^2 x\,dx = x\ln^2 x - 2x\ln x + 2x$$

14.531 $$\int \frac{\ln^n x\,dx}{x} = \frac{\ln^{n+1} x}{n+1}$$ [If $n=-1$ see 14.532.]

14.532 $$\int \frac{dx}{x\ln x} = \ln(\ln x)$$

14.533 $$\int \frac{dx}{\ln x} = \ln(\ln x) + \ln x + \frac{\ln^2 x}{2\cdot 2!} + \frac{\ln^3 x}{3\cdot 3!} + \cdots$$

14.534 $$\int \frac{x^m\,dx}{\ln x} = \ln(\ln x) + (m+1)\ln x + \frac{(m+1)^2\ln^2 x}{2\cdot 2!} + \frac{(m+1)^3\ln^3 x}{3\cdot 3!} + \cdots$$

14.535 $$\int \ln^n x\,dx = x\ln^n x - n\int \ln^{n-1} x\,dx$$

14.536 $$\int x^m \ln^n x\,dx = \frac{x^{m+1}\ln^n x}{m+1} - \frac{n}{m+1}\int x^m \ln^{n-1} x\,dx$$

If $m=-1$ see 14.531.

14.537 $$\int \ln(x^2+a^2)\,dx = x\ln(x^2+a^2) - 2x + 2a\tan^{-1}\frac{x}{a}$$

14.538 $$\int \ln(x^2-a^2)\,dx = x\ln(x^2-a^2) - 2x + a\ln\left(\frac{x+a}{x-a}\right)$$

14.539 $$\int x^m \ln(x^2\pm a^2)\,dx = \frac{x^{m+1}\ln(x^2\pm a^2)}{m+1} - \frac{2}{m+1}\int \frac{x^{m+2}}{x^2\pm a^2}\,dx$$

INTEGRALS INVOLVING $\sinh ax$

14.540 $$\int \sinh ax\,dx = \frac{\cosh ax}{a}$$

14.541 $$\int x\sinh ax\,dx = \frac{x\cosh ax}{a} - \frac{\sinh ax}{a^2}$$

14.542 $$\int x^2\sinh ax\,dx = \left(\frac{x^2}{a} + \frac{2}{a^3}\right)\cosh ax - \frac{2x}{a^2}\sinh ax$$

14.543 $\displaystyle\int \frac{\sinh ax}{x}\,dx = ax + \frac{(ax)^3}{3\cdot 3!} + \frac{(ax)^5}{5\cdot 5!} + \cdots$

14.544 $\displaystyle\int \frac{\sinh ax}{x^2}\,dx = -\frac{\sinh ax}{x} + a\int \frac{\cosh ax}{x}\,dx$ [See 14.565]

14.545 $\displaystyle\int \frac{dx}{\sinh ax} = \frac{1}{a}\ln\tanh\frac{ax}{2}$

14.546 $\displaystyle\int \frac{x\,dx}{\sinh ax} = \frac{1}{a^2}\left\{ax - \frac{(ax)^3}{18} + \frac{7(ax)^5}{1800} - \cdots + \frac{2(-1)^n(2^{2n}-1)B_n(ax)^{2n+1}}{(2n+1)!} + \cdots\right\}$

14.547 $\displaystyle\int \sinh^2 ax\,dx = \frac{\sinh ax\cosh ax}{2a} - \frac{x}{2}$

14.548 $\displaystyle\int x\sinh^2 ax\,dx = \frac{x\sinh 2ax}{4a} - \frac{\cosh 2ax}{8a^2} - \frac{x^2}{4}$

14.549 $\displaystyle\int \frac{dx}{\sinh^2 ax} = -\frac{\coth ax}{a}$

14.550 $\displaystyle\int \sinh ax\sinh px\,dx = \frac{\sinh(a+p)x}{2(a+p)} - \frac{\sinh(a-p)x}{2(a-p)}$

For $a = \pm p$ see 14.547.

14.551 $\displaystyle\int \sinh ax\sin px\,dx = \frac{a\cosh ax\sin px - p\sinh ax\cos px}{a^2+p^2}$

14.552 $\displaystyle\int \sinh ax\cos px\,dx = \frac{a\cosh ax\cos px + p\sinh ax\sin px}{a^2+p^2}$

14.553 $\displaystyle\int \frac{dx}{p+q\sinh ax} = \frac{1}{a\sqrt{p^2+q^2}}\ln\left(\frac{qe^{ax}+p-\sqrt{p^2+q^2}}{qe^{ax}+p+\sqrt{p^2+q^2}}\right)$

14.554 $\displaystyle\int \frac{dx}{(p+q\sinh ax)^2} = \frac{-q\cosh ax}{a(p^2+q^2)(p+q\sinh ax)} + \frac{p}{p^2+q^2}\int \frac{dx}{p+q\sinh ax}$

14.555 $\displaystyle\int \frac{dx}{p^2+q^2\sinh^2 ax} = \begin{cases} \dfrac{1}{ap\sqrt{q^2-p^2}}\tan^{-1}\dfrac{\sqrt{q^2-p^2}\tanh ax}{p} \\[2ex] \dfrac{1}{2ap\sqrt{p^2-q^2}}\ln\left(\dfrac{p+\sqrt{p^2-q^2}\tanh ax}{p-\sqrt{p^2-q^2}\tanh ax}\right) \end{cases}$

14.556 $\displaystyle\int \frac{dx}{p^2-q^2\sinh^2 ax} = \frac{1}{2ap\sqrt{p^2+q^2}}\ln\left(\frac{p+\sqrt{p^2+q^2}\tanh ax}{p-\sqrt{p^2+q^2}\tanh ax}\right)$

14.557 $\displaystyle\int x^m\sinh ax\,dx = \frac{x^m\cosh ax}{a} - \frac{m}{a}\int x^{m-1}\cosh ax\,dx$ [See 14.585]

14.558 $\displaystyle\int \sinh^n ax\,dx = \frac{\sinh^{n-1}ax\cosh ax}{an} - \frac{n-1}{n}\int \sinh^{n-2}ax\,dx$

14.559 $\displaystyle\int \frac{\sinh ax}{x^n}\,dx = \frac{-\sinh ax}{(n-1)x^{n-1}} + \frac{a}{n-1}\int \frac{\cosh ax}{x^{n-1}}\,dx$ [See 14.587]

14.560 $\displaystyle\int \frac{dx}{\sinh^n ax} = \frac{-\cosh ax}{a(n-1)\sinh^{n-1}ax} - \frac{n-2}{n-1}\int \frac{dx}{\sinh^{n-2}ax}$

14.561 $\displaystyle\int \frac{x\,dx}{\sinh^n ax} = \frac{-x\cosh ax}{a(n-1)\sinh^{n-1}ax} - \frac{1}{a^2(n-1)(n-2)\sinh^{n-2}ax} - \frac{n-2}{n-1}\int \frac{x\,dx}{\sinh^{n-2}ax}$

INTEGRALS INVOLVING $\cosh ax$

14.562 $$\int \cosh ax\, dx = \frac{\sinh ax}{a}$$

14.563 $$\int x \cosh ax\, dx = \frac{x \sinh ax}{a} - \frac{\cosh ax}{a^2}$$

14.564 $$\int x^2 \cosh ax\, dx = -\frac{2x \cosh ax}{a^2} + \left(\frac{x^2}{a} + \frac{2}{a^3}\right) \sinh ax$$

14.565 $$\int \frac{\cosh ax}{x}\, dx = \ln x + \frac{(ax)^2}{2 \cdot 2!} + \frac{(ax)^4}{4 \cdot 4!} + \frac{(ax)^6}{6 \cdot 6!} + \cdots$$

14.566 $$\int \frac{\cosh ax}{x^2}\, dx = -\frac{\cosh ax}{x} + a \int \frac{\sinh ax}{x}\, dx$$ [See 14.543]

14.567 $$\int \frac{dx}{\cosh ax} = \frac{2}{a} \tan^{-1} e^{ax}$$

14.568 $$\int \frac{x\, dx}{\cosh ax} = \frac{1}{a^2}\left\{\frac{(ax)^2}{2} - \frac{(ax)^4}{8} + \frac{5(ax)^6}{144} + \cdots + \frac{(-1)^n E_n (ax)^{2n+2}}{(2n+2)(2n)!} + \cdots\right\}$$

14.569 $$\int \cosh^2 ax\, dx = \frac{x}{2} + \frac{\sinh ax \cosh ax}{2a}$$

14.570 $$\int x \cosh^2 ax\, dx = \frac{x^2}{4} + \frac{x \sinh 2ax}{4a} - \frac{\cosh 2ax}{8a^2}$$

14.571 $$\int \frac{dx}{\cosh^2 ax} = \frac{\tanh ax}{a}$$

14.572 $$\int \cosh ax \cosh px\, dx = \frac{\sinh (a-p)x}{2(a-p)} + \frac{\sinh (a+p)x}{2(a+p)}$$

14.573 $$\int \cosh ax \sin px\, dx = \frac{a \sinh ax \sin px - p \cosh ax \cos px}{a^2 + p^2}$$

14.574 $$\int \cosh ax \cos px\, dx = \frac{a \sinh ax \cos px + p \cosh ax \sin px}{a^2 + p^2}$$

14.575 $$\int \frac{dx}{\cosh ax + 1} = \frac{1}{a} \tanh \frac{ax}{2}$$

14.576 $$\int \frac{dx}{\cosh ax - 1} = -\frac{1}{a} \coth \frac{ax}{2}$$

14.577 $$\int \frac{x\, dx}{\cosh ax + 1} = \frac{x}{a} \tanh \frac{ax}{2} - \frac{2}{a^2} \ln \cosh \frac{ax}{2}$$

14.578 $$\int \frac{x\, dx}{\cosh ax - 1} = -\frac{x}{a} \coth \frac{ax}{2} + \frac{2}{a^2} \ln \sinh \frac{ax}{2}$$

14.579 $$\int \frac{dx}{(\cosh ax + 1)^2} = \frac{1}{2a} \tanh \frac{ax}{2} - \frac{1}{6a} \tanh^3 \frac{ax}{2}$$

14.580 $$\int \frac{dx}{(\cosh ax - 1)^2} = \frac{1}{2a} \coth \frac{ax}{2} - \frac{1}{6a} \coth^3 \frac{ax}{2}$$

14.581 $$\int \frac{dx}{p + q \cosh ax} = \begin{cases} \dfrac{2}{a\sqrt{q^2 - p^2}} \tan^{-1} \dfrac{qe^{ax} + p}{\sqrt{q^2 - p^2}} \\[2ex] \dfrac{1}{a\sqrt{p^2 - q^2}} \ln \left(\dfrac{qe^{ax} + p - \sqrt{p^2 - q^2}}{qe^{ax} + p + \sqrt{p^2 - q^2}}\right) \end{cases}$$

14.582 $$\int \frac{dx}{(p + q \cosh ax)^2} = \frac{q \sinh ax}{a(q^2 - p^2)(p + q \cosh ax)} - \frac{p}{q^2 - p^2} \int \frac{dx}{p + q \cosh ax}$$

14.583 $\displaystyle\int \frac{dx}{p^2 - q^2\cosh^2 ax} = \begin{cases} \dfrac{1}{2ap\sqrt{p^2-q^2}} \ln\left(\dfrac{p\tanh ax + \sqrt{p^2-q^2}}{p\tanh ax - \sqrt{p^2-q^2}}\right) \\ \dfrac{-1}{ap\sqrt{q^2-p^2}} \tan^{-1} \dfrac{p\tanh ax}{\sqrt{q^2-p^2}} \end{cases}$

14.584 $\displaystyle\int \frac{dx}{p^2 + q^2\cosh^2 ax} = \begin{cases} \dfrac{1}{2ap\sqrt{p^2+q^2}} \ln\left(\dfrac{p\tanh ax + \sqrt{p^2+q^2}}{p\tanh ax - \sqrt{p^2+q^2}}\right) \\ \dfrac{1}{ap\sqrt{p^2+q^2}} \tan^{-1} \dfrac{p\tanh ax}{\sqrt{p^2+q^2}} \end{cases}$

14.585 $\displaystyle\int x^m \cosh ax\, dx = \frac{x^m \sinh ax}{a} - \frac{m}{a}\int x^{m-1}\sinh ax\, dx$ [See 14.557]

14.586 $\displaystyle\int \cosh^n ax\, dx = \frac{\cosh^{n-1} ax \sinh ax}{an} + \frac{n-1}{n}\int \cosh^{n-2} ax\, dx$

14.587 $\displaystyle\int \frac{\cosh ax}{x^n}\, dx = \frac{-\cosh ax}{(n-1)x^{n-1}} + \frac{a}{n-1}\int \frac{\sinh ax}{x^{n-1}}\, dx$ [See 14.559]

14.588 $\displaystyle\int \frac{dx}{\cosh^n ax} = \frac{\sinh ax}{a(n-1)\cosh^{n-1} ax} + \frac{n-2}{n-1}\int \frac{dx}{\cosh^{n-2} ax}$

14.589 $\displaystyle\int \frac{x\, dx}{\cosh^n ax} = \frac{x \sinh ax}{a(n-1)\cosh^{n-1} ax} + \frac{1}{(n-1)(n-2)a^2 \cosh^{n-2} ax} + \frac{n-2}{n-1}\int \frac{x\, dx}{\cosh^{n-2} ax}$

INTEGRALS INVOLVING sinh ax AND cosh ax

14.590 $\displaystyle\int \sinh ax \cosh ax\, dx = \frac{\sinh^2 ax}{2a}$

14.591 $\displaystyle\int \sinh px \cosh qx\, dx = \frac{\cosh (p+q)x}{2(p+q)} + \frac{\cosh (p-q)x}{2(p-q)}$

14.592 $\displaystyle\int \sinh^n ax \cosh ax\, dx = \frac{\sinh^{n+1} ax}{(n+1)a}$ [If $n = -1$, see 14.615.]

14.593 $\displaystyle\int \cosh^n ax \sinh ax\, dx = \frac{\cosh^{n+1} ax}{(n+1)a}$ [If $n = -1$, see 14.604.]

14.594 $\displaystyle\int \sinh^2 ax \cosh^2 ax\, dx = \frac{\sinh 4ax}{32a} - \frac{x}{8}$

14.595 $\displaystyle\int \frac{dx}{\sinh ax \cosh ax} = \frac{1}{a}\ln \tanh ax$

14.596 $\displaystyle\int \frac{dx}{\sinh^2 ax \cosh ax} = -\frac{1}{a}\tan^{-1}\sinh ax - \frac{\operatorname{csch} ax}{a}$

14.597 $\displaystyle\int \frac{dx}{\sinh ax \cosh^2 ax} = \frac{\operatorname{sech} ax}{a} + \frac{1}{a}\ln \tanh \frac{ax}{2}$

14.598 $\displaystyle\int \frac{dx}{\sinh^2 ax \cosh^2 ax} = -\frac{2\coth 2ax}{a}$

14.599 $\displaystyle\int \frac{\sinh^2 ax}{\cosh ax}\, dx = \frac{\sinh ax}{a} - \frac{1}{a}\tan^{-1}\sinh ax$

14.600 $\displaystyle\int \frac{\cosh^2 ax}{\sinh ax}\, dx = \frac{\cosh ax}{a} + \frac{1}{a}\ln \tanh \frac{ax}{2}$

14.601 $\displaystyle\int \frac{dx}{\cosh ax\,(1+\sinh ax)} = \frac{1}{2a}\ln\left(\frac{1+\sinh ax}{\cosh ax}\right) + \frac{1}{a}\tan^{-1} e^{ax}$

14.602 $$\int \frac{dx}{\sinh ax\,(\cosh ax + 1)} = \frac{1}{2a}\ln\tanh\frac{ax}{2} + \frac{1}{2a(\cosh ax + 1)}$$

14.603 $$\int \frac{dx}{\sinh ax\,(\cosh ax - 1)} = -\frac{1}{2a}\ln\tanh\frac{ax}{2} - \frac{1}{2a(\cosh ax - 1)}$$

INTEGRALS INVOLVING tanh ax

14.604 $$\int \tanh ax\,dx = \frac{1}{a}\ln\cosh ax$$

14.605 $$\int \tanh^2 ax\,dx = x - \frac{\tanh ax}{a}$$

14.606 $$\int \tanh^3 ax\,dx = \frac{1}{a}\ln\cosh ax - \frac{\tanh^2 ax}{2a}$$

14.607 $$\int \tanh^n ax\,\mathrm{sech}^2 ax\,dx = \frac{\tanh^{n+1} ax}{(n+1)a}$$

14.608 $$\int \frac{\mathrm{sech}^2 ax}{\tanh ax}\,dx = \frac{1}{a}\ln\tanh ax$$

14.609 $$\int \frac{dx}{\tanh ax} = \frac{1}{a}\ln\sinh ax$$

14.610 $$\int x\tanh ax\,dx = \frac{1}{a^2}\left\{\frac{(ax)^3}{3} - \frac{(ax)^5}{15} + \frac{2(ax)^7}{105} - \cdots \frac{(-1)^{n-1}2^{2n}(2^{2n}-1)B_n(ax)^{2n+1}}{(2n+1)!} + \cdots\right\}$$

14.611 $$\int x\tanh^2 ax\,dx = \frac{x^2}{2} - \frac{x\tanh ax}{a} + \frac{1}{a^2}\ln\cosh ax$$

14.612 $$\int \frac{\tanh ax}{x}\,dx = ax - \frac{(ax)^3}{9} + \frac{2(ax)^5}{75} - \cdots \frac{(-1)^{n-1}2^{2n}(2^{2n}-1)B_n(ax)^{2n-1}}{(2n-1)(2n)!} + \cdots$$

14.613 $$\int \frac{dx}{p + q\tanh ax} = \frac{px}{p^2 - q^2} - \frac{q}{a(p^2 - q^2)}\ln(q\sinh ax + p\cosh ax)$$

14.614 $$\int \tanh^n ax\,dx = \frac{-\tanh^{n-1} ax}{a(n-1)} + \int \tanh^{n-2} ax\,dx$$

INTEGRALS INVOLVING coth ax

14.615 $$\int \coth ax\,dx = \frac{1}{a}\ln\sinh ax$$

14.616 $$\int \coth^2 ax\,dx = x - \frac{\coth ax}{a}$$

14.617 $$\int \coth^3 ax\,dx = \frac{1}{a}\ln\sinh ax - \frac{\coth^2 ax}{2a}$$

14.618 $$\int \coth^n ax\,\mathrm{csch}^2 ax\,dx = -\frac{\coth^{n+1} ax}{(n+1)a}$$

14.619 $$\int \frac{\mathrm{csch}^2 ax}{\coth ax}\,dx = -\frac{1}{a}\ln\coth ax$$

14.620 $$\int \frac{dx}{\coth ax} = \frac{1}{a}\ln\cosh ax$$

14.621 $$\int x \coth ax\,dx = \frac{1}{a^2}\left\{ax + \frac{(ax)^3}{9} - \frac{(ax)^5}{225} + \cdots \frac{(-1)^{n-1}2^{2n}B_n(ax)^{2n+1}}{(2n+1)!} + \cdots\right\}$$

14.622 $$\int x \coth^2 ax\,dx = \frac{x^2}{2} - \frac{x \coth ax}{a} + \frac{1}{a^2}\ln \sinh ax$$

14.623 $$\int \frac{\coth ax}{x}\,dx = -\frac{1}{ax} + \frac{ax}{3} - \frac{(ax)^3}{135} + \cdots \frac{(-1)^n 2^{2n}B_n(ax)^{2n-1}}{(2n-1)(2n)!} + \cdots$$

14.624 $$\int \frac{dx}{p + q\coth ax} = \frac{px}{p^2 - q^2} - \frac{q}{a(p^2-q^2)}\ln(p \sinh ax + q\cosh ax)$$

14.625 $$\int \coth^n ax\,dx = -\frac{\coth^{n-1} ax}{a(n-1)} + \int \coth^{n-2} ax\,dx$$

INTEGRALS INVOLVING sech ax

14.626 $$\int \operatorname{sech} ax\,dx = \frac{2}{a}\tan^{-1} e^{ax}$$

14.627 $$\int \operatorname{sech}^2 ax\,dx = \frac{\tanh ax}{a}$$

14.628 $$\int \operatorname{sech}^3 ax\,dx = \frac{\operatorname{sech} ax \tanh ax}{2a} + \frac{1}{2a}\tan^{-1}\sinh ax$$

14.629 $$\int \operatorname{sech}^n ax \tanh ax\,dx = -\frac{\operatorname{sech}^n ax}{na}$$

14.630 $$\int \frac{dx}{\operatorname{sech} ax} = \frac{\sinh ax}{a}$$

14.631 $$\int x \operatorname{sech} ax\,dx = \frac{1}{a^2}\left\{\frac{(ax)^2}{2} - \frac{(ax)^4}{8} + \frac{5(ax)^6}{144} + \cdots \frac{(-1)^n E_n(ax)^{2n+2}}{(2n+2)(2n)!} + \cdots\right\}$$

14.632 $$\int x \operatorname{sech}^2 ax\,dx = \frac{x\tanh ax}{a} - \frac{1}{a^2}\ln\cosh ax$$

14.633 $$\int \frac{\operatorname{sech} ax}{x}\,dx = \ln x - \frac{(ax)^2}{4} + \frac{5(ax)^4}{96} - \frac{61(ax)^6}{4320} + \cdots \frac{(-1)^n E_n(ax)^{2n}}{2n(2n)!} + \cdots$$

14.634 $$\int \frac{dx}{q + p\operatorname{sech} ax} = \frac{x}{q} - \frac{p}{q}\int \frac{dx}{p + q\cosh ax}$$ [See 14.581]

14.635 $$\int \operatorname{sech}^n ax\,dx = \frac{\operatorname{sech}^{n-2} ax \tanh ax}{a(n-1)} + \frac{n-2}{n-1}\int \operatorname{sech}^{n-2} ax\,dx$$

INTEGRALS INVOLVING csch ax

14.636 $$\int \operatorname{csch} ax\,dx = \frac{1}{a}\ln\tanh\frac{ax}{2}$$

14.637 $$\int \operatorname{csch}^2 ax\,dx = -\frac{\coth ax}{a}$$

14.638 $$\int \operatorname{csch}^3 ax\,dx = -\frac{\operatorname{csch} ax \coth ax}{2a} - \frac{1}{2a}\ln\tanh\frac{ax}{2}$$

14.639 $$\int \operatorname{csch}^n ax \coth ax\,dx = -\frac{\operatorname{csch}^n ax}{na}$$

14.640 $\displaystyle\int \frac{dx}{\operatorname{csch} ax} = \frac{1}{a}\cosh ax$

14.641 $\displaystyle\int x \operatorname{csch} ax\, dx = \frac{1}{a^2}\left\{ax - \frac{(ax)^3}{18} + \frac{7(ax)^5}{1800} + \cdots + \frac{2(-1)^n(2^{2n-1}-1)B_n(ax)^{2n+1}}{(2n+1)!} + \cdots\right\}$

14.642 $\displaystyle\int x \operatorname{csch}^2 ax\, dx = -\frac{x \coth ax}{a} + \frac{1}{a^2}\ln \sinh ax$

14.643 $\displaystyle\int \frac{\operatorname{csch} ax}{x}\, dx = -\frac{1}{ax} - \frac{ax}{6} + \frac{7(ax)^3}{1080} + \cdots \frac{(-1)^n 2(2^{2n-1}-1)B_n(ax)^{2n-1}}{(2n-1)(2n)!} + \cdots$

14.644 $\displaystyle\int \frac{dx}{q + p \operatorname{csch} ax} = \frac{x}{q} - \frac{p}{q}\int \frac{dx}{p + q \sinh ax}$ [See 14.553]

14.645 $\displaystyle\int \operatorname{csch}^n ax\, dx = \frac{-\operatorname{csch}^{n-2} ax \coth ax}{a(n-1)} - \frac{n-2}{n-1}\int \operatorname{csch}^{n-2} ax\, dx$

INTEGRALS INVOLVING INVERSE HYPERBOLIC FUNCTIONS

14.646 $\displaystyle\int \sinh^{-1}\frac{x}{a}\, dx = x \sinh^{-1}\frac{x}{a} - \sqrt{x^2+a^2}$

14.647 $\displaystyle\int x \sinh^{-1}\frac{x}{a}\, dx = \left(\frac{x^2}{2} + \frac{a^2}{4}\right)\sinh^{-1}\frac{x}{a} - \frac{x\sqrt{x^2+a^2}}{4}$

14.648 $\displaystyle\int x^2 \sinh^{-1}\frac{x}{a}\, dx = \frac{x^3}{3}\sinh^{-1}\frac{x}{a} + \frac{(2a^2-x^2)\sqrt{x^2+a^2}}{9}$

14.649 $$\int \frac{\sinh^{-1}(x/a)}{x}\, dx = \begin{cases} \dfrac{x}{a} - \dfrac{(x/a)^3}{2\cdot 3\cdot 3} + \dfrac{1\cdot 3(x/a)^5}{2\cdot 4\cdot 5\cdot 5} - \dfrac{1\cdot 3\cdot 5(x/a)^7}{2\cdot 4\cdot 6\cdot 7\cdot 7} + \cdots & |x| < a \\ \dfrac{\ln^2(2x/a)}{2} - \dfrac{(a/x)^2}{2\cdot 2\cdot 2} + \dfrac{1\cdot 3(a/x)^4}{2\cdot 4\cdot 4\cdot 4} - \dfrac{1\cdot 3\cdot 5(a/x)^6}{2\cdot 4\cdot 6\cdot 6\cdot 6} + \cdots & x > a \\ -\dfrac{\ln^2(-2x/a)}{2} + \dfrac{(a/x)^2}{2\cdot 2\cdot 2} - \dfrac{1\cdot 3(a/x)^4}{2\cdot 4\cdot 4\cdot 4} + \dfrac{1\cdot 3\cdot 5(a/x)^6}{2\cdot 4\cdot 6\cdot 6\cdot 6} - \cdots & x < -a \end{cases}$$

14.650 $\displaystyle\int \frac{\sinh^{-1}(x/a)}{x^2}\, dx = -\frac{\sinh^{-1}(x/a)}{x} - \frac{1}{a}\ln\left(\frac{a+\sqrt{x^2+a^2}}{x}\right)$

14.651 $$\int \cosh^{-1}\frac{x}{a}\, dx = \begin{cases} x\cosh^{-1}(x/a) - \sqrt{x^2-a^2}, & \cosh^{-1}(x/a) > 0 \\ x\cosh^{-1}(x/a) + \sqrt{x^2-a^2}, & \cosh^{-1}(x/a) < 0 \end{cases}$$

14.652 $$\int x\cosh^{-1}\frac{x}{a}\, dx = \begin{cases} \tfrac{1}{4}(2x^2-a^2)\cosh^{-1}(x/a) - \tfrac{1}{4}x\sqrt{x^2-a^2}, & \cosh^{-1}(x/a) > 0 \\ \tfrac{1}{4}(2x^2-a^2)\cosh^{-1}(x/a) + \tfrac{1}{4}x\sqrt{x^2-a^2}, & \cosh^{-1}(x/a) < 0 \end{cases}$$

14.653 $$\int x^2\cosh^{-1}\frac{x}{a}\, dx = \begin{cases} \tfrac{1}{3}x^3\cosh^{-1}(x/a) - \tfrac{1}{9}(x^2+2a^2)\sqrt{x^2-a^2}, & \cosh^{-1}(x/a) > 0 \\ \tfrac{1}{3}x^3\cosh^{-1}(x/a) + \tfrac{1}{9}(x^2+2a^2)\sqrt{x^2-a^2}, & \cosh^{-1}(x/a) < 0 \end{cases}$$

14.654 $\displaystyle\int \frac{\cosh^{-1}(x/a)}{x}\, dx = \pm\left[\frac{1}{2}\ln^2(2x/a) + \frac{(a/x)^2}{2\cdot 2\cdot 2} + \frac{1\cdot 3(a/x)^4}{2\cdot 4\cdot 4\cdot 4} + \frac{1\cdot 3\cdot 5(a/x)^6}{2\cdot 4\cdot 6\cdot 6\cdot 6} + \cdots\right]$

$+$ if $\cosh^{-1}(x/a) > 0$, $-$ if $\cosh^{-1}(x/a) < 0$

14.655 $\displaystyle\int \frac{\cosh^{-1}(x/a)}{x^2}\, dx = -\frac{\cosh^{-1}(x/a)}{x} \mp \frac{1}{a}\ln\left(\frac{a+\sqrt{x^2+a^2}}{x}\right)$ [$-$ if $\cosh^{-1}(x/a) > 0$, $+$ if $\cosh^{-1}(x/a) < 0$]

14.656 $\displaystyle\int \tanh^{-1}\frac{x}{a}\, dx = x\tanh^{-1}\frac{x}{a} + \frac{a}{2}\ln(a^2-x^2)$

14.657 $\displaystyle\int x\tanh^{-1}\frac{x}{a}\, dx = \frac{ax}{2} + \tfrac{1}{2}(x^2-a^2)\tanh^{-1}\frac{x}{a}$

14.658 $\displaystyle\int x^2\tanh^{-1}\frac{x}{a}\, dx = \frac{ax^2}{6} + \frac{x^3}{3}\tanh^{-1}\frac{x}{a} + \frac{a^3}{6}\ln(a^2-x^2)$

14.659 $\displaystyle\int \frac{\tanh^{-1}(x/a)}{x}\,dx = \frac{x}{a} + \frac{(x/a)^3}{3^2} + \frac{(x/a)^5}{5^2} + \cdots$

14.660 $\displaystyle\int \frac{\tanh^{-1}(x/a)}{x^2}\,dx = -\frac{\tanh^{-1}(x/a)}{x} + \frac{1}{2a}\ln\left(\frac{x^2}{a^2-x^2}\right)$

14.661 $\displaystyle\int \coth^{-1}\frac{x}{a}\,dx = x\coth^{-1}x + \frac{a}{2}\ln(x^2-a^2)$

14.662 $\displaystyle\int x\coth^{-1}\frac{x}{a}\,dx = \frac{ax}{2} + \tfrac{1}{2}(x^2-a^2)\coth^{-1}\frac{x}{a}$

14.663 $\displaystyle\int x^2\coth^{-1}\frac{x}{a}\,dx = \frac{ax^2}{6} + \frac{x^3}{3}\coth^{-1}\frac{x}{a} + \frac{a^3}{6}\ln(x^2-a^2)$

14.664 $\displaystyle\int \frac{\coth^{-1}(x/a)}{x}\,dx = -\left(\frac{a}{x} + \frac{(a/x)^3}{3^2} + \frac{(a/x)^5}{5^2} + \cdots\right)$

14.665 $\displaystyle\int \frac{\coth^{-1}(x/a)}{x^2}\,dx = -\frac{\coth^{-1}(x/a)}{x} + \frac{1}{2a}\ln\left(\frac{x^2}{x^2-a^2}\right)$

14.666 $\displaystyle\int \operatorname{sech}^{-1}\frac{x}{a}\,dx = \begin{cases} x\operatorname{sech}^{-1}(x/a) + a\sin^{-1}(x/a), & \operatorname{sech}^{-1}(x/a) > 0 \\ x\operatorname{sech}^{-1}(x/a) - a\sin^{-1}(x/a), & \operatorname{sech}^{-1}(x/a) < 0 \end{cases}$

14.667 $\displaystyle\int x\operatorname{sech}^{-1}\frac{x}{a}\,dx = \begin{cases} \tfrac{1}{2}x^2\operatorname{sech}^{-1}(x/a) - \tfrac{1}{2}a\sqrt{a^2-x^2}, & \operatorname{sech}^{-1}(x/a) > 0 \\ \tfrac{1}{2}x^2\operatorname{sech}^{-1}(x/a) + \tfrac{1}{2}a\sqrt{a^2-x^2}, & \operatorname{sech}^{-1}(x/a) < 0 \end{cases}$

14.668 $\displaystyle\int \frac{\operatorname{sech}^{-1}(x/a)}{x}\,dx = \begin{cases} -\tfrac{1}{2}\ln(a/x)\ln(4a/x) - \dfrac{(x/a)^2}{2\cdot 2\cdot 2} - \dfrac{1\cdot 3(x/a)^4}{2\cdot 4\cdot 4\cdot 4} - \cdots, & \operatorname{sech}^{-1}(x/a) > 0 \\ \tfrac{1}{2}\ln(a/x)\ln(4a/x) + \dfrac{(x/a)^2}{2\cdot 2\cdot 2} + \dfrac{1\cdot 3(x/a)^4}{2\cdot 4\cdot 4\cdot 4} + \cdots, & \operatorname{sech}^{-1}(x/a) < 0 \end{cases}$

14.669 $\displaystyle\int \operatorname{csch}^{-1}\frac{x}{a}\,dx = x\operatorname{csch}^{-1}\frac{x}{a} \pm a\sinh^{-1}\frac{x}{a} \qquad [+ \text{ if } x > 0,\ - \text{ if } x < 0]$

14.670 $\displaystyle\int x\operatorname{csch}^{-1}\frac{x}{a}\,dx = \frac{x^2}{2}\operatorname{csch}^{-1}\frac{x}{a} \pm \frac{a\sqrt{x^2+a^2}}{2} \qquad [+ \text{ if } x > 0,\ - \text{ if } x < 0]$

14.671 $\displaystyle\int \frac{\operatorname{csch}^{-1}(x/a)}{x}\,dx = \begin{cases} \tfrac{1}{2}\ln(x/a)\ln(4a/x) + \dfrac{1(x/a)^2}{2\cdot 2\cdot 2} - \dfrac{1\cdot 3(x/a)^4}{2\cdot 4\cdot 4\cdot 4} + \cdots & 0 < x < a \\ \tfrac{1}{2}\ln(-x/a)\ln(-x/4a) - \dfrac{(x/a)^2}{2\cdot 2\cdot 2} + \dfrac{1\cdot 3(x/a)^4}{2\cdot 4\cdot 4\cdot 4} - \cdots & -a < x < 0 \\ -\dfrac{a}{x} + \dfrac{(a/x)^3}{2\cdot 3\cdot 3} - \dfrac{1\cdot 3(a/x)^5}{2\cdot 4\cdot 5\cdot 5} + \cdots & |x| > a \end{cases}$

14.672 $\displaystyle\int x^m\sinh^{-1}\frac{x}{a}\,dx = \frac{x^{m+1}}{m+1}\sinh^{-1}\frac{x}{a} - \frac{1}{m+1}\int\frac{x^{m+1}}{\sqrt{x^2+a^2}}\,dx$

14.673 $\displaystyle\int x^m\cosh^{-1}\frac{x}{a}\,dx = \begin{cases} \dfrac{x^{m+1}}{m+1}\cosh^{-1}\dfrac{x}{a} - \dfrac{1}{m+1}\displaystyle\int\frac{x^{m+1}}{\sqrt{x^2-a^2}}\,dx & \cosh^{-1}(x/a) > 0 \\ \dfrac{x^{m+1}}{m+1}\cosh^{-1}\dfrac{x}{a} + \dfrac{1}{m+1}\displaystyle\int\frac{x^{m+1}}{\sqrt{x^2-a^2}}\,dx & \cosh^{-1}(x/a) < 0 \end{cases}$

14.674 $\displaystyle\int x^m\tanh^{-1}\frac{x}{a}\,dx = \frac{x^{m+1}}{m+1}\tanh^{-1}\frac{x}{a} - \frac{a}{m+1}\int\frac{x^{m+1}}{a^2-x^2}\,dx$

14.675 $\displaystyle\int x^m\coth^{-1}\frac{x}{a}\,dx = \frac{x^{m+1}}{m+1}\coth^{-1}\frac{x}{a} - \frac{a}{m+1}\int\frac{x^{m+1}}{a^2-x^2}\,dx$

14.676 $\displaystyle\int x^m\operatorname{sech}^{-1}\frac{x}{a}\,dx = \begin{cases} \dfrac{x^{m+1}}{m+1}\operatorname{sech}^{-1}\dfrac{x}{a} + \dfrac{a}{m+1}\displaystyle\int\frac{x^m\,dx}{\sqrt{a^2-x^2}} & \operatorname{sech}^{-1}(x/a) > 0 \\ \dfrac{x^{m+1}}{m+1}\operatorname{sech}^{-1}\dfrac{x}{a} - \dfrac{a}{m+1}\displaystyle\int\frac{x^m\,dx}{\sqrt{a^2-x^2}} & \operatorname{sech}^{-1}(x/a) < 0 \end{cases}$

14.677 $\displaystyle\int x^m\operatorname{csch}^{-1}\frac{x}{a}\,dx = \frac{x^{m+1}}{m+1}\operatorname{csch}^{-1}\frac{x}{a} \pm \frac{a}{m+1}\int\frac{x^m\,dx}{\sqrt{x^2+a^2}} \qquad [+ \text{ if } x > 0,\ - \text{ if } x < 0]$

15 DEFINITE INTEGRALS

DEFINITION OF A DEFINITE INTEGRAL

Let $f(x)$ be defined in an interval $a \leqq x \leqq b$. Divide the interval into n equal parts of length $\Delta x = (b-a)/n$. Then the definite integral of $f(x)$ between $x = a$ and $x = b$ is defined as

15.1 $$\int_a^b f(x)\,dx = \lim_{n\to\infty} \{f(a)\,\Delta x + f(a+\Delta x)\,\Delta x + f(a+2\Delta x)\,\Delta x + \cdots + f(a+(n-1)\,\Delta x)\,\Delta x\}$$

The limit will certainly exist if $f(x)$ is piecewise continuous.

If $f(x) = \dfrac{d}{dx}g(x)$, then by the fundamental theorem of the integral calculus the above definite integral can be evaluated by using the result

15.2 $$\int_a^b f(x)\,dx = \int_a^b \frac{d}{dx}g(x)\,dx = g(x)\Big|_a^b = g(b) - g(a)$$

If the interval is infinite or if $f(x)$ has a singularity at some point in the interval, the definite integral is called an *improper integral* and can be defined by using appropriate limiting procedures. For example,

15.3 $$\int_a^\infty f(x)\,dx = \lim_{b\to\infty} \int_a^b f(x)\,dx$$

15.4 $$\int_{-\infty}^\infty f(x)\,dx = \lim_{\substack{a\to-\infty \\ b\to\infty}} \int_a^b f(x)\,dx$$

15.5 $$\int_a^b f(x)\,dx = \lim_{\epsilon\to 0} \int_a^{b-\epsilon} f(x)\,dx \qquad \text{if } b \text{ is a singular point}$$

15.6 $$\int_a^b f(x)\,dx = \lim_{\epsilon\to 0} \int_{a+\epsilon}^{b} f(x)\,dx \qquad \text{if } a \text{ is a singular point}$$

GENERAL FORMULAS INVOLVING DEFINITE INTEGRALS

15.7 $$\int_a^b \{f(x) \pm g(x) \pm h(x) \pm \cdots\}\,dx = \int_a^b f(x)\,dx \pm \int_a^b g(x)\,dx \pm \int_a^b h(x)\,dx \pm \cdots$$

15.8 $$\int_a^b c\,f(x)\,dx = c\int_a^b f(x)\,dx \qquad \text{where } c \text{ is any constant}$$

15.9 $$\int_a^a f(x)\,dx = 0$$

15.10 $$\int_a^b f(x)\,dx = -\int_b^a f(x)\,dx$$

15.11 $$\int_a^b f(x)\,dx = \int_a^c f(x)\,dx + \int_c^b f(x)\,dx$$

15.12 $$\int_a^b f(x)\,dx = (b-a)\,f(c) \qquad \text{where } c \text{ is between } a \text{ and } b$$

This is called the *mean value theorem* for definite integrals and is valid if $f(x)$ is continuous in $a \leqq x \leqq b$.

15.13 $$\int_a^b f(x)\,g(x)\,dx = f(c)\int_a^b g(x)\,dx \qquad \text{where } c \text{ is between } a \text{ and } b$$

This is a generalization of 15.12 and is valid if $f(x)$ and $g(x)$ are continuous in $a \leqq x \leqq b$ and $g(x) \geqq 0$.

LEIBNITZ'S RULE FOR DIFFERENTIATION OF INTEGRALS

15.14 $$\frac{d}{d\alpha}\int_{\phi_1(\alpha)}^{\phi_2(\alpha)} F(x,\alpha)\,dx = \int_{\phi_1(\alpha)}^{\phi_2(\alpha)} \frac{\partial F}{\partial \alpha}\,dx + F(\phi_2,\alpha)\frac{d\phi_1}{d\alpha} - F(\phi_1,\alpha)\frac{d\phi_2}{d\alpha}$$

APPROXIMATE FORMULAS FOR DEFINITE INTEGRALS

In the following the interval from $x = a$ to $x = b$ is subdivided into n equal parts by the points $a = x_0$, $x_1, x_2, \ldots, x_{n-1}, x_n = b$ and we let $y_0 = f(x_0)$, $y_1 = f(x_1)$, $y_2 = f(x_2)$, $\ldots$, $y_n = f(x_n)$, $h = (b-a)/n$.

Rectangular formula

15.15 $$\int_a^b f(x)\,dx \approx h(y_0 + y_1 + y_2 + \cdots + y_{n-1})$$

Trapezoidal formula

15.16 $$\int_a^b f(x)\,dx \approx \frac{h}{2}(y_0 + 2y_1 + 2y_2 + \cdots + 2y_{n-1} + y_n)$$

Simpson's formula (or parabolic formula) for n even

15.17 $$\int_a^b f(x)\,dx \approx \frac{h}{3}(y_0 + 4y_1 + 2y_2 + 4y_3 + \cdots + 2y_{n-2} + 4y_{n-1} + y_n)$$

DEFINITE INTEGRALS INVOLVING RATIONAL OR IRRATIONAL EXPRESSIONS

15.18 $$\int_0^\infty \frac{dx}{x^2 + a^2} = \frac{\pi}{2a}$$

15.19 $$\int_0^\infty \frac{x^{p-1}\,dx}{1+x} = \frac{\pi}{\sin p\pi}, \qquad 0 < p < 1$$

15.20 $$\int_0^\infty \frac{x^m\,dx}{x^n + a^n} = \frac{\pi a^{m+1-n}}{n \sin[(m+1)\pi/n]}, \qquad 0 < m+1 < n$$

15.21 $$\int_0^\infty \frac{x^m\,dx}{1 + 2x\cos\beta + x^2} = \frac{\pi}{\sin m\pi}\,\frac{\sin m\beta}{\sin\beta}$$

15.22 $$\int_0^a \frac{dx}{\sqrt{a^2 - x^2}} = \frac{\pi}{2}$$

15.23 $$\int_0^a \sqrt{a^2 - x^2}\,dx = \frac{\pi a^2}{4}$$

15.24 $$\int_0^a x^m(a^n - x^n)^p\,dx = \frac{a^{m+1+np}\,\Gamma[(m+1)/n]\,\Gamma(p+1)}{n\Gamma[(m+1)/n + p + 1]}$$

15.25 $$\int_0^\infty \frac{x^m\,dx}{(x^n + a^n)^r} = \frac{(-1)^{r-1}\pi a^{m+1-nr}\,\Gamma[(m+1)/n]}{n\sin[(m+1)\pi/n](r-1)!\,\Gamma[(m+1)/n - r + 1]}, \qquad 0 < m+1 < nr$$

DEFINITE INTEGRALS INVOLVING TRIGONOMETRIC FUNCTIONS

All letters are considered positive unless otherwise indicated.

15.26 $$\int_0^{\pi} \sin mx \sin nx \, dx = \begin{cases} 0 & m, n \text{ integers and } m \neq n \\ \pi/2 & m, n \text{ integers and } m = n \end{cases}$$

15.27 $$\int_0^{\pi} \cos mx \cos nx \, dx = \begin{cases} 0 & m, n \text{ integers and } m \neq n \\ \pi/2 & m, n \text{ integers and } m = n \end{cases}$$

15.28 $$\int_0^{\pi} \sin mx \cos nx \, dx = \begin{cases} 0 & m, n \text{ integers and } m + n \text{ odd} \\ 2m/(m^2 - n^2) & m, n \text{ integers and } m + n \text{ even} \end{cases}$$

15.29 $$\int_0^{\pi/2} \sin^2 x \, dx = \int_0^{\pi/2} \cos^2 x \, dx = \frac{\pi}{4}$$

15.30 $$\int_0^{\pi/2} \sin^{2m} x \, dx = \int_0^{\pi/2} \cos^{2m} x \, dx = \frac{1 \cdot 3 \cdot 5 \cdots 2m-1}{2 \cdot 4 \cdot 6 \cdots 2m} \frac{\pi}{2}, \qquad m = 1, 2, \ldots$$

15.31 $$\int_0^{\pi/2} \sin^{2m+1} x \, dx = \int_0^{\pi/2} \cos^{2m+1} x \, dx = \frac{2 \cdot 4 \cdot 6 \cdots 2m}{1 \cdot 3 \cdot 5 \cdots 2m+1}, \qquad m = 1, 2, \ldots$$

15.32 $$\int_0^{\pi/2} \sin^{2p-1} x \cos^{2q-1} x \, dx = \frac{\Gamma(p)\,\Gamma(q)}{2\,\Gamma(p+q)}$$

15.33 $$\int_0^{\infty} \frac{\sin px}{x} \, dx = \begin{cases} \pi/2 & p > 0 \\ 0 & p = 0 \\ -\pi/2 & p < 0 \end{cases}$$

15.34 $$\int_0^{\infty} \frac{\sin px \cos qx}{x} \, dx = \begin{cases} 0 & p > q > 0 \\ \pi/2 & 0 < p < q \\ \pi/4 & p = q > 0 \end{cases}$$

15.35 $$\int_0^{\infty} \frac{\sin px \sin qx}{x^2} \, dx = \begin{cases} \pi p/2 & 0 < p \leqq q \\ \pi q/2 & p \geqq q > 0 \end{cases}$$

15.36 $$\int_0^{\infty} \frac{\sin^2 px}{x^2} \, dx = \frac{\pi p}{2}$$

15.37 $$\int_0^{\infty} \frac{1 - \cos px}{x^2} \, dx = \frac{\pi p}{2}$$

15.38 $$\int_0^{\infty} \frac{\cos px - \cos qx}{x} \, dx = \ln \frac{q}{p}$$

15.39 $$\int_0^{\infty} \frac{\cos px - \cos qx}{x^2} \, dx = \frac{\pi(q-p)}{2}$$

15.40 $$\int_0^{\infty} \frac{\cos mx}{x^2 + a^2} \, dx = \frac{\pi}{2a} e^{-ma}$$

15.41 $$\int_0^{\infty} \frac{x \sin mx}{x^2 + a^2} \, dx = \frac{\pi}{2} e^{-ma}$$

15.42 $$\int_0^{\infty} \frac{\sin mx}{x(x^2 + a^2)} \, dx = \frac{\pi}{2a^2}(1 - e^{-ma})$$

15.43 $$\int_0^{2\pi} \frac{dx}{a + b \sin x} = \frac{2\pi}{\sqrt{a^2 - b^2}}$$

15.44 $$\int_0^{2\pi} \frac{dx}{a + b \cos x} = \frac{2\pi}{\sqrt{a^2 - b^2}}$$

15.45 $$\int_0^{\pi/2} \frac{dx}{a + b \cos x} = \frac{\cos^{-1}(b/a)}{\sqrt{a^2 - b^2}}$$

15.46 $$\int_0^{2\pi} \frac{dx}{(a + b \sin x)^2} = \int_0^{2\pi} \frac{dx}{(a + b \cos x)^2} = \frac{2\pi a}{(a^2 - b^2)^{3/2}}$$

15.47 $$\int_0^{2\pi} \frac{dx}{1 - 2a \cos x + a^2} = \frac{2\pi}{1 - a^2}, \quad 0 < a < 1$$

15.48 $$\int_0^{\pi} \frac{x \sin x \, dx}{1 - 2a \cos x + a^2} = \begin{cases} (\pi/a) \ln (1 + a) & |a| < 1 \\ \pi \ln (1 + 1/a) & |a| > 1 \end{cases}$$

15.49 $$\int_0^{\pi} \frac{\cos mx \, dx}{1 - 2a \cos x + a^2} = \frac{\pi a^m}{1 - a^2}, \quad a^2 < 1, \quad m = 0, 1, 2, \ldots$$

15.50 $$\int_0^{\infty} \sin ax^2 \, dx = \int_0^{\infty} \cos ax^2 \, dx = \frac{1}{2}\sqrt{\frac{\pi}{2a}}$$

15.51 $$\int_0^{\infty} \sin ax^n \, dx = \frac{1}{na^{1/n}} \Gamma(1/n) \sin \frac{\pi}{2n}, \quad n > 1$$

15.52 $$\int_0^{\infty} \cos ax^n \, dx = \frac{1}{na^{1/n}} \Gamma(1/n) \cos \frac{\pi}{2n}, \quad n > 1$$

15.53 $$\int_0^{\infty} \frac{\sin x}{\sqrt{x}} \, dx = \int_0^{\infty} \frac{\cos x}{\sqrt{x}} \, dx = \sqrt{\frac{\pi}{2}}$$

15.54 $$\int_0^{\infty} \frac{\sin x}{x^p} \, dx = \frac{\pi}{2\Gamma(p) \sin (p\pi/2)}, \quad 0 < p < 1$$

15.55 $$\int_0^{\infty} \frac{\cos x}{x^p} \, dx = \frac{\pi}{2\Gamma(p) \cos (p\pi/2)}, \quad 0 < p < 1$$

15.56 $$\int_0^{\infty} \sin ax^2 \cos 2bx \, dx = \frac{1}{2}\sqrt{\frac{\pi}{2a}} \left(\cos \frac{b^2}{a} - \sin \frac{b^2}{a} \right)$$

15.57 $$\int_0^{\infty} \cos ax^2 \cos 2bx \, dx = \frac{1}{2}\sqrt{\frac{\pi}{2a}} \left(\cos \frac{b^2}{a} + \sin \frac{b^2}{a} \right)$$

15.58 $$\int_0^{\infty} \frac{\sin^3 x}{x^3} \, dx = \frac{3\pi}{8}$$

15.59 $$\int_0^{\infty} \frac{\sin^4 x}{x^4} \, dx = \frac{\pi}{3}$$

15.60 $$\int_0^{\infty} \frac{\tan x}{x} \, dx = \frac{\pi}{2}$$

15.61 $$\int_0^{\pi/2} \frac{dx}{1 + \tan^m x} = \frac{\pi}{4}$$

15.62 $$\int_0^{\pi/2} \frac{x}{\sin x} \, dx = 2 \left\{ \frac{1}{1^2} - \frac{1}{3^2} + \frac{1}{5^2} - \frac{1}{7^2} + \cdots \right\}$$

15.63 $$\int_0^{1} \frac{\tan^{-1} x}{x} \, dx = \frac{1}{1^2} - \frac{1}{3^2} + \frac{1}{5^2} - \frac{1}{7^2} + \cdots$$

15.64 $$\int_0^{1} \frac{\sin^{-1} x}{x} \, dx = \frac{\pi}{2} \ln 2$$

15.65 $$\int_0^{1} \frac{1 - \cos x}{x} \, dx - \int_1^{\infty} \frac{\cos x}{x} \, dx = \gamma$$

15.66 $$\int_0^{\infty} \left(\frac{1}{1 + x^2} - \cos x \right) \frac{dx}{x} = \gamma$$

15.67 $$\int_0^{\infty} \frac{\tan^{-1} px - \tan^{-1} qx}{x} \, dx = \frac{\pi}{2} \ln \frac{p}{q}$$

DEFINITE INTEGRALS INVOLVING EXPONENTIAL FUNCTIONS

15.68 $\int_0^\infty e^{-ax}\cos bx\,dx = \dfrac{a}{a^2+b^2}$

15.69 $\int_0^\infty e^{-ax}\sin bx\,dx = \dfrac{b}{a^2+b^2}$

15.70 $\int_0^\infty \dfrac{e^{-ax}\sin bx}{x}\,dx = \tan^{-1}\dfrac{b}{a}$

15.71 $\int_0^\infty \dfrac{e^{-ax}-e^{-bx}}{x}\,dx = \ln\dfrac{b}{a}$

15.72 $\int_0^\infty e^{-ax^2}\,dx = \dfrac{1}{2}\sqrt{\dfrac{\pi}{a}}$

15.73 $\int_0^\infty e^{-ax^2}\cos bx\,dx = \dfrac{1}{2}\sqrt{\dfrac{\pi}{a}}\,e^{-b^2/4a}$

15.74 $\int_0^\infty e^{-(ax^2+bx+c)}\,dx = \dfrac{1}{2}\sqrt{\dfrac{\pi}{a}}\,e^{(b^2-4ac)/4a}\,\operatorname{erfc}\dfrac{b}{2\sqrt{a}}$

where $\operatorname{erfc}(p) = \dfrac{2}{\sqrt{\pi}}\int_p^\infty e^{-x^2}\,dx$

15.75 $\int_{-\infty}^\infty e^{-(ax^2+bx+c)}\,dx = \sqrt{\dfrac{\pi}{a}}\,e^{(b^2-4ac)/4a}$

15.76 $\int_0^\infty x^n e^{-ax}\,dx = \dfrac{\Gamma(n+1)}{a^{n+1}}$

15.77 $\int_0^\infty x^m e^{-ax^2}\,dx = \dfrac{\Gamma[(m+1)/2]}{2a^{(m+1)/2}}$

15.78 $\int_0^\infty e^{-(ax^2+b/x^2)}\,dx = \dfrac{1}{2}\sqrt{\dfrac{\pi}{a}}\,e^{-2\sqrt{ab}}$

15.79 $\int_0^\infty \dfrac{x\,dx}{e^x-1} = \dfrac{1}{1^2}+\dfrac{1}{2^2}+\dfrac{1}{3^2}+\dfrac{1}{4^2}+\cdots = \dfrac{\pi^2}{6}$

15.80 $\int_0^\infty \dfrac{x^{n-1}}{e^x-1}\,dx = \Gamma(n)\left(\dfrac{1}{1^n}+\dfrac{1}{2^n}+\dfrac{1}{3^n}+\cdots\right)$

For even n this can be summed in terms of Bernoulli numbers [see pages 108-109 and 114-115].

15.81 $\int_0^\infty \dfrac{x\,dx}{e^x+1} = \dfrac{1}{1^2}-\dfrac{1}{2^2}+\dfrac{1}{3^2}-\dfrac{1}{4^2}+\cdots = \dfrac{\pi^2}{12}$

15.82 $\int_0^\infty \dfrac{x^{n-1}}{e^x+1}\,dx = \Gamma(n)\left(\dfrac{1}{1^n}-\dfrac{1}{2^n}+\dfrac{1}{3^n}-\cdots\right)$

For some positive integer values of n the series can be summed [see pages 108-109 and 114-115].

15.83 $\int_0^\infty \dfrac{\sin mx}{e^{2\pi x}-1}\,dx = \dfrac{1}{4}\coth\dfrac{m}{2}-\dfrac{1}{2m}$

15.84 $\int_0^\infty \left(\dfrac{1}{1+x}-e^{-x}\right)\dfrac{dx}{x} = \gamma$

15.85 $\int_0^\infty \dfrac{e^{-x^2}-e^{-x}}{x}\,dx = \tfrac{1}{2}\gamma$

15.86 $\int_0^\infty \left(\dfrac{1}{e^x-1}-\dfrac{e^{-x}}{x}\right)dx = \gamma$

15.87 $$\int_0^\infty \frac{e^{-ax} - e^{-bx}}{x \sec px}\,dx = \frac{1}{2}\ln\left(\frac{b^2+p^2}{a^2+p^2}\right)$$

15.88 $$\int_0^\infty \frac{e^{-ax} - e^{-bx}}{x \csc px}\,dx = \tan^{-1}\frac{b}{p} - \tan^{-1}\frac{a}{p}$$

15.89 $$\int_0^\infty \frac{e^{-ax}(1-\cos x)}{x^2}\,dx = \cot^{-1} a - \frac{a}{2}\ln(a^2+1)$$

DEFINITE INTEGRALS INVOLVING LOGARITHMIC FUNCTIONS

15.90 $$\int_0^1 x^m (\ln x)^n\,dx = \frac{(-1)^n n!}{(m+1)^{n+1}} \qquad m > -1,\ n = 0, 1, 2, \ldots$$

If $n \neq 0, 1, 2, \ldots$ replace $n!$ by $\Gamma(n+1)$.

15.91 $$\int_0^1 \frac{\ln x}{1+x}\,dx = -\frac{\pi^2}{12}$$

15.92 $$\int_0^1 \frac{\ln x}{1-x}\,dx = -\frac{\pi^2}{6}$$

15.93 $$\int_0^1 \frac{\ln(1+x)}{x}\,dx = \frac{\pi^2}{12}$$

15.94 $$\int_0^1 \frac{\ln(1-x)}{x}\,dx = -\frac{\pi^2}{6}$$

15.95 $$\int_0^1 \ln x \ln(1+x)\,dx = 2 - 2\ln 2 - \frac{\pi^2}{12}$$

15.96 $$\int_0^1 \ln x \ln(1-x)\,dx = 2 - \frac{\pi^2}{6}$$

15.97 $$\int_0^\infty \frac{x^{p-1}\ln x}{1+x}\,dx = -\pi^2 \csc p\pi \cot p\pi \qquad 0 < p < 1$$

15.98 $$\int_0^1 \frac{x^m - x^n}{\ln x}\,dx = \ln\frac{m+1}{n+1}$$

15.99 $$\int_0^\infty e^{-x}\ln x\,dx = -\gamma$$

15.100 $$\int_0^\infty e^{-x^2}\ln x\,dx = -\frac{\sqrt{\pi}}{4}(\gamma + 2\ln 2)$$

15.101 $$\int_0^\infty \ln\left(\frac{e^x+1}{e^x-1}\right)dx = \frac{\pi^2}{4}$$

15.102 $$\int_0^{\pi/2} \ln \sin x\,dx = \int_0^{\pi/2} \ln\cos x\,dx = -\frac{\pi}{2}\ln 2$$

15.103 $$\int_0^{\pi/2} (\ln\sin x)^2\,dx = \int_0^{\pi/2} (\ln\cos x)^2\,dx = \frac{\pi}{2}(\ln 2)^2 + \frac{\pi^3}{24}$$

15.104 $$\int_0^{\pi} x\ln\sin x\,dx = -\frac{\pi^2}{2}\ln 2$$

15.105 $$\int_0^{\pi/2} \sin x \ln\sin x\,dx = \ln 2 - 1$$

15.106 $$\int_0^{2\pi} \ln(a + b\sin x)\,dx = \int_0^{2\pi} \ln(a + b\cos x)\,dx = 2\pi\ln(a + \sqrt{a^2-b^2})$$

15.107 $$\int_0^{\pi} \ln(a + b\cos x)\,dx = \pi \ln\left(\frac{a + \sqrt{a^2 - b^2}}{2}\right)$$

15.108 $$\int_0^{\pi} \ln(a^2 - 2ab\cos x + b^2)\,dx = \begin{cases} 2\pi \ln a, & a \geqq b > 0 \\ 2\pi \ln b, & b \geqq a > 0 \end{cases}$$

15.109 $$\int_0^{\pi/4} \ln(1 + \tan x)\,dx = \frac{\pi}{8}\ln 2$$

15.110 $$\int_0^{\pi/2} \sec x \ln\left(\frac{1 + b\cos x}{1 + a\cos x}\right) dx = \tfrac{1}{2}\{(\cos^{-1} a)^2 - (\cos^{-1} b)^2\}$$

15.111 $$\int_0^{a} \ln\left(2\sin\frac{x}{2}\right) dx = -\left(\frac{\sin a}{1^2} + \frac{\sin 2a}{2^2} + \frac{\sin 3a}{3^2} + \cdots\right)$$

See also 15.102.

DEFINITE INTEGRALS INVOLVING HYPERBOLIC FUNCTIONS

15.112 $$\int_0^{\infty} \frac{\sin ax}{\sinh bx}\,dx = \frac{\pi}{2b}\tanh\frac{a\pi}{2b}$$

15.113 $$\int_0^{\infty} \frac{\cos ax}{\cosh bx}\,dx = \frac{\pi}{2b}\operatorname{sech}\frac{a\pi}{2b}$$

15.114 $$\int_0^{\infty} \frac{x\,dx}{\sinh ax} = \frac{\pi^2}{4a^2}$$

15.115 $$\int_0^{\infty} \frac{x^n\,dx}{\sinh ax} = \frac{2^{n+1} - 1}{2^n a^{n+1}}\Gamma(n+1)\left\{\frac{1}{1^{n+1}} + \frac{1}{2^{n+1}} + \frac{1}{3^{n+1}} + \cdots\right\}$$

If n is an odd positive integer, the series can be summed [see page 108].

15.116 $$\int_0^{\infty} \frac{\sinh ax}{e^{bx} + 1}\,dx = \frac{\pi}{2b}\csc\frac{a\pi}{b} - \frac{1}{2a}$$

15.117 $$\int_0^{\infty} \frac{\sinh ax}{e^{bx} - 1}\,dx = \frac{1}{2a} - \frac{\pi}{2b}\cot\frac{a\pi}{b}$$

MISCELLANEOUS DEFINITE INTEGRALS

15.118 $$\int_0^{\infty} \frac{f(ax) - f(bx)}{x}\,dx = \{f(0) - f(\infty)\}\ln\frac{b}{a}$$

This is called *Frullani's integral*. It holds if $f'(x)$ is continuous and $\int_1^{\infty} \frac{f(x) - f(\infty)}{x}\,dx$ converges.

15.119 $$\int_0^{1} \frac{dx}{x^x} = \frac{1}{1^1} + \frac{1}{2^2} + \frac{1}{3^3} + \cdots$$

15.120 $$\int_{-a}^{a} (a + x)^{m-1}(a - x)^{n-1}\,dx = (2a)^{m+n-1}\frac{\Gamma(m)\,\Gamma(n)}{\Gamma(m + n)}$$

16 THE GAMMA FUNCTION

DEFINITION OF THE GAMMA FUNCTION $\Gamma(n)$ FOR $n > 0$

16.1 $$\Gamma(n) = \int_0^\infty t^{n-1}e^{-t}\,dt \qquad n > 0$$

RECURSION FORMULA

16.2 $$\Gamma(n+1) = n\,\Gamma(n)$$

16.3 $$\Gamma(n+1) = n! \qquad \text{if } n = 0, 1, 2, \ldots \text{ where } 0! = 1$$

THE GAMMA FUNCTION FOR $n < 0$

For $n < 0$ the gamma function can be defined by using 16.2, i.e.

16.4 $$\Gamma(n) = \frac{\Gamma(n+1)}{n}$$

GRAPH OF THE GAMMA FUNCTION

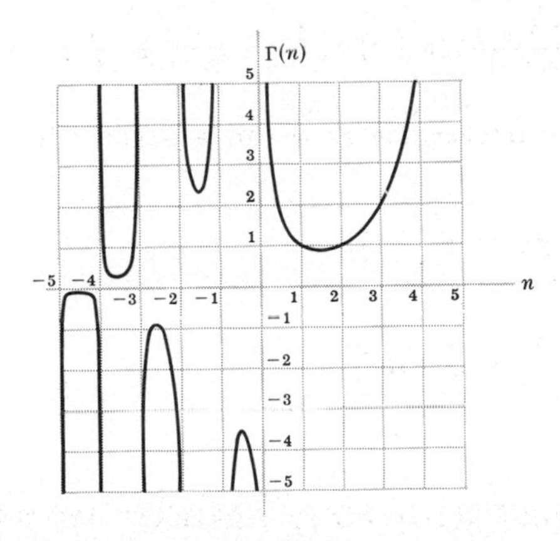

Fig. 16-1

SPECIAL VALUES FOR THE GAMMA FUNCTION

16.5 $$\Gamma(\tfrac{1}{2}) = \sqrt{\pi}$$

16.6 $$\Gamma(m+\tfrac{1}{2}) = \frac{1 \cdot 3 \cdot 5 \cdots (2m-1)}{2^m}\sqrt{\pi} \qquad m = 1, 2, 3, \ldots$$

16.7 $$\Gamma(-m+\tfrac{1}{2}) = \frac{(-1)^m 2^m \sqrt{\pi}}{1 \cdot 3 \cdot 5 \cdots (2m-1)} \qquad m = 1, 2, 3, \ldots$$

RELATIONSHIPS AMONG GAMMA FUNCTIONS

16.8 $$\Gamma(p)\,\Gamma(1-p) = \frac{\pi}{\sin p\pi}$$

16.9 $$2^{2x-1}\,\Gamma(x)\,\Gamma(x+\tfrac{1}{2}) = \sqrt{\pi}\,\Gamma(2x)$$

This is called the *duplication formula.*

16.10 $$\Gamma(x)\,\Gamma\left(x+\frac{1}{m}\right)\Gamma\left(x+\frac{2}{m}\right)\cdots\Gamma\left(x+\frac{m-1}{m}\right) = m^{1/2-mx}\,(2\pi)^{(m-1)/2}\,\Gamma(mx)$$

For $m = 2$ this reduces to 16.9.

OTHER DEFINITIONS OF THE GAMMA FUNCTION

16.11 $$\Gamma(x+1) = \lim_{k\to\infty} \frac{1\cdot 2\cdot 3\cdots k}{(x+1)(x+2)\cdots(x+k)}\,k^x$$

16.12 $$\frac{1}{\Gamma(x)} = xe^{\gamma x}\prod_{m=1}^{\infty}\left\{\left(1+\frac{x}{m}\right)e^{-x/m}\right\}$$

This is an infinite product representation for the gamma function where γ is Euler's constant.

DERIVATIVES OF THE GAMMA FUNCTION

16.13 $$\Gamma'(1) = \int_0^\infty e^{-x}\ln x\,dx = -\gamma$$

16.14 $$\frac{\Gamma'(x)}{\Gamma(x)} = -\gamma + \left(\frac{1}{1}-\frac{1}{x}\right) + \left(\frac{1}{2}-\frac{1}{x+1}\right) + \cdots + \left(\frac{1}{n}-\frac{1}{x+n-1}\right) + \cdots$$

ASYMPTOTIC EXPANSIONS FOR THE GAMMA FUNCTION

16.15 $$\Gamma(x+1) = \sqrt{2\pi x}\,x^x e^{-x}\left\{1 + \frac{1}{12x} + \frac{1}{288x^2} - \frac{139}{51{,}840x^3} + \cdots\right\}$$

This is called *Stirling's asymptotic series.*

If we let $x = n$ a positive integer in 16.15, then a useful approximation for $n!$ where n is large [e.g. $n > 10$] is given by *Stirling's formula*

16.16 $$n! \sim \sqrt{2\pi n}\,n^n e^{-n}$$

where $\sim$ is used to indicate that the ratio of the terms on each side approaches 1 as $n \to \infty$.

MISCELLANEOUS RESULTS

16.17 $$|\Gamma(ix)|^2 = \frac{\pi}{x\sinh \pi x}$$

17 THE BETA FUNCTION

DEFINITION OF THE BETA FUNCTION $B(m, n)$

17.1
$$B(m, n) = \int_0^1 t^{m-1}(1-t)^{n-1}\,dt \qquad m > 0,\ n > 0$$

RELATIONSHIP OF BETA FUNCTION TO GAMMA FUNCTION

17.2
$$B(m, n) = \frac{\Gamma(m)\,\Gamma(n)}{\Gamma(m+n)}$$

Extensions of $B(m, n)$ to $m < 0,\ n < 0$ is provided by using 16.4, page 101.

SOME IMPORTANT RESULTS

17.3
$$B(m, n) = B(n, m)$$

17.4
$$B(m, n) = 2\int_0^{\pi/2} \sin^{2m-1}\theta\,\cos^{2n-1}\theta\,d\theta$$

17.5
$$B(m, n) = \int_0^\infty \frac{t^{m-1}}{(1+t)^{m+n}}\,dt$$

17.6
$$B(m, n) = r^n(r+1)^m \int_0^1 \frac{t^{m-1}(1-t)^{n-1}}{(r+t)^{m+n}}\,dt$$

18 BASIC DIFFERENTIAL EQUATIONS and SOLUTIONS

DIFFERENTIAL EQUATION	SOLUTION
18.1 Separation of variables $f_1(x)\,g_1(y)\,dx + f_2(x)\,g_2(y)\,dy = 0$	$\int \frac{f_1(x)}{f_2(x)}\,dx + \int \frac{g_2(y)}{g_1(y)}\,dy = c$
18.2 Linear first order equation $\frac{dy}{dx} + P(x)y = Q(x)$	$y\,e^{\int P\,dx} = \int Q\,e^{\int P\,dx}\,dx + c$
18.3 Bernoulli's equation $\frac{dy}{dx} + P(x)y = Q(x)y^n$	$v\,e^{(1-n)\int P\,dx} = (1-n)\int Q\,e^{(1-n)\int P\,dx}\,dx + c$ where $v = y^{1-n}$. If $n = 1$, the solution is $\ln y = \int (Q-P)\,dx + c$
18.4 Exact equation $M(x,y)\,dx + N(x,y)\,dy = 0$ where $\partial M/\partial y = \partial N/\partial x$.	$\int M\,\partial x + \int\left(N - \frac{\partial}{\partial y}\int M\,\partial x\right)dy = c$ where ∂x indicates that the integration is to be performed with respect to x keeping y constant.
18.5 Homogeneous equation $\frac{dy}{dx} = F\left(\frac{y}{x}\right)$	$\ln x = \int \frac{dv}{F(v) - v} + c$ where $v = y/x$. If $F(v) = v$, the solution is $y = cx$.

DIFFERENTIAL EQUATION	SOLUTION
18.6 $y\,F(xy)\,dx + x\,G(xy)\,dy = 0$	$\ln x = \int \frac{G(v)\,dv}{v\{G(v)-F(v)\}} + c$ where $v = xy$. If $G(v) = F(v)$, the solution is $xy = c$.
18.7 Linear, homogeneous second order equation $\frac{d^2y}{dx^2} + a\frac{dy}{dx} + by = 0$ a, b are real constants.	Let m_1, m_2 be the roots of $m^2 + am + b = 0$. Then there are 3 cases. Case 1. m_1, m_2 real and distinct: $y = c_1e^{m_1x} + c_2e^{m_2x}$ Case 2. m_1, m_2 real and equal: $y = c_1e^{m_1x} + c_2xe^{m_1x}$ Case 3. $m_1 = p + qi$, $m_2 = p - qi$: $y = e^{px}(c_1 \cos qx + c_2 \sin qx)$ where $p = -a/2$, $q = \sqrt{b - a^2/4}$.
18.8 Linear, nonhomogeneous second order equation $\frac{d^2y}{dx^2} + a\frac{dy}{dx} + by = R(x)$ a, b are real constants.	There are 3 cases corresponding to those of entry 18.7 above. Case 1. $y = c_1e^{m_1x} + c_2e^{m_2x} + \frac{e^{m_1x}}{m_1 - m_2}\int e^{-m_1x}R(x)\,dx + \frac{e^{m_2x}}{m_2 - m_1}\int e^{-m_2x}R(x)\,dx$ Case 2. $y = c_1e^{m_1x} + c_2xe^{m_1x} + xe^{m_1x}\int e^{-m_1x}R(x)\,dx - e^{m_1x}\int xe^{-m_1x}R(x)\,dx$ Case 3. $y = e^{px}(c_1 \cos qx + c_2 \sin qx) + \frac{e^{px}\sin qx}{q}\int e^{-px}R(x)\cos qx\,dx - \frac{e^{px}\cos qx}{q}\int e^{-px}R(x)\sin qx\,dx$
18.9 Euler or Cauchy equation $x^2\frac{d^2y}{dx^2} + ax\frac{dy}{dx} + by = S(x)$	Putting $x = e^t$, the equation becomes $\frac{d^2y}{dt^2} + (a-1)\frac{dy}{dt} + by = S(e^t)$ and can then be solved as in entries 18.7 and 18.8 above.

DIFFERENTIAL EQUATION	SOLUTION
18.10 Bessel's equation	
$x^2 \dfrac{d^2y}{dx^2} + x\dfrac{dy}{dx} + (\lambda^2x^2 - n^2)y = 0$	$y = c_1 J_n(\lambda x) + c_2 Y_n(x)$ See pages 136-137.
18.11 Transformed Bessel's equation	
$x^2 \dfrac{d^2y}{dx^2} + (2p+1)x\dfrac{dy}{dx} + (\alpha^2x^{2r} + \beta^2)y = 0$	$y = x^{-p}\left\{c_1 J_{q/r}\left(\dfrac{\alpha}{r}x^r\right) + c_2 Y_{q/r}\left(\dfrac{\alpha}{r}x^r\right)\right\}$ where $q = \sqrt{p^2 - \beta^2}$.
18.12 Legendre's equation	
$(1-x^2)\dfrac{d^2y}{dx^2} - 2x\dfrac{dy}{dx} + n(n+1)y = 0$	$y = c_1 P_n(x) + c_2 Q_n(x)$ See pages 146-148.

19 SERIES of CONSTANTS

ARITHMETIC SERIES

19.1 $$a + (a+d) + (a+2d) + \cdots + \{a + (n-1)d\} = \tfrac{1}{2}n\{2a + (n-1)d\} = \tfrac{1}{2}n(a+l)$$

where $l = a + (n-1)d$ is the last term.

Some special cases are

19.2 $$1 + 2 + 3 + \cdots + n = \tfrac{1}{2}n(n+1)$$

19.3 $$1 + 3 + 5 + \cdots + (2n-1) = n^2$$

GEOMETRIC SERIES

19.4 $$a + ar + ar^2 + ar^3 + \cdots + ar^{n-1} = \frac{a(1-r^n)}{1-r} = \frac{a-rl}{1-r}$$

where $l = ar^{n-1}$ is the last term and $r \neq 1$.

If $-1 < r < 1$, then

19.5 $$a + ar + ar^2 + ar^3 + \cdots = \frac{a}{1-r}$$

ARITHMETIC-GEOMETRIC SERIES

19.6 $$a + (a+d)r + (a+2d)r^2 + \cdots + \{a + (n-1)d\}r^{n-1} = \frac{a(1-r^n)}{1-r} + \frac{rd\{1 - nr^{n-1} + (n-1)r^n\}}{(1-r)^2}$$

where $r \neq 1$.

If $-1 < r < 1$, then

19.7 $$a + (a+d)r + (a+2d)r^2 + \cdots = \frac{a}{1-r} + \frac{rd}{(1-r)^2}$$

SUMS OF POWERS OF POSITIVE INTEGERS

19.8 $$1^p + 2^p + 3^p + \cdots + n^p = \frac{n^{p+1}}{p+1} + \tfrac{1}{2}n^p + \frac{B_1 p n^{p-1}}{2!} - \frac{B_2 p(p-1)(p-2)n^{p-3}}{4!} + \cdots$$

where the series terminates at n^2 or n according as p is odd or even, and B_k are the *Bernoulli numbers* [see page 114].

Some special cases are

19.9 $1 + 2 + 3 + \cdots + n = \dfrac{n(n+1)}{2}$

19.10 $1^2 + 2^2 + 3^2 + \cdots + n^2 = \dfrac{n(n+1)(2n+1)}{6}$

19.11 $1^3 + 2^3 + 3^3 + \cdots + n^3 = \dfrac{n^2(n+1)^2}{4} = (1+2+3+\cdots+n)^2$

19.12 $1^4 + 2^4 + 3^4 + \cdots + n^4 = \dfrac{n(n+1)(2n+1)(3n^2+3n-1)}{30}$

If $S_k = 1^k + 2^k + 3^k + \cdots + n^k$ where k and n are positive integers, then

19.13 $\binom{k+1}{1}S_1 + \binom{k+1}{2}S_2 + \cdots + \binom{k+1}{k}S_k = (n+1)^{k+1} - (n+1)$

SERIES INVOLVING RECIPROCALS OF POWERS OF POSITIVE INTEGERS

19.14 $1 - \frac{1}{2} + \frac{1}{3} - \frac{1}{4} + \frac{1}{5} - \cdots = \ln 2$

19.15 $1 - \frac{1}{3} + \frac{1}{5} - \frac{1}{7} + \frac{1}{9} - \cdots = \frac{\pi}{4}$

19.16 $1 - \frac{1}{4} + \frac{1}{7} - \frac{1}{10} + \frac{1}{13} - \cdots = \frac{\pi\sqrt{3}}{9} + \frac{1}{3}\ln 2$

19.17 $1 - \frac{1}{5} + \frac{1}{9} - \frac{1}{13} + \frac{1}{17} - \cdots = \frac{\pi\sqrt{2}}{8} + \frac{\sqrt{2}\ln(1+\sqrt{2})}{4}$

19.18 $\frac{1}{2} - \frac{1}{5} + \frac{1}{8} - \frac{1}{11} + \frac{1}{14} - \cdots = \frac{\pi\sqrt{3}}{9} - \frac{1}{3}\ln 2$

19.19 $\frac{1}{1^2} + \frac{1}{2^2} + \frac{1}{3^2} + \frac{1}{4^2} + \cdots = \frac{\pi^2}{6}$

19.20 $\frac{1}{1^4} + \frac{1}{2^4} + \frac{1}{3^4} + \frac{1}{4^4} + \cdots = \frac{\pi^4}{90}$

19.21 $\frac{1}{1^6} + \frac{1}{2^6} + \frac{1}{3^6} + \frac{1}{4^6} + \cdots = \frac{\pi^6}{945}$

19.22 $\frac{1}{1^2} - \frac{1}{2^2} + \frac{1}{3^2} - \frac{1}{4^2} + \cdots = \frac{\pi^2}{12}$

19.23 $\frac{1}{1^4} - \frac{1}{2^4} + \frac{1}{3^4} - \frac{1}{4^4} + \cdots = \frac{7\pi^4}{720}$

19.24 $\frac{1}{1^6} - \frac{1}{2^6} + \frac{1}{3^6} - \frac{1}{4^6} + \cdots = \frac{31\pi^6}{30{,}240}$

19.25 $\frac{1}{1^2} + \frac{1}{3^2} + \frac{1}{5^2} + \frac{1}{7^2} + \cdots = \frac{\pi^2}{8}$

19.26 $\frac{1}{1^4} + \frac{1}{3^4} + \frac{1}{5^4} + \frac{1}{7^4} + \cdots = \frac{\pi^4}{96}$

19.27 $\frac{1}{1^6} + \frac{1}{3^6} + \frac{1}{5^6} + \frac{1}{7^6} + \cdots = \frac{\pi^6}{960}$

19.28 $\frac{1}{1^3} - \frac{1}{3^3} + \frac{1}{5^3} - \frac{1}{7^3} + \cdots = \frac{\pi^3}{32}$

19.29 $\frac{1}{1^3} + \frac{1}{3^3} - \frac{1}{5^3} - \frac{1}{7^3} + \cdots = \frac{3\pi^3\sqrt{2}}{128}$

19.30 $\frac{1}{1\cdot 3} + \frac{1}{3\cdot 5} + \frac{1}{5\cdot 7} + \frac{1}{7\cdot 9} + \cdots = \frac{1}{2}$

19.31 $\frac{1}{1\cdot 3} + \frac{1}{2\cdot 4} + \frac{1}{3\cdot 5} + \frac{1}{4\cdot 6} + \cdots = \frac{3}{4}$

19.32 $\frac{1}{1^2\cdot 3^2} + \frac{1}{3^2\cdot 5^2} + \frac{1}{5^2\cdot 7^2} + \frac{1}{7^2\cdot 9^2} + \cdots = \frac{\pi^2 - 8}{16}$

19.33 $$\frac{1}{1^2\cdot 2^2\cdot 3^2}+\frac{1}{2^2\cdot 3^2\cdot 4^2}+\frac{1}{3^2\cdot 4^2\cdot 5^2}+\cdots = \frac{4\pi^2-39}{16}$$

19.34 $$\frac{1}{a}-\frac{1}{a+d}+\frac{1}{a+2d}-\frac{1}{a+3d}+\cdots = \int_0^1 \frac{u^{a-1}\,du}{1+u^d}$$

19.35 $$\frac{1}{1^{2p}}+\frac{1}{2^{2p}}+\frac{1}{3^{2p}}+\frac{1}{4^{2p}}+\cdots = \frac{2^{2p-1}\pi^{2p}B_p}{(2p)!}$$

19.36 $$\frac{1}{1^{2p}}+\frac{1}{3^{2p}}+\frac{1}{5^{2p}}+\frac{1}{7^{2p}}+\cdots = \frac{(2^{2p}-1)\pi^{2p}B_p}{2(2p)!}$$

19.37 $$\frac{1}{1^{2p}}-\frac{1}{2^{2p}}+\frac{1}{3^{2p}}-\frac{1}{4^{2p}}+\cdots = \frac{(2^{2p-1}-1)\pi^{2p}B_p}{(2p)!}$$

19.38 $$\frac{1}{1^{2p+1}}-\frac{1}{3^{2p+1}}+\frac{1}{5^{2p+1}}-\frac{1}{7^{2p+1}}+\cdots = \frac{\pi^{2p+1}E_p}{2^{2p+2}(2p)!}$$

MISCELLANEOUS SERIES

19.39 $$\frac{1}{2}+\cos\alpha+\cos 2\alpha+\cdots+\cos n\alpha = \frac{\sin(n+\frac{1}{2})\alpha}{2\sin(\alpha/2)}$$

19.40 $$\sin\alpha+\sin 2\alpha+\sin 3\alpha+\cdots+\sin n\alpha = \frac{\sin[\frac{1}{2}(n+1)]\alpha\,\sin\frac{1}{2}n\alpha}{\sin(\alpha/2)}$$

19.41 $$1+r\cos\alpha+r^2\cos 2\alpha+r^3\cos 3\alpha+\cdots = \frac{1-r\cos\alpha}{1-2r\cos\alpha+r^2}, \quad |r|<1$$

19.42 $$r\sin\alpha+r^2\sin 2\alpha+r^3\sin 3\alpha+\cdots = \frac{r\sin\alpha}{1-2r\cos\alpha+r^2}, \quad |r|<1$$

19.43 $$1+r\cos\alpha+r^2\cos 2\alpha+\cdots+r^n\cos n\alpha = \frac{r^{n+2}\cos n\alpha-r^{n+1}\cos(n+1)\alpha-r\cos\alpha+1}{1-2r\cos\alpha+r^2}$$

19.44 $$r\sin\alpha+r^2\sin 2\alpha+\cdots+r^n\sin n\alpha = \frac{r\sin\alpha-r^{n+1}\sin(n+1)\alpha+r^{n+2}\sin n\alpha}{1-2r\cos\alpha+r^2}$$

THE EULER-MACLAURIN SUMMATION FORMULA

19.45 $$\begin{aligned}\sum_{k=1}^{n-1}F(k) &= \int_0^n F(k)\,dk - \frac{1}{2}\{F(0)+F(n)\} \\ &\quad + \frac{1}{12}\{F'(n)-F'(0)\} - \frac{1}{720}\{F'''(n)-F'''(0)\} \\ &\quad + \frac{1}{30{,}240}\{F^{(v)}(n)-F^{(v)}(0)\} - \frac{1}{1{,}209{,}600}\{F^{(vii)}(n)-F^{(vii)}(0)\} \\ &\quad + \cdots (-1)^{p-1}\frac{B_p}{(2p)!}\{F^{(2p-1)}(n)-F^{(2p-1)}(0)\} + \cdots\end{aligned}$$

THE POISSON SUMMATION FORMULA

19.46 $$\sum_{k=-\infty}^{\infty}F(k) = \sum_{m=-\infty}^{\infty}\left\{\int_{-\infty}^{\infty} e^{2\pi imx}F(x)\,dx\right\}$$

20 TAYLOR SERIES

TAYLOR SERIES FOR FUNCTIONS OF ONE VARIABLE

20.1 $$f(x) = f(a) + f'(a)(x-a) + \frac{f''(a)(x-a)^2}{2!} + \cdots + \frac{f^{(n-1)}(a)(x-a)^{n-1}}{(n-1)!} + R_n$$

where R_n, the remainder after n terms, is given by either of the following forms:

20.2 Lagrange's form $$R_n = \frac{f^{(n)}(\xi)(x-a)^n}{n!}$$

20.3 Cauchy's form $$R_n = \frac{f^{(n)}(\xi)(x-\xi)^{n-1}(x-a)}{(n-1)!}$$

The value ξ, which may be different in the two forms, lies between a and x. The result holds if $f(x)$ has continuous derivatives of order n at least.

If $\lim_{n\to\infty} R_n = 0$, the infinite series obtained is called the *Taylor series* for $f(x)$ about $x = a$. If $a = 0$ the series is often called a *Maclaurin series*. These series, often called power series, generally converge for all values of x in some interval called the *interval of convergence* and diverge for all x outside this interval.

BINOMIAL SERIES

20.4 $$(a+x)^n = a^n + na^{n-1}x + \frac{n(n-1)}{2!}a^{n-2}x^2 + \frac{n(n-1)(n-2)}{3!}a^{n-3}x^3 + \cdots$$
$$= a^n + \binom{n}{1}a^{n-1}x + \binom{n}{2}a^{n-2}x^2 + \binom{n}{3}a^{n-3}x^3 + \cdots$$

Special cases are

20.5 $$(a+x)^2 = a^2 + 2ax + x^2$$

20.6 $$(a+x)^3 = a^3 + 3a^2x + 3ax^2 + x^3$$

20.7 $$(a+x)^4 = a^4 + 4a^3x + 6a^2x^2 + 4ax^3 + x^4$$

20.8 $$(1+x)^{-1} = 1 - x + x^2 - x^3 + x^4 - \cdots \qquad -1 < x < 1$$

20.9 $$(1+x)^{-2} = 1 - 2x + 3x^2 - 4x^3 + 5x^4 - \cdots \qquad -1 < x < 1$$

20.10 $$(1+x)^{-3} = 1 - 3x + 6x^2 - 10x^3 + 15x^4 - \cdots \qquad -1 < x < 1$$

20.11 $$(1+x)^{-1/2} = 1 - \frac{1}{2}x + \frac{1\cdot 3}{2\cdot 4}x^2 - \frac{1\cdot 3\cdot 5}{2\cdot 4\cdot 6}x^3 + \cdots \qquad -1 < x \leqq 1$$

20.12 $$(1+x)^{1/2} = 1 + \frac{1}{2}x - \frac{1}{2\cdot 4}x^2 + \frac{1\cdot 3}{2\cdot 4\cdot 6}x^3 - \cdots \qquad -1 < x \leqq 1$$

20.13 $$(1+x)^{-1/3} = 1 - \frac{1}{3}x + \frac{1\cdot 4}{3\cdot 6}x^2 - \frac{1\cdot 4\cdot 7}{3\cdot 6\cdot 9}x^3 + \cdots \qquad -1 < x \leqq 1$$

20.14 $$(1+x)^{1/3} = 1 + \frac{1}{3}x - \frac{2}{3\cdot 6}x^2 + \frac{2\cdot 5}{3\cdot 6\cdot 9}x^3 - \cdots \qquad -1 < x \leqq 1$$

SERIES FOR EXPONENTIAL AND LOGARITHMIC FUNCTIONS

20.15 $e^x = 1 + x + \frac{x^2}{2!} + \frac{x^3}{3!} + \cdots \qquad -\infty < x < \infty$

20.16 $a^x = e^{x \ln a} = 1 + x \ln a + \frac{(x \ln a)^2}{2!} + \frac{(x \ln a)^3}{3!} + \cdots \qquad -\infty < x < \infty$

20.17 $\ln(1+x) = x - \frac{x^2}{2} + \frac{x^3}{3} - \frac{x^4}{4} + \cdots \qquad -1 < x \leqq 1$

20.18 $\frac{1}{2}\ln\left(\frac{1+x}{1-x}\right) = x + \frac{x^3}{3} + \frac{x^5}{5} + \frac{x^7}{7} + \cdots \qquad -1 < x < 1$

20.19 $\ln x = 2\left\{\left(\frac{x-1}{x+1}\right) + \frac{1}{3}\left(\frac{x-1}{x+1}\right)^3 + \frac{1}{5}\left(\frac{x-1}{x+1}\right)^5 + \cdots\right\} \qquad x > 0$

20.20 $\ln x = \left(\frac{x-1}{x}\right) + \frac{1}{2}\left(\frac{x-1}{x}\right)^2 + \frac{1}{3}\left(\frac{x-1}{x}\right)^3 + \cdots \qquad x \geqq \frac{1}{2}$

SERIES FOR TRIGONOMETRIC FUNCTIONS

20.21 $\sin x = x - \frac{x^3}{3!} + \frac{x^5}{5!} - \frac{x^7}{7!} + \cdots \qquad -\infty < x < \infty$

20.22 $\cos x = 1 - \frac{x^2}{2!} + \frac{x^4}{4!} - \frac{x^6}{6!} + \cdots \qquad -\infty < x < \infty$

20.23 $\tan x = x + \frac{x^3}{3} + \frac{2x^5}{15} + \frac{17x^7}{315} + \cdots + \frac{2^{2n}(2^{2n}-1)B_n x^{2n-1}}{(2n)!} + \cdots \qquad |x| < \frac{\pi}{2}$

20.24 $\cot x = \frac{1}{x} - \frac{x}{3} - \frac{x^3}{45} - \frac{2x^5}{945} - \cdots - \frac{2^{2n}B_n x^{2n-1}}{(2n)!} - \cdots \qquad 0 < |x| < \pi$

20.25 $\sec x = 1 + \frac{x^2}{2} + \frac{5x^4}{24} + \frac{61x^6}{720} + \cdots + \frac{E_n x^{2n}}{(2n)!} + \cdots \qquad |x| < \frac{\pi}{2}$

20.26 $\csc x = \frac{1}{x} + \frac{x}{6} + \frac{7x^3}{360} + \frac{31x^5}{15{,}120} + \cdots + \frac{2(2^{2n-1}-1)B_n x^{2n-1}}{(2n)!} + \cdots \qquad 0 < |x| < \pi$

20.27 $\sin^{-1} x = x + \frac{1}{2}\frac{x^3}{3} + \frac{1 \cdot 3}{2 \cdot 4}\frac{x^5}{5} + \frac{1 \cdot 3 \cdot 5}{2 \cdot 4 \cdot 6}\frac{x^7}{7} + \cdots \qquad |x| < 1$

20.28 $\cos^{-1} x = \frac{\pi}{2} - \sin^{-1} x = \frac{\pi}{2} - \left(x + \frac{1}{2}\frac{x^3}{3} + \frac{1 \cdot 3}{2 \cdot 4}\frac{x^5}{5} + \cdots\right) \qquad |x| < 1$

20.29 $\tan^{-1} x = \begin{cases} x - \frac{x^3}{3} + \frac{x^5}{5} - \frac{x^7}{7} + \cdots & |x| < 1 \\ \pm\frac{\pi}{2} - \frac{1}{x} + \frac{1}{3x^3} - \frac{1}{5x^5} + \cdots & [+ \text{ if } x \geqq 1, \; - \text{ if } x \leqq -1] \end{cases}$

20.30 $\cot^{-1} x = \frac{\pi}{2} - \tan^{-1} x = \begin{cases} \frac{\pi}{2} - \left(x - \frac{x^3}{3} + \frac{x^5}{5} - \cdots\right) & |x| < 1 \\ p\pi + \frac{1}{x} - \frac{1}{3x^3} + \frac{1}{5x^5} - \cdots & [p = 0 \text{ if } x > 1, \; p = 1 \text{ if } x < -1] \end{cases}$

20.31 $\sec^{-1} x = \cos^{-1}(1/x) = \frac{\pi}{2} - \left(\frac{1}{x} + \frac{1}{2 \cdot 3x^3} + \frac{1 \cdot 3}{2 \cdot 4 \cdot 5x^5} + \cdots\right) \qquad |x| > 1$

20.32 $\csc^{-1} x = \sin^{-1}(1/x) = \frac{1}{x} + \frac{1}{2 \cdot 3x^3} + \frac{1 \cdot 3}{2 \cdot 4 \cdot 5x^5} + \cdots \qquad |x| > 1$

SERIES FOR HYPERBOLIC FUNCTIONS

20.33 $\sinh x = x + \frac{x^3}{3!} + \frac{x^5}{5!} + \frac{x^7}{7!} + \cdots \qquad -\infty < x < \infty$

20.34 $\cosh x = 1 + \frac{x^2}{2!} + \frac{x^4}{4!} + \frac{x^6}{6!} + \cdots \qquad -\infty < x < \infty$

20.35 $\tanh x = x - \frac{x^3}{3} + \frac{2x^5}{15} - \frac{17x^7}{315} + \cdots \frac{(-1)^{n-1}2^{2n}(2^{2n}-1)B_n x^{2n-1}}{(2n)!} + \cdots \qquad |x| < \frac{\pi}{2}$

20.36 $\coth x = \frac{1}{x} + \frac{x}{3} - \frac{x^3}{45} + \frac{2x^5}{945} + \cdots \frac{(-1)^{n-1}2^{2n}B_n x^{2n-1}}{(2n)!} + \cdots \qquad 0 < |x| < \pi$

20.37 $\operatorname{sech} x = 1 - \frac{x^2}{2} + \frac{5x^4}{24} - \frac{61x^6}{720} + \cdots \frac{(-1)^n E_n x^{2n}}{(2n)!} + \cdots \qquad |x| < \frac{\pi}{2}$

20.38 $\operatorname{csch} x = \frac{1}{x} - \frac{x}{6} + \frac{7x^3}{360} - \frac{31x^5}{15{,}120} + \cdots \frac{(-1)^n 2(2^{2n-1}-1)B_n x^{2n-1}}{(2n)!} + \cdots \qquad 0 < |x| < \pi$

20.39
$$\sinh^{-1} x = \begin{cases} x - \frac{x^3}{2\cdot 3} + \frac{1\cdot 3x^5}{2\cdot 4\cdot 5} - \frac{1\cdot 3\cdot 5x^7}{2\cdot 4\cdot 6\cdot 7} + \cdots & |x| < 1 \\ \pm\left(\ln|2x| + \frac{1}{2\cdot 2x^2} - \frac{1\cdot 3}{2\cdot 4\cdot 4x^4} + \frac{1\cdot 3\cdot 5}{2\cdot 4\cdot 6\cdot 6x^6} - \cdots\right) & \begin{bmatrix} + \text{ if } x \geqq 1 \\ - \text{ if } x \leqq -1 \end{bmatrix} \end{cases}$$

20.40
$$\cosh^{-1} x = \pm\left\{\ln(2x) - \left(\frac{1}{2\cdot 2x^2} + \frac{1\cdot 3}{2\cdot 4\cdot 4x^4} + \frac{1\cdot 3\cdot 5}{2\cdot 4\cdot 6\cdot 6x^6} + \cdots\right)\right\} \qquad \begin{bmatrix} + \text{ if } \cosh^{-1}x > 0,\ x \geqq 1 \\ - \text{ if } \cosh^{-1}x < 0,\ x \geqq 1 \end{bmatrix}$$

20.41 $\tanh^{-1} x = x + \frac{x^3}{3} + \frac{x^5}{5} + \frac{x^7}{7} + \cdots \qquad |x| < 1$

20.42 $\coth^{-1} x = \frac{1}{x} + \frac{1}{3x^3} + \frac{1}{5x^5} + \frac{1}{7x^7} + \cdots \qquad |x| > 1$

MISCELLANEOUS SERIES

20.43 $e^{\sin x} = 1 + x + \frac{x^2}{2} - \frac{x^4}{8} - \frac{x^5}{15} + \cdots \qquad -\infty < x < \infty$

20.44 $e^{\cos x} = e\left(1 - \frac{x^2}{2} + \frac{x^4}{6} - \frac{31x^6}{720} + \cdots\right) \qquad -\infty < x < \infty$

20.45 $e^{\tan x} = 1 + x + \frac{x^2}{2} + \frac{x^3}{2} + \frac{3x^4}{8} + \cdots \qquad |x| < \frac{\pi}{2}$

20.46 $e^x \sin x = x + x^2 + \frac{2x^3}{3} - \frac{x^5}{30} - \frac{x^6}{90} + \cdots + \frac{2^{n/2}\sin(n\pi/4)\,x^n}{n!} + \cdots \qquad -\infty < x < \infty$

20.47 $e^x \cos x = 1 + x - \frac{x^3}{3} - \frac{x^4}{6} + \cdots + \frac{2^{n/2}\cos(n\pi/4)\,x^n}{n!} + \cdots \qquad -\infty < x < \infty$

20.48 $\ln|\sin x| = \ln|x| - \frac{x^2}{6} - \frac{x^4}{180} - \frac{x^6}{2835} - \cdots - \frac{2^{2n-1}B_n x^{2n}}{n(2n)!} + \cdots \qquad 0 < |x| < \pi$

20.49 $\ln|\cos x| = -\frac{x^2}{2} - \frac{x^4}{12} - \frac{x^6}{45} - \frac{17x^8}{2520} - \cdots - \frac{2^{2n-1}(2^{2n}-1)B_n x^{2n}}{n(2n)!} + \cdots \qquad |x| < \frac{\pi}{2}$

20.50 $\ln|\tan x| = \ln|x| + \frac{x^2}{3} + \frac{7x^4}{90} + \frac{62x^6}{2835} + \cdots + \frac{2^{2n}(2^{2n-1}-1)B_n x^{2n}}{n(2n)!} + \cdots \qquad 0 < |x| < \frac{\pi}{2}$

20.51 $\frac{\ln(1+x)}{1+x} = x - (1+\frac{1}{2})x^2 + (1+\frac{1}{2}+\frac{1}{3})x^3 - \cdots \qquad |x| < 1$

REVERSION OF POWER SERIES

If

20.52 $y = c_1x + c_2x^2 + c_3x^3 + c_4x^4 + c_5x^5 + c_6x^6 + \cdots$

then

20.53 $x = C_1y + C_2y^2 + C_3y^3 + C_4y^4 + C_5y^5 + C_6y^6 + \cdots$

where

20.54 $c_1C_1 = 1$

20.55 $c_1^3C_2 = -c_2$

20.56 $c_1^5C_3 = 2c_2^2 - c_1c_3$

20.57 $c_1^7C_4 = 5c_1c_2c_3 - 5c_2^3 - c_1^2c_4$

20.58 $c_1^9C_5 = 6c_1^2c_2c_4 + 3c_1^2c_3^2 - c_1^3c_5 + 14c_2^4 - 21c_1c_2^2c_3$

20.59 $c_1^{11}C_6 = 7c_1^3c_2c_5 + 84c_1c_2^3c_3 + 7c_1^3c_3c_4 - 28c_1^2c_2c_3^2 - c_1^4c_6 - 28c_1^2c_2^2c_4 - 42c_2^5$

TAYLOR SERIES FOR FUNCTIONS OF TWO VARIABLES

20.60
$$f(x,y) = f(a,b) + (x-a)f_x(a,b) + (y-b)f_y(a,b) + \frac{1}{2!}\{(x-a)^2f_{xx}(a,b) + 2(x-a)(y-b)f_{xy}(a,b) + (y-b)^2f_{yy}(a,b)\} + \cdots$$

where $f_x(a,b), f_y(a,b), \ldots$ denote partial derivatives with respect to $x, y, \ldots$ evaluated at $x = a$, $y = b$.

21 BERNOULLI and EULER NUMBERS

DEFINITION OF BERNOULLI NUMBERS

The *Bernoulli numbers* $B_1, B_2, B_3, \ldots$ are defined by the series

21.1 $$\frac{x}{e^x - 1} = 1 - \frac{x}{2} + \frac{B_1 x^2}{2!} - \frac{B_2 x^4}{4!} + \frac{B_3 x^6}{6!} - \cdots \qquad |x| < 2\pi$$

21.2 $$1 - \frac{x}{2}\cot\frac{x}{2} = \frac{B_1 x^2}{2!} + \frac{B_2 x^4}{4!} + \frac{B_3 x^6}{6!} + \cdots \qquad |x| < \pi$$

DEFINITION OF EULER NUMBERS

The *Euler numbers* $E_1, E_2, E_3, \ldots$ are defined by the series

21.3 $$\operatorname{sech} x = 1 - \frac{E_1 x^2}{2!} + \frac{E_2 x^4}{4!} - \frac{E_3 x^6}{6!} + \cdots \qquad |x| < \frac{\pi}{2}$$

21.4 $$\sec x = 1 + \frac{E_1 x^2}{2!} + \frac{E_2 x^4}{4!} + \frac{E_3 x^6}{6!} + \cdots \qquad |x| < \frac{\pi}{2}$$

TABLE OF FIRST FEW BERNOULLI AND EULER NUMBERS

Bernoulli numbers	Euler numbers
$B_1 = 1/6$	$E_1 = 1$
$B_2 = 1/30$	$E_2 = 5$
$B_3 = 1/42$	$E_3 = 61$
$B_4 = 1/30$	$E_4 = 1385$
$B_5 = 5/66$	$E_5 = 50{,}521$
$B_6 = 691/2730$	$E_6 = 2{,}702{,}765$
$B_7 = 7/6$	$E_7 = 199{,}360{,}981$
$B_8 = 3617/510$	$E_8 = 19{,}391{,}512{,}145$
$B_9 = 43{,}867/798$	$E_9 = 2{,}404{,}879{,}675{,}441$
$B_{10} = 174{,}611/330$	$E_{10} = 370{,}371{,}188{,}237{,}525$
$B_{11} = 854{,}513/138$	$E_{11} = 69{,}348{,}874{,}393{,}137{,}901$
$B_{12} = 236{,}364{,}091/2730$	$E_{12} = 15{,}514{,}534{,}163{,}557{,}086{,}905$

RELATIONSHIPS OF BERNOULLI AND EULER NUMBERS

21.5 $$\binom{2n+1}{2}2^2B_1 - \binom{2n+1}{4}2^4B_2 + \binom{2n+1}{6}2^6B_3 - \cdots (-1)^{n-1}(2n+1)2^{2n}B_n = 2n$$

21.6 $$E_n = \binom{2n}{2}E_{n-1} - \binom{2n}{4}E_{n-2} + \binom{2n}{6}E_{n-3} - \cdots (-1)^n$$

21.7 $$B_n = \frac{2n}{2^{2n}(2^{2n}-1)}\left\{\binom{2n-1}{1}E_{n-1} - \binom{2n-1}{3}E_{n-2} + \binom{2n-1}{5}E_{n-3} - \cdots (-1)^{n-1}\right\}$$

SERIES INVOLVING BERNOULLI AND EULER NUMBERS

21.8 $$B_n = \frac{(2n)!}{2^{2n-1}\pi^{2n}}\left\{1 + \frac{1}{2^{2n}} + \frac{1}{3^{2n}} + \cdots\right\}$$

21.9 $$B_n = \frac{2(2n)!}{(2^{2n}-1)\pi^{2n}}\left\{1 + \frac{1}{3^{2n}} + \frac{1}{5^{2n}} + \cdots\right\}$$

21.10 $$B_n = \frac{(2n)!}{(2^{2n-1}-1)\pi^{2n}}\left\{1 - \frac{1}{2^{2n}} + \frac{1}{3^{2n}} - \cdots\right\}$$

21.11 $$E_n = \frac{2^{2n+2}(2n)!}{\pi^{2n+1}}\left\{1 - \frac{1}{3^{2n+1}} + \frac{1}{5^{2n+1}} - \cdots\right\}$$

ASYMPTOTIC FORMULA FOR BERNOULLI NUMBERS

21.12 $$B_n \sim 4n^{2n}(\pi e)^{-2n}\sqrt{\pi n}$$

22 FORMULAS from VECTOR ANALYSIS

VECTORS AND SCALARS

Various quantities in physics such as temperature, volume and speed can be specified by a real number. Such quantities are called *scalars*.

Other quantities such as force, velocity and momentum require for their specification a direction as well as magnitude. Such quantities are called *vectors*. A vector is represented by an arrow or directed line segment indicating direction. The magnitude of the vector is determined by the length of the arrow, using an appropriate unit.

NOTATION FOR VECTORS

A vector is denoted by a bold faced letter such as $\mathbf{A}$ [Fig. 22-1]. The magnitude is denoted by $|\mathbf{A}|$ or A. The tail end of the arrow is called the *initial point* while the head is called the *terminal point.*

FUNDAMENTAL DEFINITIONS

1. **Equality of vectors.** Two vectors are equal if they have the same magnitude and direction. Thus $\mathbf{A} = \mathbf{B}$ in Fig. 22-1.

2. **Multiplication of a vector by a scalar.** If m is any real number (scalar), then $m\mathbf{A}$ is a vector whose magnitude is $|m|$ times the magnitude of $\mathbf{A}$ and whose direction is the same as or opposite to $\mathbf{A}$ according as $m > 0$ or $m < 0$. If $m = 0$, then $m\mathbf{A} = \mathbf{0}$ is called the *zero* or *null vector.*

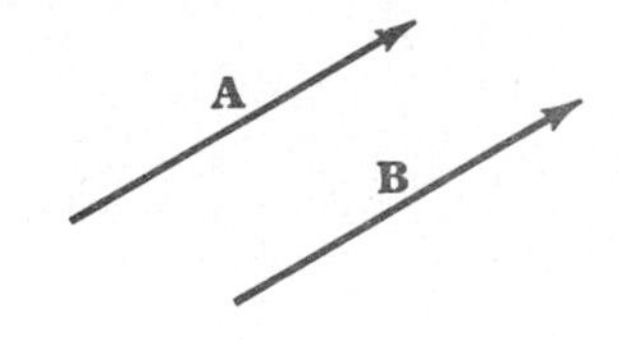

Fig. 22-1

3. **Sums of vectors.** The sum or resultant of $\mathbf{A}$ and $\mathbf{B}$ is a vector $\mathbf{C} = \mathbf{A} + \mathbf{B}$ formed by placing the initial point of $\mathbf{B}$ on the terminal point of $\mathbf{A}$ and joining the initial point of $\mathbf{A}$ to the terminal point of $\mathbf{B}$ [Fig. 22-2(b)]. This definition is equivalent to the parallelogram law for vector addition as indicated in Fig. 22-2(c). The vector $\mathbf{A} - \mathbf{B}$ is defined as $\mathbf{A} + (-\mathbf{B})$.

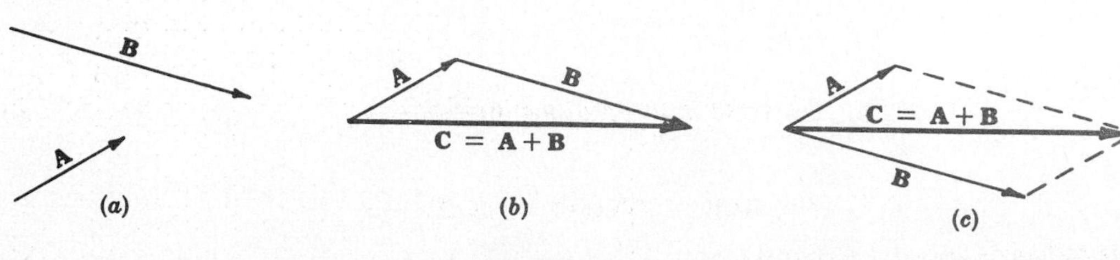

Fig. 22-2

Extensions to sums of more than two vectors are immediate. Thus Fig. 22-3 shows how to obtain the sum **E** of the vectors **A**, **B**, **C** and **D**.

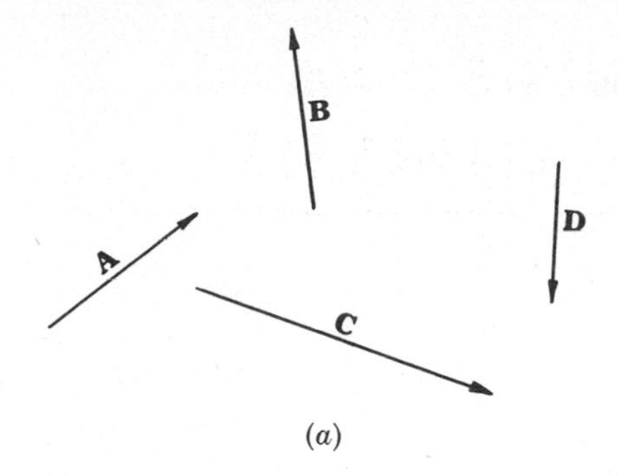

(a)

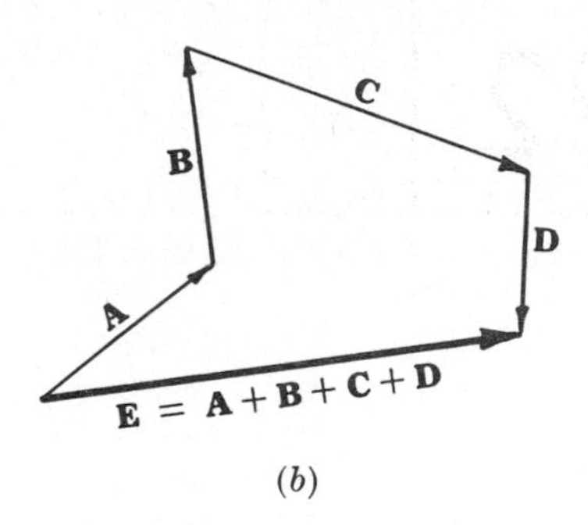

(b)

Fig. 22-3

4. **Unit vectors.** A *unit vector* is a vector with unit magnitude. If **A** is a vector, then a unit vector in the direction of **A** is $\mathbf{a} = \mathbf{A}/A$ where $A > 0$.

LAWS OF VECTOR ALGEBRA

If **A**, **B**, **C** are vectors and m, n are scalars, then

22.1 $\mathbf{A} + \mathbf{B} = \mathbf{B} + \mathbf{A}$ Commutative law for addition

22.2 $\mathbf{A} + (\mathbf{B} + \mathbf{C}) = (\mathbf{A} + \mathbf{B}) + \mathbf{C}$ Associative law for addition

22.3 $m(n\mathbf{A}) = (mn)\mathbf{A} = n(m\mathbf{A})$ Associative law for scalar multiplication

22.4 $(m + n)\mathbf{A} = m\mathbf{A} + n\mathbf{A}$ Distributive law

22.5 $m(\mathbf{A} + \mathbf{B}) = m\mathbf{A} + m\mathbf{B}$ Distributive law

COMPONENTS OF A VECTOR

A vector **A** can be represented with initial point at the origin of a rectangular coordinate system. If **i**, **j**, **k** are unit vectors in the directions of the positive x, y, z axes, then

22.6
$$\mathbf{A} = A_1\mathbf{i} + A_2\mathbf{j} + A_3\mathbf{k}$$

where $A_1\mathbf{i}, A_2\mathbf{j}, A_3\mathbf{k}$ are called *component vectors* of **A** in the **i**, **j**, **k** directions and A_1, A_2, A_3 are called the *components* of **A**.

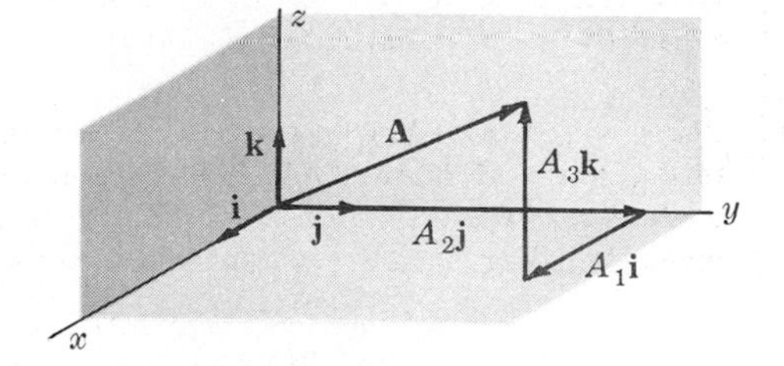

Fig. 22-4

DOT OR SCALAR PRODUCT

22.7
$$\mathbf{A} \cdot \mathbf{B} = AB \cos\theta \qquad 0 \leqq \theta \leqq \pi$$

where θ is the angle between **A** and **B**.

Fundamental results are

22.8 $$\mathbf{A}\cdot\mathbf{B} = \mathbf{B}\cdot\mathbf{A}$$ Commutative law

22.9 $$\mathbf{A}\cdot(\mathbf{B}+\mathbf{C}) = \mathbf{A}\cdot\mathbf{B} + \mathbf{A}\cdot\mathbf{C}$$ Distributive law

22.10 $$\mathbf{A}\cdot\mathbf{B} = A_1B_1 + A_2B_2 + A_3B_3$$

where $\mathbf{A} = A_1\mathbf{i} + A_2\mathbf{j} + A_3\mathbf{k}$, $\mathbf{B} = B_1\mathbf{i} + B_2\mathbf{j} + B_3\mathbf{k}$.

CROSS OR VECTOR PRODUCT

22.11 $$\mathbf{A}\times\mathbf{B} = AB\sin\theta\,\mathbf{u} \qquad 0 \leqq \theta \leqq \pi$$

where θ is the angle between $\mathbf{A}$ and $\mathbf{B}$ and $\mathbf{u}$ is a unit vector perpendicular to the plane of $\mathbf{A}$ and $\mathbf{B}$ such that $\mathbf{A}, \mathbf{B}, \mathbf{u}$ form a *right-handed system* [i.e. a right-threaded screw rotated through an angle less than 180° from $\mathbf{A}$ to $\mathbf{B}$ will advance in the direction of $\mathbf{u}$ as in Fig. 22-5].

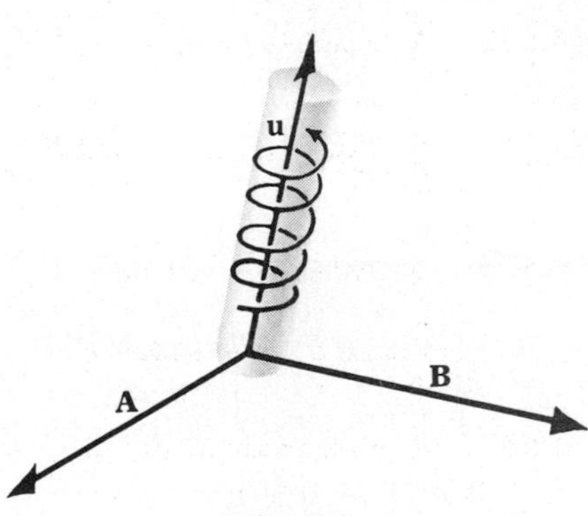

Fig. 22-5

Fundamental results are

22.12 $$\mathbf{A}\times\mathbf{B} = \begin{vmatrix} \mathbf{i} & \mathbf{j} & \mathbf{k} \\ A_1 & A_2 & A_3 \\ B_1 & B_2 & B_3 \end{vmatrix}$$
$$= (A_2B_3 - A_3B_2)\mathbf{i} + (A_3B_1 - A_1B_3)\mathbf{j} + (A_1B_2 - A_2B_1)\mathbf{k}$$

22.13 $$\mathbf{A}\times\mathbf{B} = -\mathbf{B}\times\mathbf{A}$$

22.14 $$\mathbf{A}\times(\mathbf{B}+\mathbf{C}) = \mathbf{A}\times\mathbf{B} + \mathbf{A}\times\mathbf{C}$$

22.15 $$|\mathbf{A}\times\mathbf{B}| = \text{area of parallelogram having sides } \mathbf{A} \text{ and } \mathbf{B}$$

MISCELLANEOUS FORMULAS INVOLVING DOT AND CROSS PRODUCTS

22.16 $$\mathbf{A}\cdot(\mathbf{B}\times\mathbf{C}) = \begin{vmatrix} A_1 & A_2 & A_3 \\ B_1 & B_2 & B_3 \\ C_1 & C_2 & C_3 \end{vmatrix} = A_1B_2C_3 + A_2B_3C_1 + A_3B_1C_2 - A_3B_2C_1 - A_2B_1C_3 - A_1B_3C_2$$

22.17 $$|\mathbf{A}\cdot(\mathbf{B}\times\mathbf{C})| = \text{volume of parallelepiped with sides } \mathbf{A}, \mathbf{B}, \mathbf{C}$$

22.18 $$\mathbf{A}\times(\mathbf{B}\times\mathbf{C}) = \mathbf{B}(\mathbf{A}\cdot\mathbf{C}) - \mathbf{C}(\mathbf{A}\cdot\mathbf{B})$$

22.19 $$(\mathbf{A}\times\mathbf{B})\times\mathbf{C} = \mathbf{B}(\mathbf{A}\cdot\mathbf{C}) - \mathbf{A}(\mathbf{B}\cdot\mathbf{C})$$

22.20 $$(\mathbf{A}\times\mathbf{B})\cdot(\mathbf{C}\times\mathbf{D}) = (\mathbf{A}\cdot\mathbf{C})(\mathbf{B}\cdot\mathbf{D}) - (\mathbf{A}\cdot\mathbf{D})(\mathbf{B}\cdot\mathbf{C})$$

22.21 $$(\mathbf{A}\times\mathbf{B})\times(\mathbf{C}\times\mathbf{D}) = \mathbf{C}\{\mathbf{A}\cdot(\mathbf{B}\times\mathbf{D})\} - \mathbf{D}\{\mathbf{A}\cdot(\mathbf{B}\times\mathbf{C})\}$$
$$= \mathbf{B}\{\mathbf{A}\cdot(\mathbf{C}\times\mathbf{D})\} - \mathbf{A}\{\mathbf{B}\cdot(\mathbf{C}\times\mathbf{D})\}$$

DERIVATIVES OF VECTORS

The derivative of a vector function $\mathbf{A}(u) = A_1(u)\mathbf{i} + A_2(u)\mathbf{j} + A_3(u)\mathbf{k}$ of the scalar variable u is given by

22.22 $$\frac{d\mathbf{A}}{du} = \lim_{\Delta u \to 0} \frac{\mathbf{A}(u+\Delta u) - \mathbf{A}(u)}{\Delta u} = \frac{dA_1}{du}\mathbf{i} + \frac{dA_2}{du}\mathbf{j} + \frac{dA_3}{du}\mathbf{k}$$

Partial derivatives of a vector function $\mathbf{A}(x, y, z)$ are similarly defined. We assume that all derivatives exist unless otherwise specified.

FORMULAS INVOLVING DERIVATIVES

22.23 $$\frac{d}{du}(\mathbf{A}\cdot\mathbf{B}) = \mathbf{A}\cdot\frac{d\mathbf{B}}{du} + \frac{d\mathbf{A}}{du}\cdot\mathbf{B}$$

22.24 $$\frac{d}{du}(\mathbf{A}\times\mathbf{B}) = \mathbf{A}\times\frac{d\mathbf{B}}{du} + \frac{d\mathbf{A}}{du}\times\mathbf{B}$$

22.25 $$\frac{d}{du}\{\mathbf{A}\cdot(\mathbf{B}\times\mathbf{C})\} = \frac{d\mathbf{A}}{du}\cdot(\mathbf{B}\times\mathbf{C}) + \mathbf{A}\cdot\left(\frac{d\mathbf{B}}{du}\times\mathbf{C}\right) + \mathbf{A}\cdot\left(\mathbf{B}\times\frac{d\mathbf{C}}{du}\right)$$

22.26 $$\mathbf{A}\cdot\frac{d\mathbf{A}}{du} = A\frac{dA}{du}$$

22.27 $$\mathbf{A}\cdot\frac{d\mathbf{A}}{du} = 0 \quad \text{if } |\mathbf{A}| \text{ is a constant}$$

THE DEL OPERATOR

The operator *del* is defined by

22.28 $$\nabla = \mathbf{i}\frac{\partial}{\partial x} + \mathbf{j}\frac{\partial}{\partial y} + \mathbf{k}\frac{\partial}{\partial z}$$

In the results below we assume that $U = U(x, y, z)$, $V = V(x, y, z)$, $\mathbf{A} = \mathbf{A}(x, y, z)$ and $\mathbf{B} = \mathbf{B}(x, y, z)$ have partial derivatives.

THE GRADIENT

22.29 $$\text{Gradient of } U = \text{grad } U = \nabla U = \left(\mathbf{i}\frac{\partial}{\partial x} + \mathbf{j}\frac{\partial}{\partial y} + \mathbf{k}\frac{\partial}{\partial z}\right)U = \frac{\partial U}{\partial x}\mathbf{i} + \frac{\partial U}{\partial y}\mathbf{j} + \frac{\partial U}{\partial z}\mathbf{k}$$

THE DIVERGENCE

22.30 $$\text{Divergence of } \mathbf{A} = \text{div } \mathbf{A} = \nabla\cdot\mathbf{A} = \left(\mathbf{i}\frac{\partial}{\partial x} + \mathbf{j}\frac{\partial}{\partial y} + \mathbf{k}\frac{\partial}{\partial z}\right)\cdot(A_1\mathbf{i} + A_2\mathbf{j} + A_3\mathbf{k})$$

$$= \frac{\partial A_1}{\partial x} + \frac{\partial A_2}{\partial y} + \frac{\partial A_3}{\partial z}$$

THE CURL

22.31 Curl of $\mathbf{A}$ = curl $\mathbf{A}$ = $\nabla \times \mathbf{A}$

$$= \left(\mathbf{i}\frac{\partial}{\partial x} + \mathbf{j}\frac{\partial}{\partial y} + \mathbf{k}\frac{\partial}{\partial z}\right) \times (A_1\mathbf{i} + A_2\mathbf{j} + A_3\mathbf{k})$$

$$= \begin{vmatrix} \mathbf{i} & \mathbf{j} & \mathbf{k} \\ \dfrac{\partial}{\partial x} & \dfrac{\partial}{\partial y} & \dfrac{\partial}{\partial z} \\ A_1 & A_2 & A_3 \end{vmatrix}$$

$$= \left(\frac{\partial A_3}{\partial y} - \frac{\partial A_2}{\partial z}\right)\mathbf{i} + \left(\frac{\partial A_1}{\partial z} - \frac{\partial A_3}{\partial x}\right)\mathbf{j} + \left(\frac{\partial A_2}{\partial x} - \frac{\partial A_1}{\partial y}\right)\mathbf{k}$$

THE LAPLACIAN

22.32 Laplacian of U = $\nabla^2 U = \nabla \cdot (\nabla U) = \dfrac{\partial^2 U}{\partial x^2} + \dfrac{\partial^2 U}{\partial y^2} + \dfrac{\partial^2 U}{\partial z^2}$

22.33 Laplacian of $\mathbf{A}$ = $\nabla^2 \mathbf{A} = \dfrac{\partial^2 \mathbf{A}}{\partial x^2} + \dfrac{\partial^2 \mathbf{A}}{\partial y^2} + \dfrac{\partial^2 \mathbf{A}}{\partial z^2}$

THE BIHARMONIC OPERATOR

22.34 Biharmonic operator on U = $\nabla^4 U = \nabla^2(\nabla^2 U)$

$$= \frac{\partial^4 U}{\partial x^4} + \frac{\partial^4 U}{\partial y^4} + \frac{\partial^4 U}{\partial z^4} + 2\frac{\partial^4 U}{\partial x^2\,\partial y^2} + 2\frac{\partial^4 U}{\partial y^2\,\partial z^2} + 2\frac{\partial^4 U}{\partial x^2\,\partial z^2}$$

MISCELLANEOUS FORMULAS INVOLVING ∇

22.35 $\nabla(U + V) = \nabla U + \nabla V$

22.36 $\nabla \cdot (\mathbf{A} + \mathbf{B}) = \nabla \cdot \mathbf{A} + \nabla \cdot \mathbf{B}$

22.37 $\nabla \times (\mathbf{A} + \mathbf{B}) = \nabla \times \mathbf{A} + \nabla \times \mathbf{B}$

22.38 $\nabla \cdot (U\mathbf{A}) = (\nabla U) \cdot \mathbf{A} + U(\nabla \cdot \mathbf{A})$

22.39 $\nabla \times (U\mathbf{A}) = (\nabla U) \times \mathbf{A} + U(\nabla \times \mathbf{A})$

22.40 $\nabla \cdot (\mathbf{A} \times \mathbf{B}) = \mathbf{B} \cdot (\nabla \times \mathbf{A}) - \mathbf{A} \cdot (\nabla \times \mathbf{B})$

22.41 $\nabla \times (\mathbf{A} \times \mathbf{B}) = (\mathbf{B} \cdot \nabla)\mathbf{A} - \mathbf{B}(\nabla \cdot \mathbf{A}) - (\mathbf{A} \cdot \nabla)\mathbf{B} + \mathbf{A}(\nabla \cdot \mathbf{B})$

22.42 $\nabla(\mathbf{A} \cdot \mathbf{B}) = (\mathbf{B} \cdot \nabla)\mathbf{A} + (\mathbf{A} \cdot \nabla)\mathbf{B} + \mathbf{B} \times (\nabla \times \mathbf{A}) + \mathbf{A} \times (\nabla \times \mathbf{B})$

22.43 $\nabla \times (\nabla U) = \mathbf{0}$, i.e. the curl of the gradient of U is zero.

22.44 $\nabla \cdot (\nabla \times \mathbf{A}) = 0$, i.e. the divergence of the curl of $\mathbf{A}$ is zero.

22.45 $\nabla \times (\nabla \times \mathbf{A}) = \nabla(\nabla \cdot \mathbf{A}) - \nabla^2 \mathbf{A}$

INTEGRALS INVOLVING VECTORS

If $\mathbf{A}(u) = \frac{d}{du}\mathbf{B}(u)$, then the *indefinite integral* of $\mathbf{A}(u)$ is

22.46 $$\int \mathbf{A}(u)\,du = \mathbf{B}(u) + \mathbf{c} \qquad \mathbf{c} = \text{constant vector}$$

The *definite integral* of $\mathbf{A}(u)$ from $u = a$ to $u = b$ in this case is given by

22.47 $$\int_a^b \mathbf{A}(u)\,du = \mathbf{B}(b) - \mathbf{B}(a)$$

The definite integral can be defined as on page 94.

LINE INTEGRALS

Consider a space curve C joining two points $P_1(a_1, a_2, a_3)$ and $P_2(b_1, b_2, b_3)$ as in Fig. 22-6. Divide the curve into n parts by points of subdivision $(x_1, y_1, z_1), \ldots, (x_{n-1}, y_{n-1}, z_{n-1})$. Then the *line integral* of a vector $\mathbf{A}(x, y, z)$ along C is defined as

22.48 $$\int_C \mathbf{A}\cdot d\mathbf{r} = \int_{P_1}^{P_2} \mathbf{A}\cdot d\mathbf{r} = \lim_{n\to\infty} \sum_{p=1}^{n} \mathbf{A}(x_p, y_p, z_p)\cdot \Delta\mathbf{r}_p$$

where $\Delta\mathbf{r}_p = \Delta x_p\,\mathbf{i} + \Delta y_p\,\mathbf{j} + \Delta z_p\,\mathbf{k}$, $\Delta x_p = x_{p+1} - x_p$, $\Delta y_p = y_{p+1} - y_p$, $\Delta z_p = z_{p+1} - z_p$ and where it is assumed that as $n \to \infty$ the largest of the magnitudes $|\Delta\mathbf{r}_p|$ approaches zero. The result 22.48 is a generalization of the ordinary definite integral [page 94].

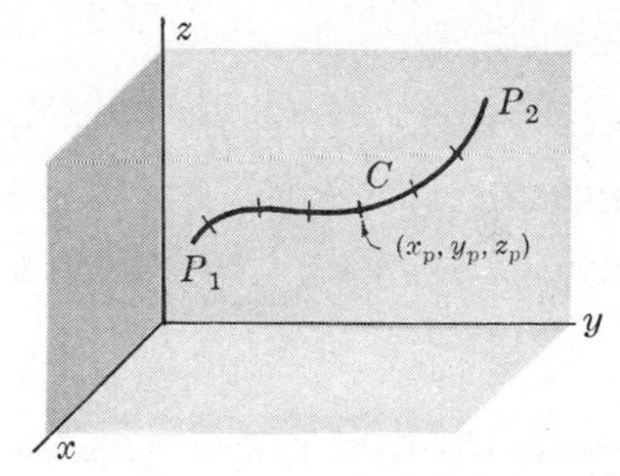

Fig. 22-6

The line integral 22.48 can also be written

22.49 $$\int_C \mathbf{A}\cdot d\mathbf{r} = \int_C (A_1\,dx + A_2\,dy + A_3\,dz)$$

using $\mathbf{A} = A_1\mathbf{i} + A_2\mathbf{j} + A_3\mathbf{k}$ and $d\mathbf{r} = dx\,\mathbf{i} + dy\,\mathbf{j} + dz\,\mathbf{k}$.

PROPERTIES OF LINE INTEGRALS

22.50 $$\int_{P_1}^{P_2} \mathbf{A}\cdot d\mathbf{r} = -\int_{P_2}^{P_1} \mathbf{A}\cdot d\mathbf{r}$$

22.51 $$\int_{P_1}^{P_2} \mathbf{A}\cdot d\mathbf{r} = \int_{P_1}^{P_3} \mathbf{A}\cdot d\mathbf{r} + \int_{P_3}^{P_2} \mathbf{A}\cdot d\mathbf{r}$$

INDEPENDENCE OF THE PATH

In general a line integral has a value which depends on the particular path C joining points P_1 and P_2 in a region $\mathcal{R}$. However, in case $\mathbf{A} = \nabla\phi$ or $\nabla \times \mathbf{A} = \mathbf{0}$ where ϕ and its partial derivatives are continuous in $\mathcal{R}$, the line integral $\int_C \mathbf{A}\cdot d\mathbf{r}$ is independent of the path. In such case

22.52 $$\int_C \mathbf{A}\cdot d\mathbf{r} = \int_{P_1}^{P_2} \mathbf{A}\cdot d\mathbf{r} = \phi(P_2) - \phi(P_1)$$

where $\phi(P_1)$ and $\phi(P_2)$ denote the values of ϕ at P_1 and P_2 respectively. In particular if C is a closed curve,

22.53 $$\int_C \mathbf{A}\cdot d\mathbf{r} = \oint_C \mathbf{A}\cdot d\mathbf{r} = 0$$

where the circle on the integral sign is used to emphasize that C is closed.

MULTIPLE INTEGRALS

Let $F(x,y)$ be a function defined in a region $\mathcal{R}$ of the xy plane as in Fig. 22-7. Subdivide the region into n parts by lines parallel to the x and y axes as indicated. Let $\Delta A_p = \Delta x_p\,\Delta y_p$ denote an area of one of these parts. Then the integral of $F(x,y)$ over $\mathcal{R}$ is defined as

22.54 $$\int_{\mathcal{R}} F(x,y)\,dA = \lim_{n\to\infty}\sum_{p=1}^{n} F(x_p,y_p)\,\Delta A_p$$

provided this limit exists.

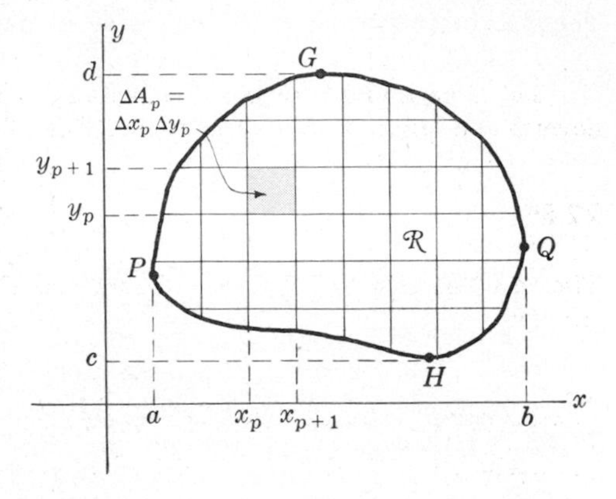

Fig. 22-7

In such case the integral can also be written as

22.55 $$\int_{x=a}^{b}\int_{y=f_1(x)}^{f_2(x)} F(x,y)\,dy\,dx = \int_{x=a}^{b}\left\{\int_{y=f_1(x)}^{f_2(x)} F(x,y)\,dy\right\}dx$$

where $y = f_1(x)$ and $y = f_2(x)$ are the equations of curves PHQ and PGQ respectively and a and b are the x coordinates of points P and Q. The result can also be written as

22.56 $$\int_{y=c}^{d}\int_{x=g_1(y)}^{g_2(y)} F(x,y)\,dx\,dy = \int_{y=c}^{d}\left\{\int_{x=g_1(y)}^{g_2(y)} F(x,y)\,dx\right\}dy$$

where $x = g_1(y)$, $x = g_2(y)$ are the equations of curves HPG and HQG respectively and c and d are the y coordinates of H and G.

These are called *double integrals* or *area integrals*. The ideas can be similarly extended to *triple* or *volume integrals* or to higher *multiple integrals*.

SURFACE INTEGRALS

Subdivide the surface S [see Fig. 22-8] into n elements of area ΔS_p, $p = 1,2,\ldots,n$. Let $\mathbf{A}(x_p,y_p,z_p) = \mathbf{A}_p$ where (x_p,y_p,z_p) is a point P in ΔS_p. Let $\mathbf{N}_p$ be a unit normal to ΔS_p at P. Then the surface integral of the normal component of $\mathbf{A}$ over S is defined as

22.57 $$\int_S \mathbf{A}\cdot\mathbf{N}\,dS = \lim_{n\to\infty}\sum_{p=1}^{n}\mathbf{A}_p\cdot\mathbf{N}_p\,\Delta S_p$$

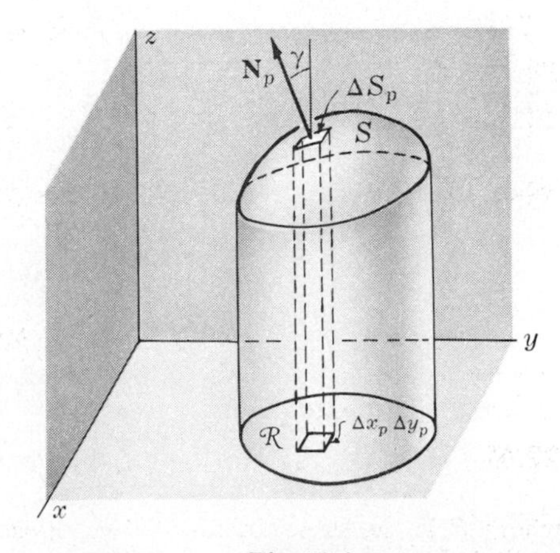

Fig. 22-8

RELATION BETWEEN SURFACE AND DOUBLE INTEGRALS

If $\mathcal{R}$ is the projection of S on the xy plane, then [see Fig. 22-8]

22.58
$$\int_S \mathbf{A}\cdot\mathbf{N}\,dS = \iint_{\mathcal{R}} \mathbf{A}\cdot\mathbf{N}\,\frac{dx\,dy}{|\mathbf{N}\cdot\mathbf{k}|}$$

THE DIVERGENCE THEOREM

Let S be a closed surface bounding a region of volume V; then if $\mathbf{N}$ is the positive (outward drawn) normal and $d\mathbf{S} = \mathbf{N}\,dS$, we have [see Fig. 22-9]

22.59
$$\int_V \nabla\cdot\mathbf{A}\,dV = \int_S \mathbf{A}\cdot d\mathbf{S}$$

The result is also called *Gauss' theorem* or *Green's theorem.*

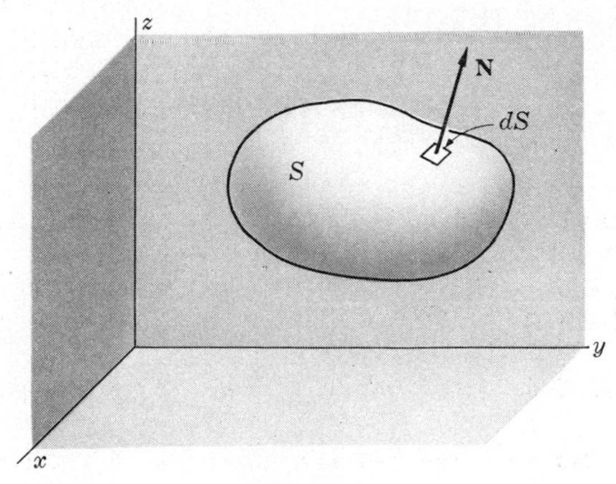

Fig. 22-9

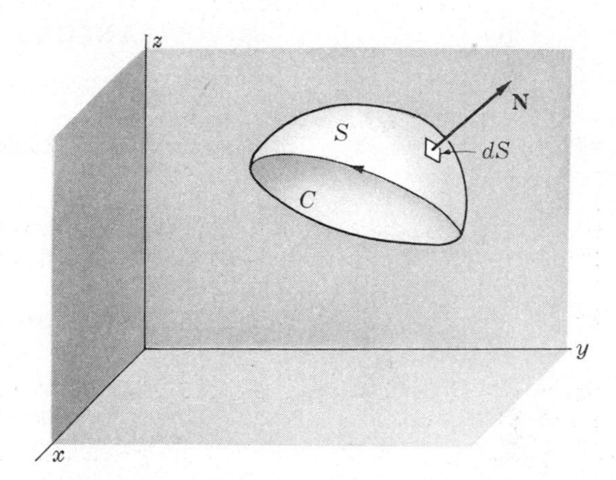

Fig. 22-10

STOKE'S THEOREM

Let S be an open two-sided surface bounded by a closed non-intersecting curve C [simple closed curve] as in Fig. 22-10. Then

22.60
$$\oint_C \mathbf{A}\cdot d\mathbf{r} = \int_S (\nabla\times\mathbf{A})\cdot d\mathbf{S}$$

where the circle on the integral is used to emphasize that C is closed.

GREEN'S THEOREM IN THE PLANE

22.61
$$\oint_C (P\,dx + Q\,dy) = \int_R \left(\frac{\partial Q}{\partial x} - \frac{\partial P}{\partial y}\right) dx\,dy$$

where R is the area bounded by the closed curve C. This result is a special case of the divergence theorem or Stoke's theorem.

GREEN'S FIRST IDENTITY

22.62 $$\int_V \{\phi \nabla^2\psi + (\nabla\phi)\cdot(\nabla\psi)\}\,dV = \int (\phi\nabla\psi)\cdot d\mathbf{S}$$

where ϕ and ψ are scalar functions.

GREEN'S SECOND IDENTITY

22.63 $$\int_V (\phi\nabla^2\psi - \psi\nabla^2\phi)\,dV = \int_S (\phi\nabla\psi - \psi\nabla\phi)\cdot d\mathbf{S}$$

MISCELLANEOUS INTEGRAL THEOREMS

22.64 $$\int_V \nabla\times\mathbf{A}\,dV = \int_S d\mathbf{S}\times\mathbf{A}$$

22.65 $$\int_C \phi\,d\mathbf{r} = \int_S d\mathbf{S}\times\nabla\phi$$

CURVILINEAR COORDINATES

A point P in space [see Fig. 22-11] can be located by rectangular coordinates (x, y, z) or curvilinear coordinates (u_1, u_2, u_3) where the transformation equations from one set of coordinates to the other are given by

22.66 $$x = x(u_1, u_2, u_3)$$
$$y = y(u_1, u_2, u_3)$$
$$z = z(u_1, u_2, u_3)$$

If u_2 and u_3 are constant, then as u_1 varies, the position vector $\mathbf{r} = x\mathbf{i} + y\mathbf{j} + z\mathbf{k}$ of P describes a curve called the u_1 coordinate curve. Similarly we define the u_2 and u_3 coordinate curves through P. The vectors $\partial\mathbf{r}/\partial u_1$, $\partial\mathbf{r}/\partial u_2$, $\partial\mathbf{r}/\partial u_3$ represent tangent vectors to the u_1, u_2, u_3 coordinate curves. Letting $\mathbf{e}_1, \mathbf{e}_2, \mathbf{e}_3$ be unit tangent vectors to these curves, we have

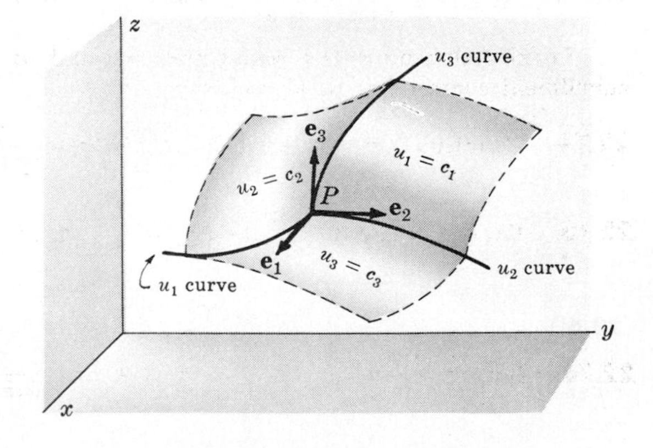

Fig. 22-11

22.67 $$\frac{\partial\mathbf{r}}{\partial u_1} = h_1\mathbf{e}_1, \quad \frac{\partial\mathbf{r}}{\partial u_2} = h_2\mathbf{e}_2, \quad \frac{\partial\mathbf{r}}{\partial u_3} = h_3\mathbf{e}_3$$

where

22.68 $$h_1 = \left|\frac{\partial\mathbf{r}}{\partial u_1}\right|, \quad h_2 = \left|\frac{\partial\mathbf{r}}{\partial u_2}\right|, \quad h_3 = \left|\frac{\partial\mathbf{r}}{\partial u_3}\right|$$

are called *scale factors*. If $\mathbf{e}_1, \mathbf{e}_2, \mathbf{e}_3$ are mutually perpendicular, the curvilinear coordinate system is called *orthogonal*.

FORMULAS INVOLVING ORTHOGONAL CURVILINEAR COORDINATES

22.69 $$d\mathbf{r} = \frac{\partial \mathbf{r}}{\partial u_1} du_1 + \frac{\partial \mathbf{r}}{\partial u_2} du_2 + \frac{\partial \mathbf{r}}{\partial u_3} du_3 = h_1\, du_1\, \mathbf{e}_1 + h_2\, du_2\, \mathbf{e}_2 + h_3\, du_3\, \mathbf{e}_3$$

22.70 $$ds^2 = d\mathbf{r} \cdot d\mathbf{r} = h_1^2\, du_1^2 + h_2^2\, du_2^2 + h_3^2\, du_3^2$$

where ds is the element of arc length.

If dV is the element of volume, then

22.71 $$\begin{aligned} dV &= |(h_1\mathbf{e}_1\, du_1) \cdot (h_2\mathbf{e}_2\, du_2) \times (h_3\mathbf{e}_3\, du_3)| = h_1h_2h_3\, du_1\, du_2\, du_3 \\ &= \left| \frac{\partial \mathbf{r}}{\partial u_1} \cdot \frac{\partial \mathbf{r}}{\partial u_2} \times \frac{\partial \mathbf{r}}{\partial u_3} \right| du_1\, du_2\, du_3 = \left| \frac{\partial(x,y,z)}{\partial(u_1,u_2,u_3)} \right| du_1\, du_2\, du_3 \end{aligned}$$

where

22.72 $$\frac{\partial(x,y,z)}{\partial(u_1,u_2,u_3)} = \begin{vmatrix} \partial x/\partial u_1 & \partial x/\partial u_2 & \partial x/\partial u_3 \\ \partial y/\partial u_1 & \partial y/\partial u_2 & \partial y/\partial u_3 \\ \partial z/\partial u_1 & \partial z/\partial u_2 & \partial z/\partial u_3 \end{vmatrix}$$

is called the *Jacobian* of the transformation.

TRANSFORMATION OF MULTIPLE INTEGRALS

The result 22.72 can be used to transform multiple integrals from rectangular to curvilinear coordinates. For example, we have

22.73 $$\iiint_{\mathcal{R}} F(x,y,z)\, dx\, dy\, dz = \iiint_{\mathcal{R}'} G(u_1,u_2,u_3) \left| \frac{\partial(x,y,z)}{\partial(u_1,u_2,u_3)} \right| du_1\, du_2\, du_3$$

where $\mathcal{R}'$ is the region into which $\mathcal{R}$ is mapped by the transformation and $G(u_1, u_2, u_3)$ is the value of $F(x, y, z)$ corresponding to the transformation.

GRADIENT, DIVERGENCE, CURL AND LAPLACIAN

In the following, Φ is a scalar function and $\mathbf{A} = A_1\mathbf{e}_1 + A_2\mathbf{e}_2 + A_3\mathbf{e}_3$ a vector function of orthogonal curvilinear coordinates u_1, u_2, u_3.

22.74 Gradient of Φ = grad Φ = $\nabla\Phi$ = $\frac{1}{h_1}\frac{\partial \Phi}{\partial u_1} + \frac{1}{h_2}\frac{\partial \Phi}{\partial u_2} + \frac{1}{h_3}\frac{\partial \Phi}{\partial u_3}$

22.75 Divergence of $\mathbf{A}$ = div $\mathbf{A}$ = $\nabla \cdot \mathbf{A}$ = $\frac{1}{h_1h_2h_3}\left[\frac{\partial}{\partial u_1}(h_2h_3A_1) + \frac{\partial}{\partial u_2}(h_3h_1A_2) + \frac{\partial}{\partial u_3}(h_1h_2A_3)\right]$

22.76 Curl of $\mathbf{A}$ = curl $\mathbf{A}$ = $\nabla \times \mathbf{A}$ = $\frac{1}{h_1h_2h_3}\begin{vmatrix} h_1\mathbf{e}_1 & h_2\mathbf{e}_2 & h_3\mathbf{e}_3 \\ \frac{\partial}{\partial u_1} & \frac{\partial}{\partial u_2} & \frac{\partial}{\partial u_3} \\ h_1A_1 & h_2A_2 & h_3A_3 \end{vmatrix}$

$$= \frac{1}{h_2h_3}\left[\frac{\partial}{\partial u_2}(h_3A_3) - \frac{\partial}{\partial u_3}(h_2A_2)\right]\mathbf{e}_1 + \frac{1}{h_1h_3}\left[\frac{\partial}{\partial u_3}(h_1A_1) - \frac{\partial}{\partial u_1}(h_3A_3)\right]\mathbf{e}_2 + \frac{1}{h_1h_2}\left[\frac{\partial}{\partial u_1}(h_2A_2) - \frac{\partial}{\partial u_2}(h_1A_1)\right]\mathbf{e}_3$$

22.77 Laplacian of Φ = $\nabla^2\Phi$ = $\frac{1}{h_1h_2h_3}\left[\frac{\partial}{\partial u_1}\left(\frac{h_2h_3}{h_1}\frac{\partial \Phi}{\partial u_1}\right) + \frac{\partial}{\partial u_2}\left(\frac{h_3h_1}{h_2}\frac{\partial \Phi}{\partial u_2}\right) + \frac{\partial}{\partial u_3}\left(\frac{h_1h_2}{h_3}\frac{\partial \Phi}{\partial u_3}\right)\right]$

Note that the biharmonic operator $\nabla^4\Phi = \nabla^2(\nabla^2\Phi)$ can be obtained from 22.77.

SPECIAL ORTHOGONAL COORDINATE SYSTEMS

Cylindrical Coordinates (r, θ, z) [See Fig. 22-12]

22.78 $$x = r\cos\theta, \quad y = r\sin\theta, \quad z = z$$

22.79 $$h_1^2 = 1, \quad h_2^2 = r^2, \quad h_3^2 = 1$$

22.80 $$\nabla^2\Phi = \frac{\partial^2\Phi}{\partial r^2} + \frac{1}{r}\frac{\partial\Phi}{\partial r} + \frac{1}{r^2}\frac{\partial^2\Phi}{\partial\theta^2} + \frac{\partial^2\Phi}{\partial z^2}$$

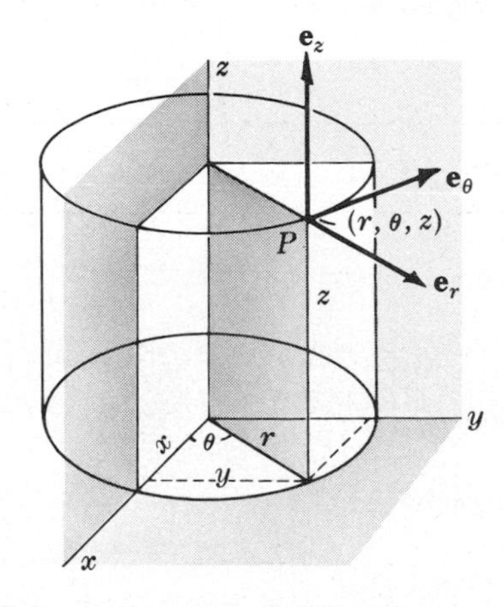

Fig. 22-12. Cylindrical coordinates.

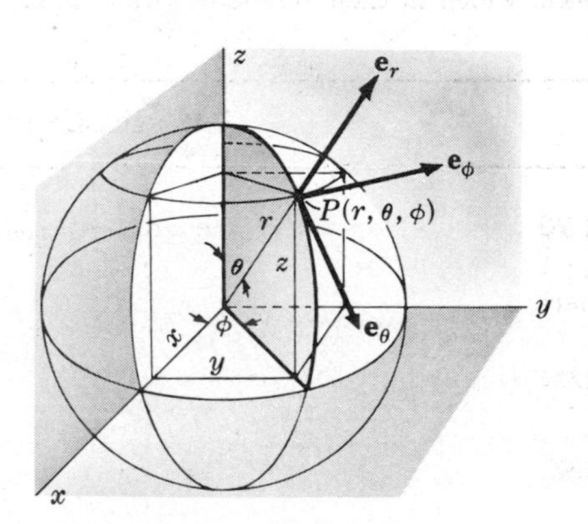

Fig. 22-13. Spherical coordinates.

Spherical Coordinates (r, θ, ϕ) [See Fig. 22-13]

22.81 $$x = r\sin\theta\cos\phi, \quad y = r\sin\theta\sin\phi, \quad z = r\cos\theta$$

22.82 $$h_1^2 = 1, \quad h_2^2 = r^2, \quad h_3^2 = r^2\sin^2\theta$$

22.83 $$\nabla^2\Phi = \frac{1}{r^2}\frac{\partial}{\partial r}\left(r^2\frac{\partial\Phi}{\partial r}\right) + \frac{1}{r^2\sin\theta}\frac{\partial}{\partial\theta}\left(\sin\theta\frac{\partial\Phi}{\partial\theta}\right) + \frac{1}{r^2\sin^2\theta}\frac{\partial^2\Phi}{\partial\phi^2}$$

Parabolic Cylindrical Coordinates (u, v, z)

22.84 $$x = \tfrac{1}{2}(u^2 - v^2), \quad y = uv, \quad z = z$$

22.85 $$h_1^2 = h_2^2 = u^2 + v^2, \quad h_3^2 = 1$$

22.86 $$\nabla^2\Phi = \frac{1}{u^2+v^2}\left(\frac{\partial^2\Phi}{\partial u^2} + \frac{\partial^2\Phi}{\partial v^2}\right) + \frac{\partial^2\Phi}{\partial z^2}$$

The traces of the coordinate surfaces on the xy plane are shown in Fig. 22-14. They are confocal parabolas with a common axis.

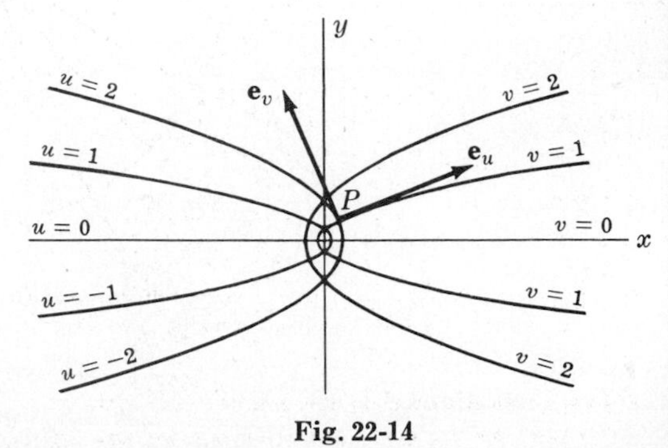

Fig. 22-14

Paraboloidal Coordinates (u, v, ϕ)

22.87 $$x = uv\cos\phi, \quad y = uv\sin\phi, \quad z = \tfrac{1}{2}(u^2 - v^2)$$

where $$u \geqq 0, \quad v \geqq 0, \quad 0 \leqq \phi < 2\pi$$

22.88 $$h_1^2 = h_2^2 = u^2 + v^2, \quad h_3^2 = u^2 v^2$$

22.89 $$\nabla^2\Phi = \frac{1}{u(u^2+v^2)}\frac{\partial}{\partial u}\left(u\frac{\partial\Phi}{\partial u}\right) + \frac{1}{v(u^2+v^2)}\frac{\partial}{\partial v}\left(v\frac{\partial\Phi}{\partial v}\right) + \frac{1}{u^2v^2}\frac{\partial^2\Phi}{\partial\phi^2}$$

Two sets of coordinate surfaces are obtained by revolving the parabolas of Fig. 22-14 about the x axis which is then relabeled the z axis.

Elliptic Cylindrical Coordinates (u, v, z)

22.90 $$x = a\cosh u\cos v, \quad y = a\sinh u\sin v, \quad z = z$$

where $$u \geqq 0, \quad 0 \leqq v < 2\pi, \quad -\infty < z < \infty$$

22.91 $$h_1^2 = h_2^2 = a^2(\sinh^2 u + \sin^2 v), \quad h_3^2 = 1$$

22.92 $$\nabla^2\Phi = \frac{1}{a^2(\sinh^2 u + \sin^2 v)}\left(\frac{\partial^2\Phi}{\partial u^2} + \frac{\partial^2\Phi}{\partial v^2}\right) + \frac{\partial^2\Phi}{\partial z^2}$$

The traces of the coordinate surfaces on the xy plane are shown in Fig. 22-15. They are confocal ellipses and hyperbolas.

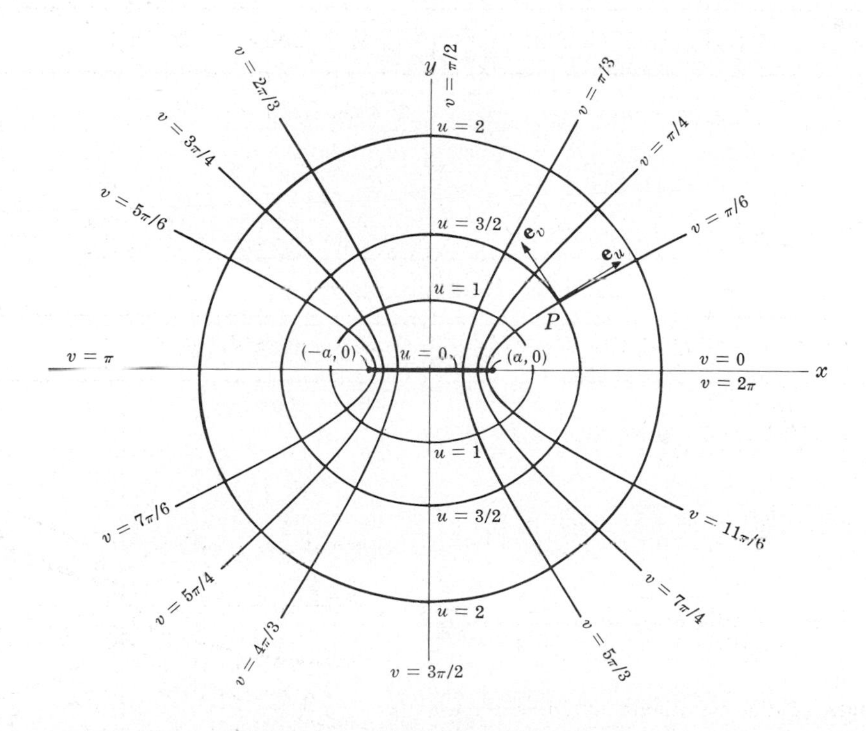

Fig. 22-15. Elliptic cylindrical coordinates.

Prolate Spheroidal Coordinates (ξ, η, ϕ)

22.93 $$x = a \sinh \xi \sin \eta \cos \phi, \quad y = a \sinh \xi \sin \eta \sin \phi, \quad z = a \cosh \xi \cos \eta$$

where $$\xi \geqq 0, \quad 0 \leqq \eta \leqq \pi, \quad 0 \leqq \phi < 2\pi$$

22.94 $$h_1^2 = h_2^2 = a^2(\sinh^2 \xi + \sin^2 \eta), \quad h_3^2 = a^2 \sinh^2 \xi \sin^2 \eta$$

22.95 $$\nabla^2\Phi = \frac{1}{a^2(\sinh^2 \xi + \sin^2 \eta) \sinh \xi} \frac{\partial}{\partial \xi}\left(\sinh \xi \frac{\partial \Phi}{\partial \xi}\right) + \frac{1}{a^2(\sinh^2 \xi + \sin^2 \eta) \sin \eta} \frac{\partial}{\partial \eta}\left(\sin \eta \frac{\partial \Phi}{\partial \eta}\right) + \frac{1}{a^2 \sinh^2 \xi \sin^2 \eta} \frac{\partial^2 \Phi}{\partial \phi^2}$$

Two sets of coordinate surfaces are obtained by revolving the curves of Fig. 22-15 about the x axis which is relabeled the z axis. The third set of coordinate surfaces consists of planes passing through this axis.

Oblate Spheroidal Coordinates (ξ, η, ϕ)

22.96 $$x = a \cosh \xi \cos \eta \cos \phi, \quad y = a \cosh \xi \cos \eta \sin \phi, \quad z = a \sinh \xi \sin \eta$$

where $$\xi \geqq 0, \quad -\pi/2 \leqq \eta \leqq \pi/2, \quad 0 \leqq \phi < 2\pi$$

22.97 $$h_1^2 = h_2^2 = a^2(\sinh^2 \xi + \sin^2 \eta), \quad h_3^2 = a^2 \cosh^2 \xi \cos^2 \eta$$

22.98 $$\nabla^2\Phi = \frac{1}{a^2(\sinh^2 \xi + \sin^2 \eta) \cosh \xi} \frac{\partial}{\partial \xi}\left(\cosh \xi \frac{\partial \Phi}{\partial \xi}\right) + \frac{1}{a^2(\sinh^2 \xi + \sin^2 \eta) \cos \eta} \frac{\partial}{\partial \eta}\left(\cos \eta \frac{\partial \Phi}{\partial \eta}\right) + \frac{1}{a^2 \cosh^2 \xi \cos^2 \eta} \frac{\partial^2 \Phi}{\partial \phi^2}$$

Two sets of coordinate surfaces are obtained by revolving the curves of Fig. 22-15 about the y axis which is relabeled the z axis. The third set of coordinate surfaces are planes passing through this axis.

Bipolar Coordinates (u, v, z)

22.99 $$x = \frac{a \sinh v}{\cosh v - \cos u}, \quad y = \frac{a \sin u}{\cosh v - \cos u}, \quad z = z$$

where $$0 \leqq u < 2\pi, \quad -\infty < v < \infty, \quad -\infty < z < \infty$$

or

22.100 $$x^2 + (y - a \cot u)^2 = a^2 \csc^2 u, \quad (x - a \coth v)^2 + y^2 = a^2 \operatorname{csch}^2 v, \quad z = z$$

22.101 $$h_1^2 = h_2^2 = \frac{a^2}{(\cosh v - \cos u)^2}, \quad h_3^2 = 1$$

22.102 $$\nabla^2\Phi = \frac{(\cosh v - \cos u)^2}{a^2}\left(\frac{\partial^2 \Phi}{\partial u^2} + \frac{\partial^2 \Phi}{\partial v^2}\right) + \frac{\partial^2 \Phi}{\partial z^2}$$

The traces of the coordinate surfaces on the xy plane are shown in Fig. 22-16 below.

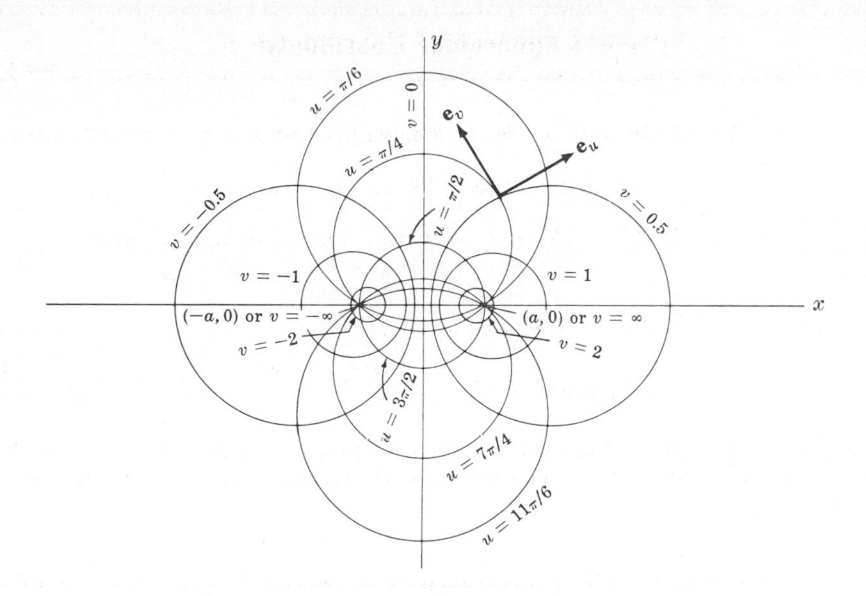

Fig. 22-16. Bipolar coordinates.

Toroidal Coordinates (u, v, ϕ)

22.103 $$x = \frac{a \sinh v \cos \phi}{\cosh v - \cos u}, \quad y = \frac{a \sinh v \sin \phi}{\cosh v - \cos u}, \quad z = \frac{a \sin u}{\cosh v - \cos u}$$

22.104 $$h_1^2 = h_2^2 = \frac{a^2}{(\cosh v - \cos u)^2}, \quad h_3^2 = \frac{a^2 \sinh^2 v}{(\cosh v - \cos u)^2}$$

22.105 $$\nabla^2\Phi = \frac{(\cosh v - \cos u)^3}{a^2} \frac{\partial}{\partial u}\left(\frac{1}{\cosh v - \cos u} \frac{\partial \Phi}{\partial u}\right)$$
$$+ \frac{(\cosh v - \cos u)^3}{a^2 \sinh v} \frac{\partial}{\partial v}\left(\frac{\sinh v}{\cosh v - \cos u} \frac{\partial \Phi}{\partial v}\right) + \frac{(\cosh v - \cos u)^2}{a^2 \sinh^2 v} \frac{\partial^2 \Phi}{\partial \phi^2}$$

The coordinate surfaces are obtained by revolving the curves of Fig. 22-16 about the y axis which is relabeled the z axis.

Conical Coordinates (λ, μ, ν)

22.106 $$x = \frac{\lambda\mu\nu}{ab}, \quad y = \frac{\lambda}{a}\sqrt{\frac{(\mu^2 - a^2)(\nu^2 - a^2)}{a^2 - b^2}}, \quad z = \frac{\lambda}{b}\sqrt{\frac{(\mu^2 - b^2)(\nu^2 - b^2)}{b^2 - a^2}}$$

22.107 $$h_1^2 = 1, \quad h_2^2 = \frac{\lambda^2(\mu^2 - \nu^2)}{(\mu^2 - a^2)(b^2 - \mu^2)}, \quad h_3^2 = \frac{\lambda^2(\mu^2 - \nu^2)}{(\nu^2 - a^2)(\nu^2 - b^2)}$$

Confocal Ellipsoidal Coordinates (λ, μ, ν)

22.108
$$\begin{cases} \dfrac{x^2}{a^2-\lambda} + \dfrac{y^2}{b^2-\lambda} + \dfrac{z^2}{c^2-\lambda} = 1 & \lambda < c^2 < b^2 < a^2 \\ \dfrac{x^2}{a^2-\mu} + \dfrac{y^2}{b^2-\mu} + \dfrac{z^2}{c^2-\mu} = 1 & c^2 < \mu < b^2 < a^2 \\ \dfrac{x^2}{a^2-\nu} + \dfrac{y^2}{b^2-\nu} + \dfrac{z^2}{c^2-\nu} = 1 & c^2 < b^2 < \nu < a^2 \end{cases}$$

or

22.109
$$\begin{cases} x^2 = \dfrac{(a^2-\lambda)(a^2-\mu)(a^2-\nu)}{(a^2-b^2)(a^2-c^2)} \\ y^2 = \dfrac{(b^2-\lambda)(b^2-\mu)(b^2-\nu)}{(b^2-a^2)(b^2-c^2)} \\ z^2 = \dfrac{(c^2-\lambda)(c^2-\mu)(c^2-\nu)}{(c^2-a^2)(c^2-b^2)} \end{cases}$$

22.110
$$\begin{cases} h_1^2 = \dfrac{(\mu-\lambda)(\nu-\lambda)}{4(a^2-\lambda)(b^2-\lambda)(c^2-\lambda)} \\ h_2^2 = \dfrac{(\nu-\mu)(\lambda-\mu)}{4(a^2-\mu)(b^2-\mu)(c^2-\mu)} \\ h_3^2 = \dfrac{(\lambda-\nu)(\mu-\nu)}{4(a^2-\nu)(b^2-\nu)(c^2-\nu)} \end{cases}$$

Confocal Paraboloidal Coordinates (λ, μ, ν)

22.111
$$\begin{cases} \dfrac{x^2}{a^2-\lambda} + \dfrac{y^2}{b^2-\lambda} = z - \lambda & -\infty < \lambda < b^2 \\ \dfrac{x^2}{a^2-\mu} + \dfrac{y^2}{b^2-\mu} = z - \mu & b^2 < \mu < a^2 \\ \dfrac{x^2}{a^2-\nu} + \dfrac{y^2}{b^2-\nu} = z - \nu & a^2 < \nu < \infty \end{cases}$$

or

22.112
$$\begin{cases} x^2 = \dfrac{(a^2-\lambda)(a^2-\mu)(a^2-\nu)}{b^2-a^2} \\ y^2 = \dfrac{(b^2-\lambda)(b^2-\mu)(b^2-\nu)}{a^2-b^2} \\ z = \lambda + \mu + \nu - a^2 - b^2 \end{cases}$$

22.113
$$\begin{cases} h_1^2 = \dfrac{(\mu-\lambda)(\nu-\lambda)}{4(a^2-\lambda)(b^2-\lambda)} \\ h_2^2 = \dfrac{(\nu-\mu)(\lambda-\mu)}{4(a^2-\mu)(b^2-\mu)} \\ h_3^2 = \dfrac{(\lambda-\nu)(\mu-\nu)}{16(a^2-\nu)(b^2-\nu)} \end{cases}$$

23 FOURIER SERIES

DEFINITION OF A FOURIER SERIES

The Fourier series corresponding to a function $f(x)$ defined in the interval $c \leqq x \leqq c+2L$ where c and $L > 0$ are constants, is defined as

23.1
$$\frac{a_0}{2} + \sum_{n=1}^{\infty}\left(a_n \cos\frac{n\pi x}{L} + b_n \sin\frac{n\pi x}{L}\right)$$

where

23.2
$$\begin{cases} a_n = \dfrac{1}{L}\displaystyle\int_c^{c+2L} f(x)\cos\frac{n\pi x}{L}\,dx \\ b_n = \dfrac{1}{L}\displaystyle\int_c^{c+2L} f(x)\sin\frac{n\pi x}{L}\,dx \end{cases}$$

If $f(x)$ and $f'(x)$ are piecewise continuous and $f(x)$ is defined by periodic extension of period $2L$, i.e. $f(x+2L) = f(x)$, then the series converges to $f(x)$ if x is a point of continuity and to $\frac{1}{2}\{f(x+0) + f(x-0)\}$ if x is a point of discontinuity.

COMPLEX FORM OF FOURIER SERIES

Assuming that the series 23.1 converges to $f(x)$, we have

23.3
$$f(x) = \sum_{n=-\infty}^{\infty} c_n e^{in\pi x/L}$$

where

23.4
$$c_n = \frac{1}{L}\int_c^{c+2L} f(x)e^{-in\pi x/L}\,dx = \begin{cases} \frac{1}{2}(a_n - ib_n) & n > 0 \\ \frac{1}{2}(a_{-n} + ib_{-n}) & n < 0 \\ \frac{1}{2}a_0 & n = 0 \end{cases}$$

PARSEVAL'S IDENTITY

23.5
$$\frac{1}{L}\int_c^{c+2L} \{f(x)\}^2\,dx = \frac{a_0^2}{2} + \sum_{n=1}^{\infty}(a_n^2 + b_n^2)$$

GENERALIZED PARSEVAL IDENTITY

23.6
$$\frac{1}{L}\int_c^{c+2L} f(x)\,g(x)\,dx = \frac{a_0 c_0}{2} + \sum_{n=1}^{\infty}(a_n c_n + b_n d_n)$$

where a_n, b_n and c_n, d_n are the Fourier coefficients corresponding to $f(x)$ and $g(x)$ respectively.

SPECIAL FOURIER SERIES AND THEIR GRAPHS

23.7

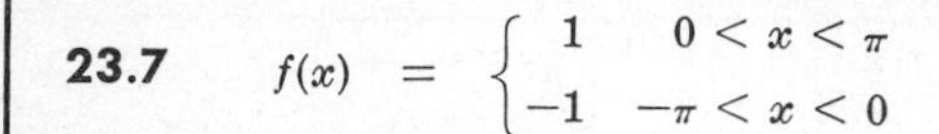

$$f(x) = \begin{cases} 1 & 0 < x < \pi \\ -1 & -\pi < x < 0 \end{cases}$$

$$\frac{4}{\pi}\left(\frac{\sin x}{1} + \frac{\sin 3x}{3} + \frac{\sin 5x}{5} + \cdots\right)$$

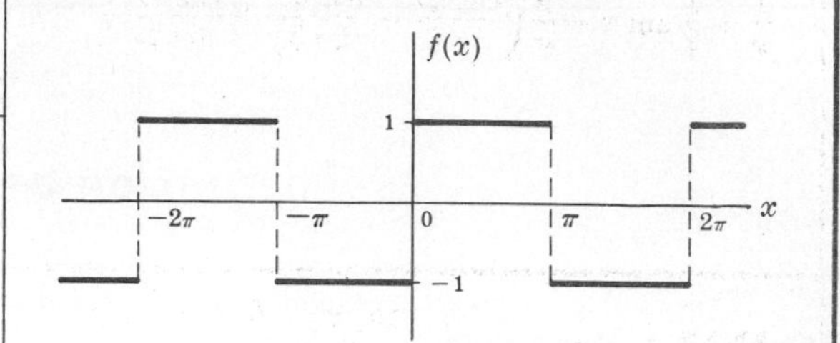

Fig. 23-1

23.8 $f(x) = |x| = \begin{cases} x & 0 < x < \pi \\ -x & -\pi < x < 0 \end{cases}$

$$\frac{\pi}{2} - \frac{4}{\pi}\left(\frac{\cos x}{1^2} + \frac{\cos 3x}{3^2} + \frac{\cos 5x}{5^2} + \cdots\right)$$

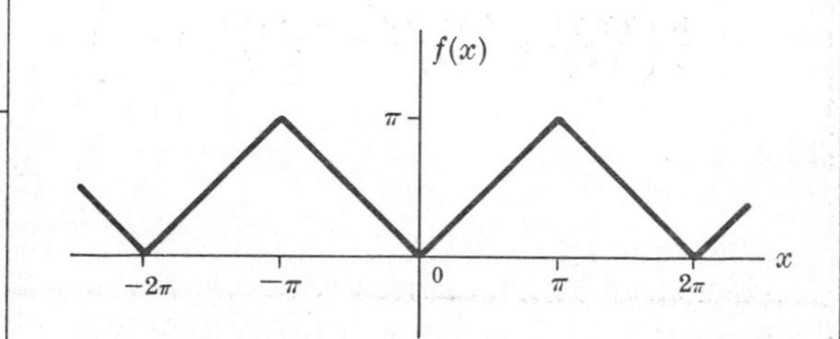

Fig. 23-2

23.9 $f(x) = x, \quad -\pi < x < \pi$

$$2\left(\frac{\sin x}{1} - \frac{\sin 2x}{2} + \frac{\sin 3x}{3} - \cdots\right)$$

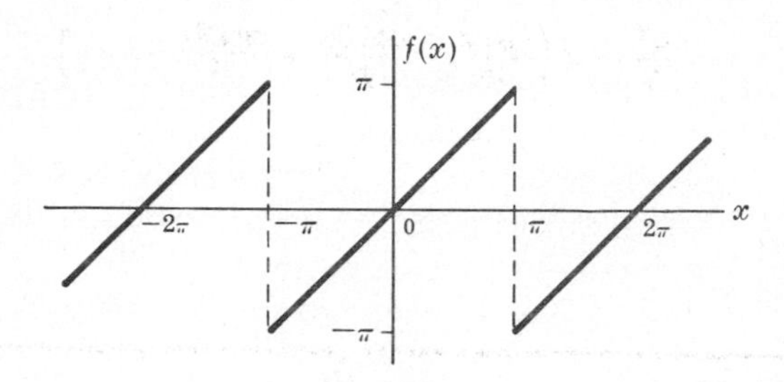

Fig. 23-3

23.10 $f(x) = x, \quad 0 < x < 2\pi$

$$\pi - 2\left(\frac{\sin x}{1} + \frac{\sin 2x}{2} + \frac{\sin 3x}{3} + \cdots\right)$$

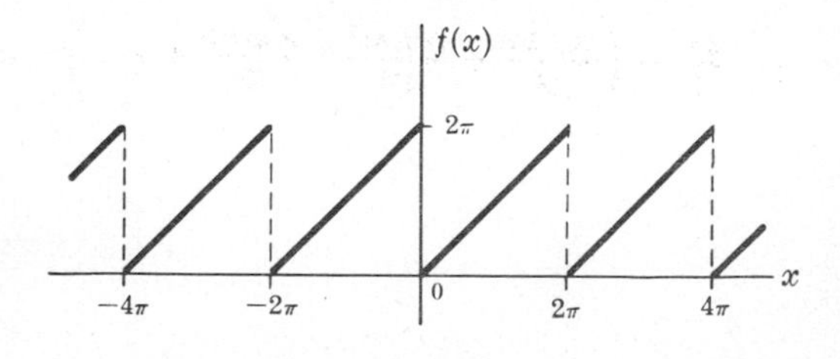

Fig. 23-4

23.11 $f(x) = |\sin x|, \quad -\pi < x < \pi$

$$\frac{2}{\pi} - \frac{4}{\pi}\left(\frac{\cos 2x}{1 \cdot 3} + \frac{\cos 4x}{3 \cdot 5} + \frac{\cos 6x}{5 \cdot 7} + \cdots\right)$$

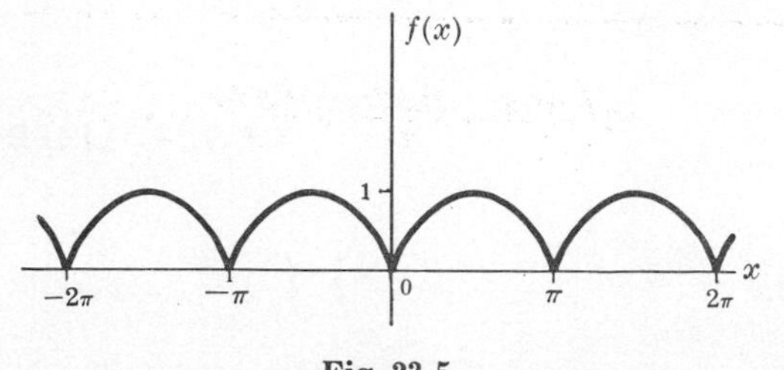

Fig. 23-5

23.12 $f(x) = \begin{cases} \sin x & 0 < x < \pi \\ 0 & \pi < x < 2\pi \end{cases}$

$$\frac{1}{\pi} + \frac{1}{2}\sin x - \frac{2}{\pi}\left(\frac{\cos 2x}{1 \cdot 3} + \frac{\cos 4x}{3 \cdot 5} + \frac{\cos 6x}{5 \cdot 7} + \cdots\right)$$

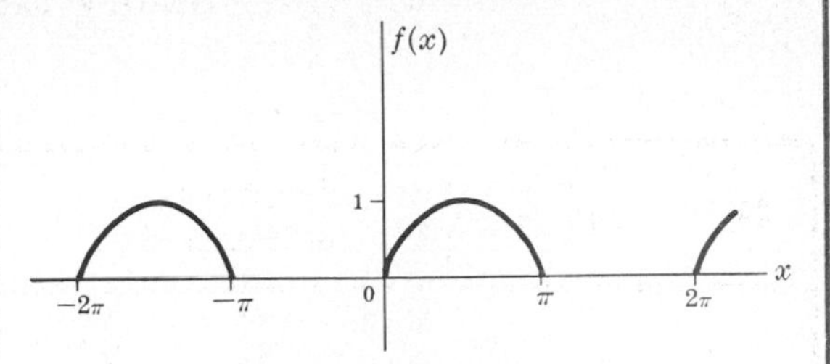

Fig. 23-6

23.13 $f(x) = \begin{cases} \cos x & 0 < x < \pi \\ -\cos x & -\pi < x < 0 \end{cases}$

$$\frac{8}{\pi}\left(\frac{\sin 2x}{1 \cdot 3} + \frac{2 \sin 4x}{3 \cdot 5} + \frac{3 \sin 6x}{5 \cdot 7} + \cdots\right)$$

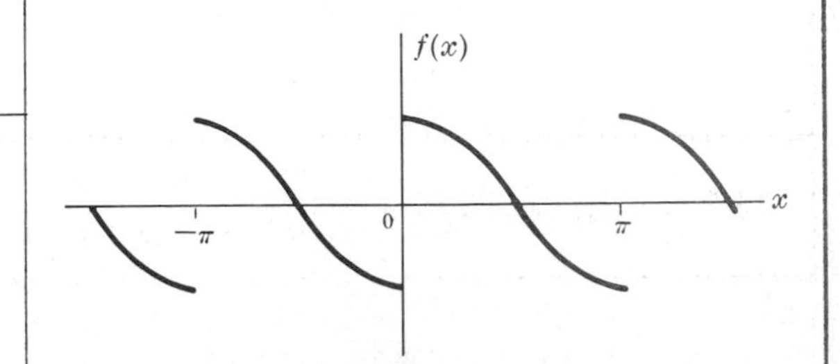

Fig. 23-7

23.14 $f(x) = x^2, \quad -\pi < x < \pi$

$$\frac{\pi^2}{3} - 4\left(\frac{\cos x}{1^2} - \frac{\cos 2x}{2^2} + \frac{\cos 3x}{3^2} - \cdots\right)$$

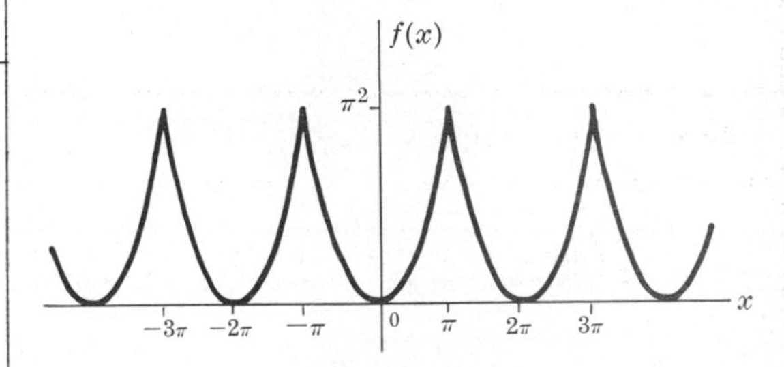

Fig. 23-8

23.15 $f(x) = x(\pi - x), \quad 0 < x < \pi$

$$\frac{\pi^2}{6} - \left(\frac{\cos 2x}{1^2} + \frac{\cos 4x}{2^2} + \frac{\cos 6x}{3^2} + \cdots\right)$$

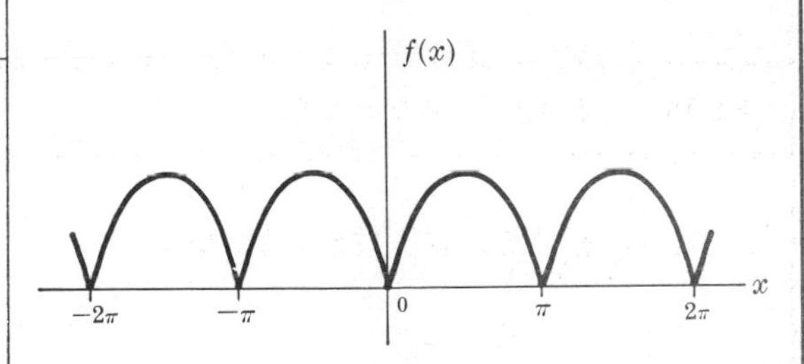

Fig. 23-9

23.16 $f(x) = x(\pi - x)(\pi + x), \quad -\pi < x < \pi$

$$12\left(\frac{\sin x}{1^3} - \frac{\sin 2x}{2^3} + \frac{\sin 3x}{3^3} - \cdots\right)$$

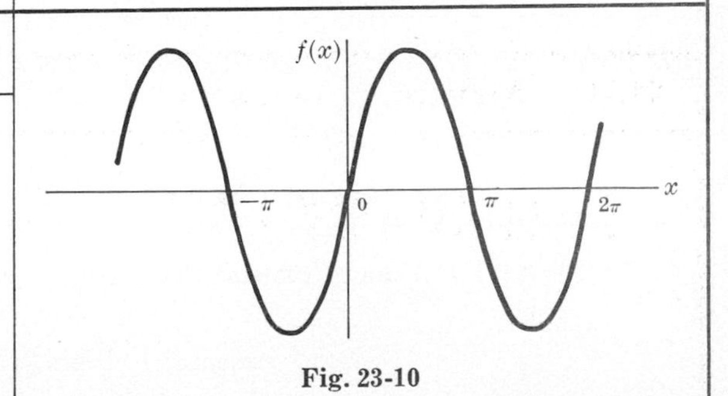

Fig. 23-10

23.17 $f(x) = \begin{cases} 0 & 0 < x < \pi - \alpha \\ 1 & \pi - \alpha < x < \pi + \alpha \\ 0 & \pi + \alpha < x < 2\pi \end{cases}$

$$\frac{\alpha}{\pi} - \frac{2}{\pi}\left(\frac{\sin \alpha \cos x}{1} - \frac{\sin 2\alpha \cos 2x}{2} + \frac{\sin 3\alpha \cos 3x}{3} - \cdots\right)$$

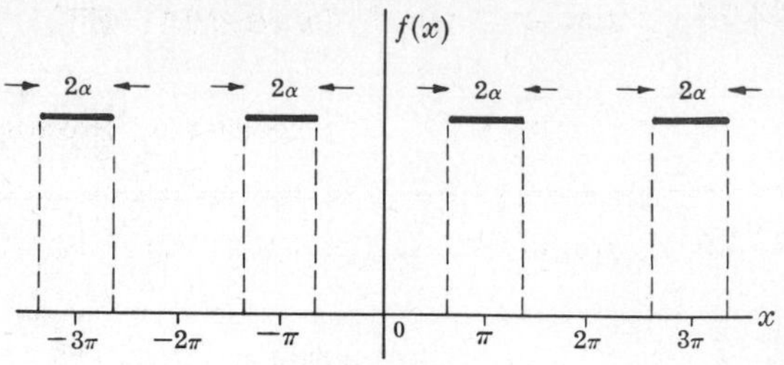

Fig. 23-11

23.18 $f(x) = \begin{cases} x(\pi - x) & 0 < x < \pi \\ -x(\pi - x) & -\pi < x < 0 \end{cases}$

$$\frac{8}{\pi}\left(\frac{\sin x}{1^3} + \frac{\sin 3x}{3^3} + \frac{\sin 5x}{5^3} + \cdots\right)$$

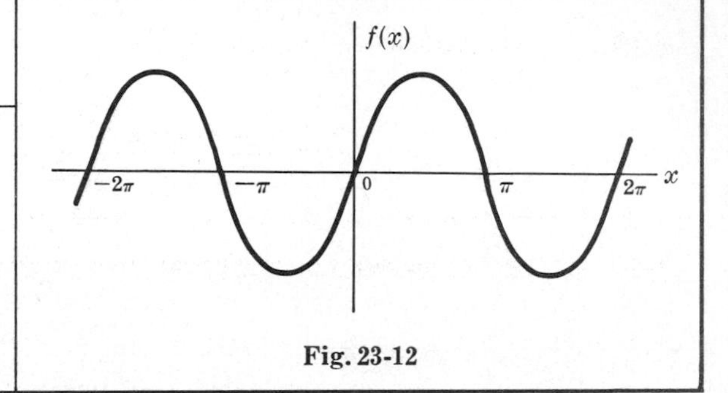

Fig. 23-12

MISCELLANEOUS FOURIER SERIES

23.19 $f(x) = \sin \mu x, \quad -\pi < x < \pi, \quad \mu \neq \text{integer}$

$$\frac{2 \sin \mu\pi}{\pi}\left(\frac{\sin x}{1^2 - \mu^2} - \frac{2 \sin 2x}{2^2 - \mu^2} + \frac{3 \sin 3x}{3^2 - \mu^2} - \cdots\right)$$

23.20 $f(x) = \cos \mu x, \quad -\pi < x < \pi, \quad \mu \neq \text{integer}$

$$\frac{2\mu \sin \mu\pi}{\pi}\left(\frac{1}{2\mu^2} + \frac{\cos x}{1^2 - \mu^2} - \frac{\cos 2x}{2^2 - \mu^2} + \frac{\cos 3x}{3^2 - \mu^2} - \cdots\right)$$

23.21 $f(x) = \tan^{-1}[(a \sin x)/(1 - a \cos x)], \quad -\pi < x < \pi, \quad |a| < 1$

$$a \sin x + \frac{a^2}{2} \sin 2x + \frac{a^3}{3} \sin 3x + \cdots$$

23.22 $f(x) = \ln(1 - 2a \cos x + a^2), \quad -\pi < x < \pi, \quad |a| < 1$

$$-2\left(a \cos x + \frac{a^2}{2} \cos 2x + \frac{a^3}{3} \cos 3x + \cdots\right)$$

23.23 $f(x) = \frac{1}{2} \tan^{-1}[(2a \sin x)/(1 - a^2)], \quad -\pi < x < \pi, \quad |a| < 1$

$$a \sin x + \frac{a^3}{3} \sin 3x + \frac{a^5}{5} \sin 5x + \cdots$$

23.24 $f(x) = \frac{1}{2}\tan^{-1}[(2a\cos x)/(1-a^2)], \quad -\pi < x < \pi, \quad |a| < 1$

$$a\cos x - \frac{a^3}{3}\cos 3x + \frac{a^5}{5}\cos 5x - \cdots$$

23.25 $f(x) = e^{\mu x}, \quad -\pi < x < \pi$

$$\frac{2\sinh \mu\pi}{\pi}\left(\frac{1}{2\mu} + \sum_{n=1}^{\infty}\frac{(-1)^n(\mu\cos nx - n\sin nx)}{\mu^2+n^2}\right)$$

23.26 $f(x) = \sinh \mu x, \quad -\pi < x < \pi$

$$\frac{2\sinh \mu\pi}{\pi}\left(\frac{\sin x}{1^2+\mu^2} - \frac{2\sin 2x}{2^2+\mu^2} + \frac{3\sin 3x}{3^2+\mu^2} - \cdots\right)$$

23.27 $f(x) = \cosh \mu x, \quad -\pi < x < \pi$

$$\frac{2\mu\sinh \mu\pi}{\pi}\left(\frac{1}{2\mu^2} - \frac{\cos x}{1^2+\mu^2} + \frac{\cos 2x}{2^2+\mu^2} - \frac{\cos 3x}{3^2+\mu^2} + \cdots\right)$$

23.28 $f(x) = \ln|\sin \tfrac{1}{2}x|, \quad 0 < x < \pi$

$$-\left(\ln 2 + \frac{\cos x}{1} + \frac{\cos 2x}{2} + \frac{\cos 3x}{3} + \cdots\right)$$

23.29 $f(x) = \ln|\cos \tfrac{1}{2}x|, \quad -\pi < x < \pi$

$$-\left(\ln 2 - \frac{\cos x}{1} + \frac{\cos 2x}{2} - \frac{\cos 3x}{3} + \cdots\right)$$

23.30 $f(x) = \tfrac{1}{6}\pi^2 - \tfrac{1}{2}\pi x + \tfrac{1}{4}x^2, \quad 0 \leqq x \leqq 2\pi$

$$\frac{\cos x}{1^2} + \frac{\cos 2x}{2^2} + \frac{\cos 3x}{3^2} + \cdots$$

23.31 $f(x) = \tfrac{1}{12}x(x-\pi)(x-2\pi), \quad 0 \leqq x \leqq 2\pi$

$$\frac{\sin x}{1^3} + \frac{\sin 2x}{2^3} + \frac{\sin 3x}{3^3} + \cdots$$

23.32 $f(x) = \tfrac{1}{90}\pi^4 - \tfrac{1}{12}\pi^2x^2 + \tfrac{1}{12}\pi x^3 - \tfrac{1}{48}x^4, \quad 0 \leqq x \leqq 2\pi$

$$\frac{\cos x}{1^4} + \frac{\cos 2x}{2^4} + \frac{\cos 3x}{3^4} + \cdots$$

24 BESSEL FUNCTIONS

BESSEL'S DIFFERENTIAL EQUATION

24.1 $$x^2y'' + xy' + (x^2 - n^2)y = 0 \qquad n \geqq 0$$

Solutions of this equation are called *Bessel functions of order n*.

BESSEL FUNCTIONS OF THE FIRST KIND OF ORDER n

24.2 $$\begin{aligned} J_n(x) &= \frac{x^n}{2^n \Gamma(n+1)}\left\{1 - \frac{x^2}{2(2n+2)} + \frac{x^4}{2 \cdot 4(2n+2)(2n+4)} - \cdots\right\} \\ &= \sum_{k=0}^{\infty} \frac{(-1)^k (x/2)^{n+2k}}{k!\,\Gamma(n+k+1)} \end{aligned}$$

24.3 $$\begin{aligned} J_{-n}(x) &= \frac{x^{-n}}{2^{-n} \Gamma(1-n)}\left\{1 - \frac{x^2}{2(2-2n)} + \frac{x^4}{2 \cdot 4(2-2n)(4-2n)} - \cdots\right\} \\ &= \sum_{k=0}^{\infty} \frac{(-1)^k (x/2)^{2k-n}}{k!\,\Gamma(k+1-n)} \end{aligned}$$

24.4 $$J_{-n}(x) = (-1)^n J_n(x) \qquad n = 0, 1, 2, \ldots$$

If $n \neq 0, 1, 2, \ldots$, $J_n(x)$ and $J_{-n}(x)$ are linearly independent.

If $n \neq 0, 1, 2, \ldots$, $J_n(x)$ is bounded at $x = 0$ while $J_{-n}(x)$ is unbounded.

For $n = 0, 1$ we have

24.5 $$J_0(x) = 1 - \frac{x^2}{2^2} + \frac{x^4}{2^2 \cdot 4^2} - \frac{x^6}{2^2 \cdot 4^2 \cdot 6^2} + \cdots$$

24.6 $$J_1(x) = \frac{x}{2} - \frac{x^3}{2^2 \cdot 4} + \frac{x^5}{2^2 \cdot 4^2 \cdot 6} - \frac{x^7}{2^2 \cdot 4^2 \cdot 6^2 \cdot 8} + \cdots$$

24.7 $$J_0'(x) = -J_1(x)$$

BESSEL FUNCTIONS OF THE SECOND KIND OF ORDER n

24.8 $$Y_n(x) = \begin{cases} \dfrac{J_n(x)\cos n\pi - J_{-n}(x)}{\sin n\pi} & n \neq 0, 1, 2, \ldots \\[2ex] \lim\limits_{p \to n} \dfrac{J_p(x)\cos p\pi - J_{-p}(x)}{\sin p\pi} & n = 0, 1, 2, \ldots \end{cases}$$

This is also called *Weber's function* or *Neumann's function* [also denoted by $N_n(x)$].

For $n = 0, 1, 2, \ldots,$ L'Hospital's rule yields

24.9
$$Y_n(x) = \frac{2}{\pi}\{\ln(x/2) + \gamma\}J_n(x) - \frac{1}{\pi}\sum_{k=0}^{n-1}\frac{(n-k-1)!}{k!}(x/2)^{2k-n}$$
$$- \frac{1}{\pi}\sum_{k=0}^{\infty}(-1)^k\{\Phi(k) + \Phi(n+k)\}\frac{(x/2)^{2k+n}}{k!\,(n+k)!}$$

where $\gamma = .5772156\ldots$ is Euler's constant [page 1] and

24.10
$$\Phi(p) = 1 + \frac{1}{2} + \frac{1}{3} + \cdots + \frac{1}{p}, \qquad \Phi(0) = 0$$

For $n = 0$,

24.11
$$Y_0(x) = \frac{2}{\pi}\{\ln(x/2) + \gamma\}J_0(x) + \frac{2}{\pi}\left\{\frac{x^2}{2^2} - \frac{x^4}{2^2 4^2}(1 + \tfrac{1}{2}) + \frac{x^6}{2^2 4^2 6^2}(1 + \tfrac{1}{2} + \tfrac{1}{3}) - \cdots\right\}$$

24.12
$$Y_{-n}(x) = (-1)^n Y_n(x) \qquad n = 0, 1, 2, \ldots$$

For any value $n \geqq 0$, $J_n(x)$ is bounded at $x = 0$ while $Y_n(x)$ is unbounded.

GENERAL SOLUTION OF BESSEL'S DIFFERENTIAL EQUATION

24.13 $y = AJ_n(x) + BJ_{-n}(x) \qquad n \neq 0, 1, 2, \ldots$

24.14 $y = AJ_n(x) + BY_n(x) \qquad$ all n

24.15 $y = AJ_n(x) + BJ_n(x)\displaystyle\int\frac{dx}{xJ_n^2(x)} \qquad$ all n

where A and B are arbitrary constants.

GENERATING FUNCTION FOR $J_n(x)$

24.16
$$e^{x(t-1/t)/2} = \sum_{n=-\infty}^{\infty} J_n(x)t^n$$

RECURRENCE FORMULAS FOR BESSEL FUNCTIONS

24.17
$$J_{n+1}(x) = \frac{2n}{x}J_n(x) - J_{n-1}(x)$$

24.18
$$J_n'(x) = \tfrac{1}{2}\{J_{n-1}(x) - J_{n+1}(x)\}$$

24.19
$$xJ_n'(x) = xJ_{n-1}(x) - nJ_n(x)$$

24.20
$$xJ_n'(x) = nJ_n(x) - xJ_{n+1}(x)$$

24.21
$$\frac{d}{dx}\{x^n J_n(x)\} = x^n J_{n-1}(x)$$

24.22
$$\frac{d}{dx}\{x^{-n} J_n(x)\} = -x^{-n} J_{n+1}(x)$$

The functions $Y_n(x)$ satisfy identical relations.

BESSEL FUNCTIONS OF ORDER EQUAL TO HALF AN ODD INTEGER

In this case the functions are expressible in terms of sines and cosines.

24.23 $J_{1/2}(x) = \sqrt{\frac{2}{\pi x}} \sin x$

24.24 $J_{-1/2}(x) = \sqrt{\frac{2}{\pi x}} \cos x$

24.25 $J_{3/2}(x) = \sqrt{\frac{2}{\pi x}} \left(\frac{\sin x}{x} - \cos x \right)$

24.26 $J_{-3/2}(x) = \sqrt{\frac{2}{\pi x}} \left(\frac{\cos x}{x} + \sin x \right)$

24.27 $J_{5/2}(x) = \sqrt{\frac{2}{\pi x}} \left\{ \left(\frac{3}{x^2} - 1 \right) \sin x - \frac{3}{x} \cos x \right\}$

24.28 $J_{-5/2}(x) = \sqrt{\frac{2}{\pi x}} \left\{ \frac{3}{x} \sin x + \left(\frac{3}{x^2} - 1 \right) \cos x \right\}$

For further results use the recurrence formula. Results for $Y_{1/2}(x), Y_{3/2}(x), \ldots$ are obtained from 24.8.

HANKEL FUNCTIONS OF FIRST AND SECOND KINDS OF ORDER n

24.29 $H_n^{(1)}(x) = J_n(x) + i Y_n(x)$

24.30 $H_n^{(2)}(x) = J_n(x) - i Y_n(x)$

BESSEL'S MODIFIED DIFFERENTIAL EQUATION

24.31
$$x^2 y'' + x y' - (x^2 + n^2) y = 0 \qquad n \geqq 0$$

Solutions of this equation are called *modified Bessel functions of order n.*

MODIFIED BESSEL FUNCTIONS OF THE FIRST KIND OF ORDER n

24.32
$$I_n(x) = i^{-n} J_n(ix) = e^{-n\pi i/2} J_n(ix)$$
$$= \frac{x^n}{2^n \Gamma(n+1)} \left\{ 1 + \frac{x^2}{2(2n+2)} + \frac{x^4}{2 \cdot 4(2n+2)(2n+4)} + \cdots \right\} = \sum_{k=0}^{\infty} \frac{(x/2)^{n+2k}}{k!\, \Gamma(n+k+1)}$$

24.33
$$I_{-n}(x) = i^n J_{-n}(ix) = e^{n\pi i/2} J_{-n}(ix)$$
$$= \frac{x^{-n}}{2^{-n} \Gamma(1-n)} \left\{ 1 + \frac{x^2}{2(2-2n)} + \frac{x^4}{2 \cdot 4(2-2n)(4-2n)} + \cdots \right\} = \sum_{k=0}^{\infty} \frac{(x/2)^{2k-n}}{k!\, \Gamma(k+1-n)}$$

24.34
$$I_{-n}(x) = I_n(x) \qquad n = 0, 1, 2, \ldots$$

If $n \neq 0, 1, 2, \ldots,$ then $I_n(x)$ and $I_{-n}(x)$ are linearly independent.

For $n = 0, 1,$ we have

24.35 $I_0(x) = 1 + \frac{x^2}{2^2} + \frac{x^4}{2^2 \cdot 4^2} + \frac{x^6}{2^2 \cdot 4^2 \cdot 6^2} + \cdots$

24.36 $I_1(x) = \frac{x}{2} + \frac{x^3}{2^2 \cdot 4} + \frac{x^5}{2^2 \cdot 4^2 \cdot 6} + \frac{x^7}{2^2 \cdot 4^2 \cdot 6^2 \cdot 8} + \cdots$

24.37 $I_0'(x) = I_1(x)$

MODIFIED BESSEL FUNCTIONS OF THE SECOND KIND OF ORDER n

24.38 $$K_n(x) = \begin{cases} \dfrac{\pi}{2\sin n\pi}\{I_{-n}(x) - I_n(x)\} & n \neq 0,1,2,\ldots \\ \lim\limits_{p\to n} \dfrac{\pi}{2\sin p\pi}\{I_{-p}(x) - I_p(x)\} & n = 0,1,2,\ldots \end{cases}$$

For $n = 0, 1, 2, \ldots$, L'Hospital's rule yields

24.39 $$K_n(x) = (-1)^{n+1}\{\ln(x/2) + \gamma\}I_n(x) + \frac{1}{2}\sum_{k=0}^{n-1}(-1)^k(n-k-1)!\,(x/2)^{2k-n} + \frac{(-1)^n}{2}\sum_{k=0}^{\infty}\frac{(x/2)^{n+2k}}{k!\,(n+k)!}\{\Phi(k) + \Phi(n+k)\}$$

where $\Phi(p)$ is given by 24.10.

For $n = 0$,

24.40 $$K_0(x) = -\{\ln(x/2) + \gamma\}I_0(x) + \frac{x^2}{2^2} + \frac{x^4}{2^2\cdot 4^2}(1+\tfrac{1}{2}) + \frac{x^6}{2^2\cdot 4^2\cdot 6^2}(1+\tfrac{1}{2}+\tfrac{1}{3}) + \cdots$$

24.41 $$K_{-n}(x) = K_n(x) \qquad n = 0,1,2,\ldots$$

GENERAL SOLUTION OF BESSEL'S MODIFIED EQUATION

24.42 $$y = AI_n(x) + BI_{-n}(x) \qquad n \neq 0,1,2,\ldots$$

24.43 $$y = AI_n(x) + BK_n(x) \qquad \text{all } n$$

24.44 $$y = AI_n(x) + BI_n(x)\int\frac{dx}{x\,I_n^2(x)} \qquad \text{all } n$$

where A and B are arbitrary constants.

GENERATING FUNCTION FOR $I_n(x)$

24.45 $$e^{x(t+1/t)/2} = \sum_{n=-\infty}^{\infty} I_n(x)t^n$$

RECURRENCE FORMULAS FOR MODIFIED BESSEL FUNCTIONS

24.46 $$I_{n+1}(x) = I_{n-1}(x) - \frac{2n}{x}I_n(x)$$

24.47 $$I_n'(x) = \tfrac{1}{2}\{I_{n-1}(x) + I_{n+1}(x)\}$$

24.48 $$xI_n'(x) = xI_{n-1}(x) - nI_n(x)$$

24.49 $$xI_n'(x) = xI_{n+1}(x) + nI_n(x)$$

24.50 $$\frac{d}{dx}\{x^nI_n(x)\} = x^nI_{n-1}(x)$$

24.51 $$\frac{d}{dx}\{x^{-n}I_n(x)\} = x^{-n}I_{n+1}(x)$$

24.52 $$K_{n+1}(x) = K_{n-1}(x) + \frac{2n}{x}K_n(x)$$

24.53 $$K_n'(x) = \tfrac{1}{2}\{K_{n-1}(x) + K_{n+1}(x)\}$$

24.54 $$xK_n'(x) = -xK_{n-1}(x) - nK_n(x)$$

24.55 $$xK_n'(x) = nK_n(x) - xK_{n+1}(x)$$

24.56 $$\frac{d}{dx}\{x^nK_n(x)\} = -x^nK_{n-1}(x)$$

24.57 $$\frac{d}{dx}\{x^{-n}K_n(x)\} = -x^{-n}K_{n+1}(x)$$

MODIFIED BESSEL FUNCTIONS OF ORDER EQUAL TO HALF AN ODD INTEGER

In this case the functions are expressible in terms of hyperbolic sines and cosines.

24.58 $$I_{1/2}(x) = \sqrt{\frac{2}{\pi x}}\sinh x$$

24.59 $$I_{-1/2}(x) = \sqrt{\frac{2}{\pi x}}\cosh x$$

24.60 $$I_{3/2}(x) = \sqrt{\frac{2}{\pi x}}\left(\cosh x - \frac{\sinh x}{x}\right)$$

24.61 $$I_{-3/2}(x) = \sqrt{\frac{2}{\pi x}}\left(\sinh x - \frac{\cosh x}{x}\right)$$

24.62 $$I_{5/2}(x) = \sqrt{\frac{2}{\pi x}}\left\{\left(\frac{3}{x^2}+1\right)\sinh x - \frac{3}{x}\cosh x\right\}$$

24.63 $$I_{-5/2}(x) = \sqrt{\frac{2}{\pi x}}\left\{\left(\frac{3}{x^2}+1\right)\cosh x - \frac{3}{x}\sinh x\right\}$$

For further results use the recurrence formula 24.46. Results for $K_{1/2}(x), K_{3/2}(x), \ldots$ are obtained from 24.38.

Ber AND Bei FUNCTIONS

The real and imaginary parts of $J_n(xe^{3\pi i/4})$ are denoted by $\text{Ber}_n(x)$ and $\text{Bei}_n(x)$ where

24.64 $$\text{Ber}_n(x) = \sum_{k=0}^{\infty} \frac{(x/2)^{2k+n}}{k!\,\Gamma(n+k+1)} \cos\frac{(3n+2k)\pi}{4}$$

24.65 $$\text{Bei}_n(x) = \sum_{k=0}^{\infty} \frac{(x/2)^{2k+n}}{k!\,\Gamma(n+k+1)} \sin\frac{(3n+2k)\pi}{4}$$

If $n = 0$,

24.66 $$\text{Ber}(x) = 1 - \frac{(x/2)^4}{2!^2} + \frac{(x/2)^8}{4!^2} - \cdots$$

24.67 $$\text{Bei}(x) = (x/2)^2 - \frac{(x/2)^6}{3!^2} + \frac{(x/2)^{10}}{5!^2} - \cdots$$

Ker AND Kei FUNCTIONS

The real and imaginary parts of $e^{-n\pi i/2} K_n(xe^{\pi i/4})$ are denoted by $\text{Ker}_n(x)$ and $\text{Kei}_n(x)$ where

24.68 $$\begin{aligned}\text{Ker}_n(x) = {} & -\{\ln(x/2)+\gamma\}\,\text{Ber}_n(x) + \tfrac{1}{4}\pi\,\text{Bei}_n(x) \\ & + \frac{1}{2}\sum_{k=0}^{n-1} \frac{(n-k-1)!\,(x/2)^{2k-n}}{k!} \cos\frac{(3n+2k)\pi}{4} \\ & + \frac{1}{2}\sum_{k=0}^{\infty} \frac{(x/2)^{n+2k}}{k!\,(n+k)!}\{\Phi(k)+\Phi(n+k)\} \cos\frac{(3n+2k)\pi}{4}\end{aligned}$$

24.69 $$\begin{aligned}\text{Kei}_n(x) = {} & -\{\ln(x/2)+\gamma\}\,\text{Bei}_n(x) - \tfrac{1}{4}\pi\,\text{Ber}_n(x) \\ & - \frac{1}{2}\sum_{k=0}^{n-1} \frac{(n-k-1)!\,(x/2)^{2k-n}}{k!} \sin\frac{(3n+2k)\pi}{4} \\ & + \frac{1}{2}\sum_{k=0}^{\infty} \frac{(x/2)^{n+2k}}{k!\,(n+k)!}\{\Phi(k)+\Phi(n+k)\} \sin\frac{(3n+2k)\pi}{4}\end{aligned}$$

and Φ is given by 24.10, page 137.

If $n = 0$,

24.70 $$\text{Ker}(x) = -\{\ln(x/2)+\gamma\}\,\text{Ber}(x) + \frac{\pi}{4}\text{Bei}(x) + 1 - \frac{(x/2)^4}{2!^2}(1+\tfrac{1}{2}) + \frac{(x/2)^8}{4!^2}(1+\tfrac{1}{2}+\tfrac{1}{3}+\tfrac{1}{4}) - \cdots$$

24.71 $$\text{Kei}(x) = -\{\ln(x/2)+\gamma\}\,\text{Bei}(x) - \frac{\pi}{4}\text{Ber}(x) + (x/2)^2 - \frac{(x/2)^6}{3!^2}(1+\tfrac{1}{2}+\tfrac{1}{3}) + \cdots$$

DIFFERENTIAL EQUATION FOR Ber, Bei, Ker, Kei FUNCTIONS

24.72 $$x^2y'' + xy' - (ix^2 + n^2)y = 0$$

The general solution of this equation is

24.73 $$y = A\{\mathrm{Ber}_n(x) + i\,\mathrm{Bei}_n(x)\} + B\{\mathrm{Ker}_n(x) + i\,\mathrm{Kei}_n(x)\}$$

GRAPHS OF BESSEL FUNCTIONS

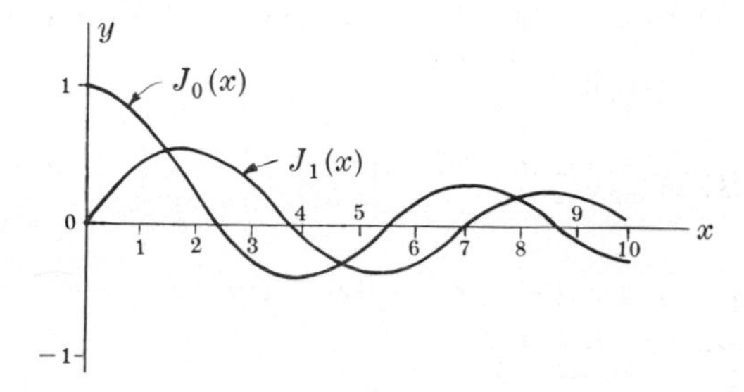

Fig. 24-1

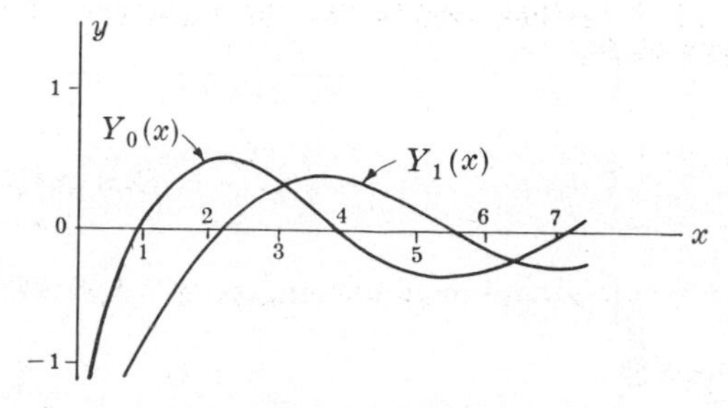

Fig. 24-2

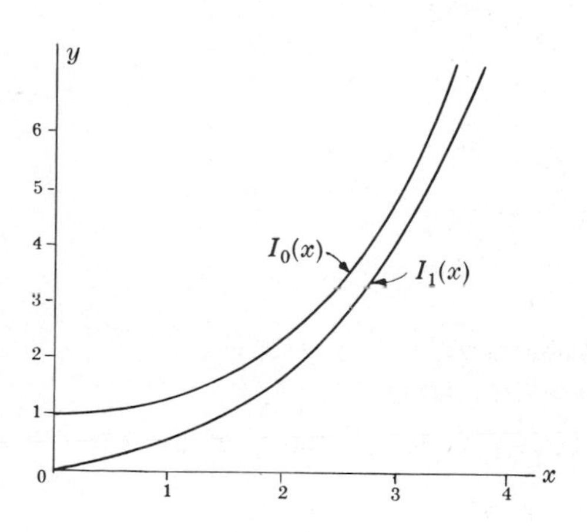

Fig. 24-3

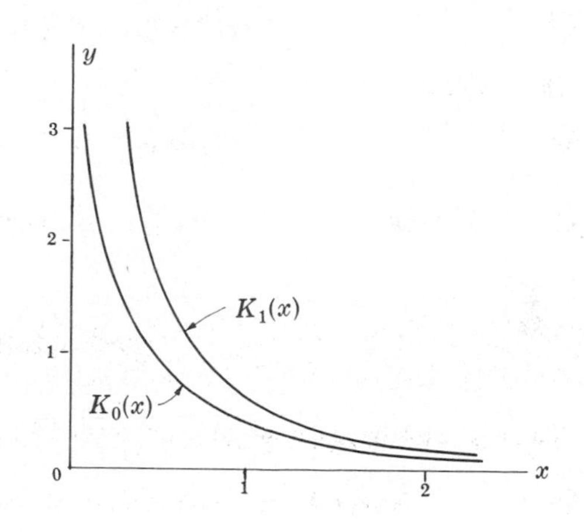

Fig. 24-4

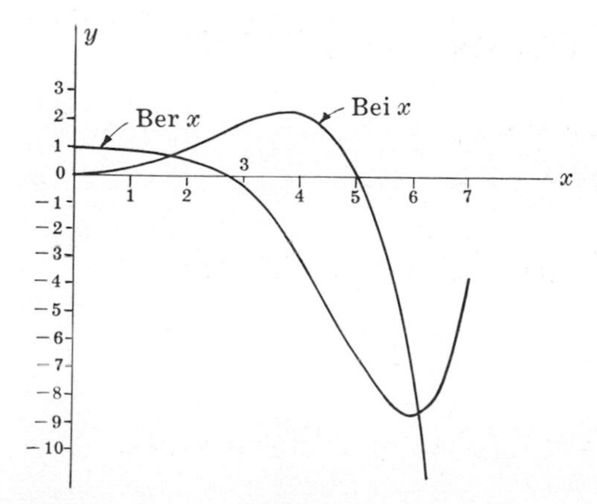

Fig. 24-5

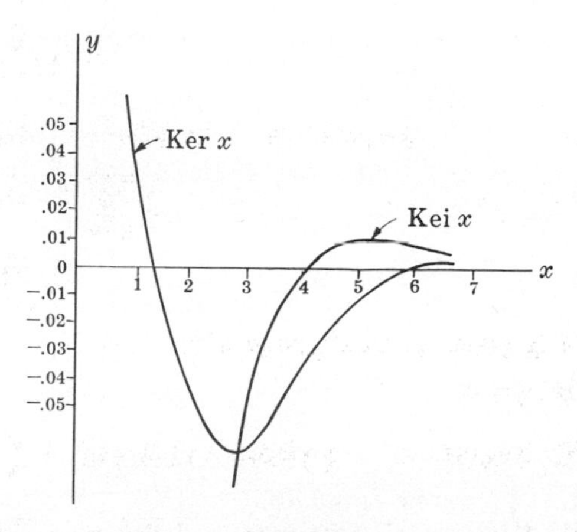

Fig. 24-6

INDEFINITE INTEGRALS INVOLVING BESSEL FUNCTIONS

24.74 $\int x\,J_0(x)\,dx = x\,J_1(x)$

24.75 $\int x^2 J_0(x)\,dx = x^2 J_1(x) + x\,J_0(x) - \int J_0(x)\,dx$

24.76 $\int x^m J_0(x)\,dx = x^m J_1(x) + (m-1)x^{m-1} J_0(x) - (m-1)^2 \int x^{m-2} J_0(x)\,dx$

24.77 $\int \frac{J_0(x)}{x^2}\,dx = J_1(x) - \frac{J_0(x)}{x} - \int J_0(x)\,dx$

24.78 $\int \frac{J_0(x)}{x^m}\,dx = \frac{J_1(x)}{(m-1)^2 x^{m-2}} - \frac{J_0(x)}{(m-1)x^{m-1}} - \frac{1}{(m-1)^2}\int \frac{J_0(x)}{x^{m-2}}\,dx$

24.79 $\int J_1(x)\,dx = -J_0(x)$

24.80 $\int x\,J_1(x)\,dx = -x\,J_0(x) + \int J_0(x)\,dx$

24.81 $\int x^m J_1(x)\,dx = -x^m J_0(x) + m \int x^{m-1} J_0(x)\,dx$

24.82 $\int \frac{J_1(x)}{x}\,dx = -J_1(x) + \int J_0(x)\,dx$

24.83 $\int \frac{J_1(x)}{x^m}\,dx = -\frac{J_1(x)}{m x^{m-1}} + \frac{1}{m}\int \frac{J_0(x)}{x^{m-1}}\,dx$

24.84 $\int x^n J_{n-1}(x)\,dx = x^n J_n(x)$

24.85 $\int x^{-n} J_{n+1}(x)\,dx = -x^{-n} J_n(x)$

24.86 $\int x^m J_n(x)\,dx = -x^m J_{n-1}(x) + (m+n-1)\int x^{m-1} J_{n-1}(x)\,dx$

24.87 $\int x\,J_n(\alpha x)\,J_n(\beta x)\,dx = \frac{x\{\alpha\,J_n(\beta x)\,J_n'(\alpha x) - \beta\,J_n(\alpha x)\,J_n'(\beta x)\}}{\beta^2 - \alpha^2}$

24.88 $\int x\,J_n^2(\alpha x)\,dx = \frac{x^2}{2}\{J_n'(\alpha x)\}^2 + \frac{x^2}{2}\left(1 - \frac{n^2}{\alpha^2 x^2}\right)\{J_n(\alpha x)\}^2$

The above results also hold if we replace $J_n(x)$ by $Y_n(x)$ or, more generally, $A\,J_n(x) + B\,Y_n(x)$ where A and B are constants.

DEFINITE INTEGRALS INVOLVING BESSEL FUNCTIONS

24.89 $\int_0^\infty e^{-ax} J_0(bx)\,dx = \frac{1}{\sqrt{a^2+b^2}}$

24.90 $\int_0^\infty e^{-ax} J_n(bx)\,dx = \frac{(\sqrt{a^2+b^2}-a)^n}{b^n\sqrt{a^2+b^2}} \qquad n > -1$

24.91 $\int_0^\infty \cos ax\,J_0(bx)\,dx = \begin{cases} \frac{1}{\sqrt{a^2-b^2}} & a > b \\ 0 & a < b \end{cases}$

24.92 $\int_0^\infty J_n(bx)\,dx = \frac{1}{b} \qquad n > -1$

24.93 $\int_0^\infty \frac{J_n(bx)}{x}\,dx = \frac{1}{n} \qquad n = 1, 2, 3, \ldots$

24.94 $\int_0^\infty e^{-ax} J_0(b\sqrt{x})\,dx = \frac{e^{-b^2/4a}}{a}$

24.95 $\int_0^1 x\,J_n(\alpha x)\,J_n(\beta x)\,dx = \frac{\alpha\,J_n(\beta)\,J_n'(\alpha) - \beta\,J_n(\alpha)\,J_n'(\beta)}{\beta^2 - \alpha^2}$

24.96 $\int_0^1 x\,J_n^2(\alpha x)\,dx = \frac{1}{2}\{J_n'(\alpha)\}^2 + \frac{1}{2}(1 - n^2/\alpha^2)\{J_n(\alpha)\}^2$

24.97 $\int_0^1 x\,J_0(\alpha x)\,I_0(\beta x)\,dx = \frac{\beta\,J_0(\alpha)\,I_0'(\beta) - \alpha\,J_0'(\alpha)\,I_0(\beta)}{\alpha^2 + \beta^2}$

INTEGRAL REPRESENTATIONS FOR BESSEL FUNCTIONS

24.98 $J_0(x) = \frac{1}{\pi}\int_0^\pi \cos(x \sin\theta)\,d\theta$

24.99 $J_n(x) = \frac{1}{\pi}\int_0^\pi \cos(n\theta - x\sin\theta)\,d\theta, \qquad n = \text{integer}$

24.100 $J_n(x) = \frac{x^n}{2^n\sqrt{\pi}\,\Gamma(n+\frac{1}{2})}\int_0^\pi \cos(x\sin\theta)\cos^{2n}\theta\,d\theta, \qquad n > -\frac{1}{2}$

24.101 $Y_0(x) = -\frac{2}{\pi}\int_0^\infty \cos(x\cosh u)\,du$

24.102 $I_0(x) = \frac{1}{\pi}\int_0^\pi \cosh(x\sin\theta)\,d\theta = \frac{1}{2\pi}\int_0^{2\pi} e^{x\sin\theta}\,d\theta$

ASYMPTOTIC EXPANSIONS

24.103 $J_n(x) \sim \sqrt{\frac{2}{\pi x}}\cos\left(x - \frac{n\pi}{2} - \frac{\pi}{4}\right)$ where x is large

24.104 $Y_n(x) \sim \sqrt{\frac{2}{\pi x}}\sin\left(x - \frac{n\pi}{2} - \frac{\pi}{4}\right)$ where x is large

24.105 $J_n(x) \sim \frac{1}{\sqrt{2\pi n}}\left(\frac{ex}{2n}\right)^n$ where n is large

24.106 $Y_n(x) \sim -\sqrt{\frac{2}{\pi n}}\left(\frac{ex}{2n}\right)^{-n}$ where n is large

24.107 $I_n(x) \sim \frac{e^x}{\sqrt{2\pi x}}$ where x is large

24.108 $K_n(x) \sim \frac{e^{-x}}{\sqrt{2\pi x}}$ where x is large

ORTHOGONAL SERIES OF BESSEL FUNCTIONS

Let $\lambda_1, \lambda_2, \lambda_3, \ldots$ be the positive roots of $R\,J_n(x) + Sx\,J_n'(x) = 0$, $n > -1$. Then the following series expansions hold under the conditions indicated.

$S = 0$, $R \neq 0$, i.e. $\lambda_1, \lambda_2, \lambda_3, \ldots$ are positive roots of $J_n(x) = 0$

24.109 $$f(x) = A_1 J_n(\lambda_1 x) + A_2 J_n(\lambda_2 x) + A_3 J_n(\lambda_3 x) + \cdots$$

where

24.110 $$A_k = \frac{2}{J_{n+1}^2(\lambda_k)} \int_0^1 x\,f(x)\,J_n(\lambda_k x)\,dx$$

In particular if $n = 0$,

24.111 $$f(x) = A_1 J_0(\lambda_1 x) + A_2 J_0(\lambda_2 x) + A_3 J_0(\lambda_3 x) + \cdots$$

where

24.112 $$A_k = \frac{2}{J_1^2(\lambda_k)} \int_0^1 x\,f(x)\,J_0(\lambda_k x)\,dx$$

$R/S > -n$

24.113 $$f(x) = A_1 J_n(\lambda_1 x) + A_2 J_n(\lambda_2 x) + A_3 J_n(\lambda_3 x) + \cdots$$

where

24.114 $$A_k = \frac{2}{J_n^2(\lambda_k) - J_{n-1}(\lambda_k)\,J_{n+1}(\lambda_k)} \int_0^1 x\,f(x)\,J_n(\lambda_k x)\,dx$$

In particular if $n = 0$,

24.115 $$f(x) = A_1 J_0(\lambda_1 x) + A_2 J_0(\lambda_2 x) + A_3 J_0(\lambda_3 x) + \cdots$$

where

24.116 $$A_k = \frac{2}{J_0^2(\lambda_k) + J_1^2(\lambda_k)} \int_0^1 x\,f(x)\,J_0(\lambda_k x)\,dx$$

$R/S = -n$

24.117 $$f(x) = A_0 x^n + A_1 J_n(\lambda_1 x) + A_2 J_n(\lambda_2 x) + \cdots$$

where

24.118 $$\begin{cases} A_0 = 2(n+1) \displaystyle\int_0^1 x^{n+1} f(x)\,dx \\ A_k = \dfrac{2}{J_n^2(\lambda_k) - J_{n-1}(\lambda_k)\,J_{n+1}(\lambda_k)} \displaystyle\int_0^1 x\,f(x)\,J_n(\lambda_k x)\,dx \end{cases}$$

In particular if $n = 0$ so that $R = 0$ [i.e. $\lambda_1, \lambda_2, \lambda_3, \ldots$ are the positive roots of $J_1(x) = 0$],

24.119 $$f(x) = A_0 + A_1 J_0(\lambda_1 x) + A_2 J_0(\lambda_2 x) + \cdots$$

where

24.120 $$\begin{cases} A_0 = 2 \displaystyle\int_0^1 x\,f(x)\,dx \\ A_k = \dfrac{2}{J_0^2(\lambda_k)} \displaystyle\int_0^1 x\,f(x)\,J_0(\lambda_k x)\,dx \end{cases}$$

$R/S < -n$

In this case there are two pure imaginary roots $\pm i\lambda_0$ as well as the positive roots $\lambda_1, \lambda_2, \lambda_3, \ldots$ and we have

24.121 $$f(x) = A_0 I_n(\lambda_0 x) + A_1 J_n(\lambda_1 x) + A_2 J_n(\lambda_2 x) + \cdots$$

where

24.122 $$\begin{cases} A_0 = \dfrac{2}{I_n^2(\lambda_0) + I_{n-1}(\lambda_0)\, I_{n+1}(\lambda_0)} \displaystyle\int_0^1 x\, f(x)\, I_n(\lambda_0 x)\, dx \\ A_k = \dfrac{2}{J_n^2(\lambda_k) - J_{n-1}(\lambda_k)\, J_{n+1}(\lambda_k)} \displaystyle\int_0^1 x\, f(x)\, J_n(\lambda_k x)\, dx \end{cases}$$

MISCELLANEOUS RESULTS

24.123 $$\cos(x \sin\theta) = J_0(x) + 2J_2(x)\cos 2\theta + 2J_4(x)\cos 4\theta + \cdots$$

24.124 $$\sin(x \sin\theta) = 2J_1(x)\sin\theta + 2J_3(x)\sin 3\theta + 2J_5(x)\sin 5\theta + \cdots$$

24.125 $$J_n(x+y) = \sum_{k=-\infty}^{\infty} J_k(x)\, J_{n-k}(y) \qquad n = 0, \pm 1, \pm 2, \ldots$$

This is called the *addition formula* for Bessel functions.

24.126 $$1 = J_0(x) + 2J_2(x) + \cdots + 2J_{2n}(x) + \cdots$$

24.127 $$x = 2\{J_1(x) + 3J_3(x) + 5J_5(x) + \cdots + (2n+1)J_{2n+1}(x) + \cdots\}$$

24.128 $$x^2 = 2\{4J_2(x) + 16J_4(x) + 36J_6(x) + \cdots + (2n)^2 J_{2n}(x) + \cdots\}$$

24.129 $$\frac{x J_1(x)}{4} = J_2(x) - 2J_4(x) + 3J_6(x) - \cdots$$

24.130 $$1 = J_0^2(x) + 2J_1^2(x) + 2J_2^2(x) + 2J_3^2(x) + \cdots$$

24.131 $$J_n''(x) = \tfrac{1}{4}\{J_{n-2}(x) - 2J_n(x) + J_{n+2}(x)\}$$

24.132 $$J_n'''(x) = \tfrac{1}{8}\{J_{n-3}(x) - 3J_{n-1}(x) + 3J_{n+1}(x) - J_{n+3}(x)\}$$

Formulas 24.131 and 24.132 can be generalized.

24.133 $$J_n'(x)\, J_{-n}(x) - J_{-n}'\, J_n(x) = \frac{2 \sin n\pi}{\pi x}$$

24.134 $$J_n(x)\, J_{-n+1}(x) + J_{-n}(x)\, J_{n-1}(x) = \frac{2 \sin n\pi}{\pi x}$$

24.135 $$J_{n+1}(x)\, Y_n(x) - J_n(x)\, Y_{n+1}(x) = J_n(x)\, Y_n'(x) - J_n'(x)\, Y_n(x) = \frac{2}{\pi x}$$

24.136 $$\sin x = 2\{J_1(x) - J_3(x) + J_5(x) - \cdots\}$$

24.137 $$\cos x = J_0(x) - 2J_2(x) + 2J_4(x) - \cdots$$

24.138 $$\sinh x = 2\{I_1(x) + I_3(x) + I_5(x) + \cdots\}$$

24.139 $$\cosh x = I_0(x) + 2\{I_2(x) + I_4(x) + I_6(x) + \cdots\}$$

25 LEGENDRE FUNCTIONS

LEGENDRE'S DIFFERENTIAL EQUATION

25.1 $$(1-x^2)y'' - 2xy' + n(n+1)y = 0$$

Solutions of this equation are called *Legendre functions of order n.*

LEGENDRE POLYNOMIALS

If $n = 0, 1, 2, \ldots,$ solutions of 25.1 are Legendre polynomials $P_n(x)$ given by *Rodrigue's formula*

25.2 $$P_n(x) = \frac{1}{2^n n!}\frac{d^n}{dx^n}(x^2-1)^n$$

SPECIAL LEGENDRE POLYNOMIALS

25.3 $P_0(x) = 1$

25.4 $P_1(x) = x$

25.5 $P_2(x) = \frac{1}{2}(3x^2-1)$

25.6 $P_3(x) = \frac{1}{2}(5x^3-3x)$

25.7 $P_4(x) = \frac{1}{8}(35x^4-30x^2+3)$

25.8 $P_5(x) = \frac{1}{8}(63x^5-70x^3+15x)$

25.9 $P_6(x) = \frac{1}{16}(231x^6-315x^4+105x^2-5)$

25.10 $P_7(x) = \frac{1}{16}(429x^7-693x^5+315x^3-35x)$

LEGENDRE POLYNOMIALS IN TERMS OF θ WHERE $x = \cos\theta$

25.11 $P_0(\cos\theta) = 1$

25.12 $P_1(\cos\theta) = \cos\theta$

25.13 $P_2(\cos\theta) = \frac{1}{4}(1+3\cos 2\theta)$

25.14 $P_3(\cos\theta) = \frac{1}{8}(3\cos\theta + 5\cos 3\theta)$

25.15 $P_4(\cos\theta) = \frac{1}{64}(9 + 20\cos 2\theta + 35\cos 4\theta)$

25.16 $P_5(\cos\theta) = \frac{1}{128}(30\cos\theta + 35\cos 3\theta + 63\cos 5\theta)$

25.17 $P_6(\cos\theta) = \frac{1}{512}(50 + 105\cos 2\theta + 126\cos 4\theta + 231\cos 6\theta)$

25.18 $P_7(\cos\theta) = \frac{1}{1024}(175\cos\theta + 189\cos 3\theta + 231\cos 5\theta + 429\cos 7\theta)$

GENERATING FUNCTION FOR LEGENDRE POLYNOMIALS

25.19 $$\frac{1}{\sqrt{1-2tx+t^2}} = \sum_{n=0}^{\infty} P_n(x)t^n$$

RECURRENCE FORMULAS FOR LEGENDRE POLYNOMIALS

25.20 $$(n+1)P_{n+1}(x) - (2n+1)xP_n(x) + nP_{n-1}(x) = 0$$

25.21 $$P'_{n+1}(x) - xP'_n(x) = (n+1)P_n(x)$$

25.22 $$xP'_n(x) - P'_{n-1}(x) = nP_n(x)$$

25.23 $$P'_{n+1}(x) - P'_{n-1}(x) = (2n+1)P_n(x)$$

25.24 $$(x^2-1)P'_n(x) = nxP_n(x) - nP_{n-1}(x)$$

ORTHOGONALITY OF LEGENDRE POLYNOMIALS

25.25 $$\int_{-1}^{1} P_m(x)P_n(x)\,dx = 0 \qquad m \neq n$$

25.26 $$\int_{-1}^{1} \{P_n(x)\}^2\,dx = \frac{2}{2n+1}$$

Because of 25.25, $P_m(x)$ and $P_n(x)$ are called *orthogonal* in $-1 \leqq x \leqq 1$.

ORTHOGONAL SERIES OF LEGENDRE POLYNOMIALS

25.27 $$f(x) = A_0P_0(x) + A_1P_1(x) + A_2P_2(x) + \cdots$$

where

25.28 $$A_k = \frac{2k+1}{2}\int_{-1}^{1} f(x)P_k(x)\,dx$$

SPECIAL RESULTS INVOLVING LEGENDRE POLYNOMIALS

25.29 $P_n(1) = 1$ **25.30** $P_n(-1) = (-1)^n$ **25.31** $P_n(-x) = (-1)^nP_n(x)$

25.32 $$P_n(0) = \begin{cases} 0 & n \text{ odd} \\ (-1)^{n/2}\dfrac{1\cdot 3\cdot 5\cdots(n-1)}{2\cdot 4\cdot 6\cdots n} & n \text{ even} \end{cases}$$

25.33 $$P_n(x) = \frac{1}{\pi}\int_0^{\pi} (x + \sqrt{x^2-1}\cos\phi)^n\,d\phi$$

25.34 $$\int P_n(x)\,dx = \frac{P_{n+1}(x) - P_{n-1}(x)}{2n+1}$$

25.35 $$|P_n(x)| \leqq 1$$

25.36 $$P_n(x) = \frac{1}{2^{n+1}\pi i}\oint_C \frac{(z^2-1)^n}{(z-x)^{n+1}}\,dz$$

where C is a simple closed curve having x as interior point.

GENERAL SOLUTION OF LEGENDRE'S EQUATION

The general solution of Legendre's equation is

25.37 $$y = A\,U_n(x) + B\,V_n(x)$$

where

25.38 $$U_n(x) = 1 - \frac{n(n+1)}{2!}x^2 + \frac{n(n-2)(n+1)(n+3)}{4!}x^4 - \cdots$$

25.39 $$V_n(x) = x - \frac{(n-1)(n+2)}{3!}x^3 + \frac{(n-1)(n-3)(n+2)(n+4)}{5!}x^5 - \cdots$$

These series converge for $-1 < x < 1$.

LEGENDRE FUNCTIONS OF THE SECOND KIND

If $n = 0, 1, 2, \ldots$ one of the series 25.38, 25.39 terminates. In such cases,

25.40 $$P_n(x) = \begin{cases} U_n(x)/U_n(1) & n = 0, 2, 4, \ldots \\ V_n(x)/V_n(1) & n = 1, 3, 5, \ldots \end{cases}$$

where

25.41 $$U_n(1) = (-1)^{n/2}\,2^n\left[\left(\frac{n}{2}\right)!\right]^2\Big/ n! \qquad n = 0, 2, 4, \ldots$$

25.42 $$V_n(1) = (-1)^{(n-1)/2}\,2^{n-1}\left[\left(\frac{n-1}{2}\right)!\right]^2\Big/ n! \qquad n = 1, 3, 5, \ldots$$

The nonterminating series in such case with a suitable multiplicative constant is denoted by $Q_n(x)$ and is called *Legendre's function of the second kind of order n.* We define

25.43 $$Q_n(x) = \begin{cases} U_n(1)\,V_n(x) & n = 0, 2, 4, \ldots \\ -V_n(1)\,U_n(x) & n = 1, 3, 5, \ldots \end{cases}$$

SPECIAL LEGENDRE FUNCTIONS OF THE SECOND KIND

25.44 $$Q_0(x) = \frac{1}{2}\ln\left(\frac{1+x}{1-x}\right)$$

25.45 $$Q_1(x) = \frac{x}{2}\ln\left(\frac{1+x}{1-x}\right) - 1$$

25.46 $$Q_2(x) = \frac{3x^2-1}{4}\ln\left(\frac{1+x}{1-x}\right) - \frac{3x}{2}$$

25.47 $$Q_3(x) = \frac{5x^3-3x}{4}\ln\left(\frac{1+x}{1-x}\right) - \frac{5x^2}{2} + \frac{2}{3}$$

The functions $Q_n(x)$ satisfy recurrence formulas exactly analogous to 25.20 through 25.24.

Using these, the general solution of Legendre's equation can also be written

25.48 $$y = A\,P_n(x) + B\,Q_n(x)$$

26 ASSOCIATED LEGENDRE FUNCTIONS

LEGENDRE'S ASSOCIATED DIFFERENTIAL EQUATION

26.1
$$(1-x^2)y'' - 2xy' + \left\{n(n+1) - \frac{m^2}{1-x^2}\right\}y = 0$$

Solutions of this equation are called *associated Legendre functions.* We restrict ourselves to the important case where m, n are nonnegative integers.

ASSOCIATED LEGENDRE FUNCTIONS OF THE FIRST KIND

26.2
$$P_n^m(x) = (1-x^2)^{m/2}\frac{d^m}{dx^m}P_n(x) = \frac{(1-x^2)^{m/2}}{2^n n!}\frac{d^{m+n}}{dx^{m+n}}(x^2-1)^n$$

where $P_n(x)$ are Legendre polynomials [page 146]. We have

26.3
$$P_n^0(x) = P_n(x)$$

26.4
$$P_n^m(x) = 0 \quad \text{if } m > n$$

SPECIAL ASSOCIATED LEGENDRE FUNCTIONS OF THE FIRST KIND

26.5 $P_1^1(x) = (1-x^2)^{1/2}$

26.6 $P_2^1(x) = 3x(1-x^2)^{1/2}$

26.7 $P_2^2(x) = 3(1-x^2)$

26.8 $P_3^1(x) = \frac{3}{2}(5x^2-1)(1-x^2)^{1/2}$

26.9 $P_3^2(x) = 15x(1-x^2)$

26.10 $P_3^3(x) = 15(1-x^2)^{3/2}$

GENERATING FUNCTION FOR $P_n^m(x)$

26.11
$$\frac{(2m)!\,(1-x^2)^{m/2}t^m}{2^m m!\,(1-2tx+t^2)^{m+1/2}} = \sum_{n=m}^{\infty} P_n^m(x)t^n$$

RECURRENCE FORMULAS

26.12
$$(n+1-m)\,P_{n+1}^m(x) - (2n+1)x\,P_n^m(x) + (n+m)\,P_{n-1}^m(x) = 0$$

26.13
$$P_n^{m+2}(x) - \frac{2(m+1)x}{(1-x^2)^{1/2}}P_n^{m+1}(x) + (n-m)(n+m+1)\,P_n^m(x) = 0$$

ORTHOGONALITY OF $P_n^m(x)$

26.14 $$\int_{-1}^{1} P_n^m(x)\,P_l^m(x)\,dx = 0 \qquad \text{if } n \neq l$$

26.15 $$\int_{-1}^{1} \{P_n^m(x)\}^2\,dx = \frac{2}{2n+1}\,\frac{(n+m)!}{(n-m)!}$$

ORTHOGONAL SERIES

26.16 $$f(x) = A_m P_m^m(x) + A_{m+1} P_{m+1}^m(x) + A_{m+2} P_{m+2}^m(x) + \cdots$$

where

26.17 $$A_k = \frac{2k+1}{2}\,\frac{(k-m)!}{(k+m)!}\int_{-1}^{1} f(x)\,P_k^m(x)\,dx$$

ASSOCIATED LEGENDRE FUNCTIONS OF THE SECOND KIND

26.18 $$Q_n^m(x) = (1-x^2)^{m/2}\,\frac{d^m}{dx^m}\,Q_n(x)$$

where $Q_n(x)$ are Legendre functions of the second kind [page 148].

These functions are unbounded at $x = \pm 1$, whereas $P_n^m(x)$ are bounded at $x = \pm 1$.

The functions $Q_n^m(x)$ satisfy the same recurrence relations as $P_n^m(x)$ [see 26.12 and 26.13].

GENERAL SOLUTION OF LEGENDRE'S ASSOCIATED EQUATION

26.19 $$y = A\,P_n^m(x) + B\,Q_n^m(x)$$

27 HERMITE POLYNOMIALS

HERMITE'S DIFFERENTIAL EQUATION

27.1
$$y'' - 2xy' + 2ny = 0$$

HERMITE POLYNOMIALS

If $n = 0, 1, 2, \ldots$ then solutions of Hermite's equation are Hermite polynomials $H_n(x)$ given by *Rodrigue's formula*

27.2
$$H_n(x) = (-1)^n e^{x^2} \frac{d^n}{dx^n}(e^{-x^2})$$

SPECIAL HERMITE POLYNOMIALS

27.3 $H_0(x) = 1$

27.4 $H_1(x) = 2x$

27.5 $H_2(x) = 4x^2 - 2$

27.6 $H_3(x) = 8x^3 - 12x$

27.7 $H_4(x) = 16x^4 - 48x^2 + 12$

27.8 $H_5(x) = 32x^5 - 160x^3 + 120x$

27.9 $H_6(x) = 64x^6 - 480x^4 + 720x^2 - 120$

27.10 $H_7(x) = 128x^7 - 1344x^5 + 3360x^3 - 1680x$

GENERATING FUNCTION

27.11
$$e^{2tx - t^2} = \sum_{n=0}^{\infty} \frac{H_n(x)\, t^n}{n!}$$

RECURRENCE FORMULAS

27.12
$$H_{n+1}(x) = 2x\, H_n(x) - 2n\, H_{n-1}(x)$$

27.13
$$H_n'(x) = 2n\, H_{n-1}(x)$$

ORTHOGONALITY OF HERMITE POLYNOMIALS

27.14 $$\int_{-\infty}^{\infty} e^{-x^2} H_m(x) H_n(x)\, dx = 0 \qquad m \neq n$$

27.15 $$\int_{-\infty}^{\infty} e^{-x^2} \{H_n(x)\}^2\, dx = 2^n n! \sqrt{\pi}$$

ORTHOGONAL SERIES

27.16 $$f(x) = A_0 H_0(x) + A_1 H_1(x) + A_2 H_2(x) + \cdots$$

where

27.17 $$A_k = \frac{1}{2^k k! \sqrt{\pi}} \int_{-\infty}^{\infty} e^{-x^2} f(x) H_k(x)\, dx$$

SPECIAL RESULTS

27.18 $$H_n(x) = (2x)^n - \frac{n(n-1)}{1!}(2x)^{n-2} + \frac{n(n-1)(n-2)(n-3)}{2!}(2x)^{n-4} - \cdots$$

27.19 $$H_n(-x) = (-1)^n H_n(x)$$

27.20 $$H_{2n-1}(0) = 0$$

27.21 $$H_{2n}(0) = (-1)^n 2^n \cdot 1 \cdot 3 \cdot 5 \cdots (2n-1)$$

27.22 $$\int_0^x H_n(t)\, dt = \frac{H_{n+1}(x)}{2(n+1)} - \frac{H_{n+1}(0)}{2(n+1)}$$

27.23 $$\frac{d}{dx}\{e^{-x^2} H_n(x)\} = -e^{-x^2} H_{n+1}(x)$$

27.24 $$\int_0^x e^{-t^2} H_n(t)\, dt = H_{n-1}(0) - e^{-x^2} H_{n-1}(x)$$

27.25 $$\int_{-\infty}^{\infty} t^n e^{-t^2} H_n(xt)\, dt = \sqrt{\pi}\, n!\, P_n(x)$$

27.26 $$H_n(x+y) = \sum_{k=0}^{n} \frac{1}{2^{n/2}} \binom{n}{k} H_k(x\sqrt{2})\, H_{n-k}(y\sqrt{2})$$

This is called the *addition formula* for Hermite polynomials.

27.27 $$\sum_{k=0}^{n} \frac{H_k(x) H_k(y)}{2^k k!} = \frac{H_{n+1}(x) H_n(y) - H_n(x) H_{n+1}(y)}{2^{n+1} n! (x-y)}$$

28 LAGUERRE POLYNOMIALS

LAGUERRE'S DIFFERENTIAL EQUATION

28.1
$$xy'' + (1-x)y' + ny = 0$$

LAGUERRE POLYNOMIALS

If $n = 0, 1, 2, \ldots$ then solutions of Laguerre's equation are Laguerre polynomials $L_n(x)$ and are given by *Rodrigue's formula.*

28.2
$$L_n(x) = e^x \frac{d^n}{dx^n}(x^n e^{-x})$$

SPECIAL LAGUERRE POLYNOMIALS

28.3 $L_0(x) = 1$

28.4 $L_1(x) = -x + 1$

28.5 $L_2(x) = x^2 - 4x + 2$

28.6 $L_3(x) = -x^3 + 9x^2 - 18x + 6$

28.7 $L_4(x) = x^4 - 16x^3 + 72x^2 - 96x + 24$

28.8 $L_5(x) = -x^5 + 25x^4 - 200x^3 + 600x^2 - 600x + 120$

28.9 $L_6(x) = x^6 - 36x^5 + 450x^4 - 2400x^3 + 5400x^2 - 4320x + 720$

28.10 $L_7(x) = -x^7 + 49x^6 - 882x^5 + 7350x^4 - 29{,}400x^3 + 52{,}920x^2 - 35{,}280x + 5040$

GENERATING FUNCTION

28.11
$$\frac{e^{-xt/1-t}}{1-t} = \sum_{n=0}^{\infty} \frac{L_n(x)\, t^n}{n!}$$

RECURRENCE FORMULAS

28.12
$$L_{n+1}(x) - (2n+1-x)\,L_n(x) + n^2\,L_{n-1}(x) = 0$$

28.13
$$L_n'(x) - n\,L_{n-1}'(x) + n\,L_{n-1}(x) = 0$$

28.14
$$x\,L_n'(x) = n\,L_n(x) - n^2\,L_{n-1}(x)$$

ORTHOGONALITY OF LAGUERRE POLYNOMIALS

28.15 $$\int_0^\infty e^{-x} L_m(x) L_n(x)\, dx = 0 \qquad m \neq n$$

28.16 $$\int_0^\infty e^{-x} \{L_n(x)\}^2\, dx = (n!)^2$$

ORTHOGONAL SERIES

28.17 $$f(x) = A_0 L_0(x) + A_1 L_1(x) + A_2 L_2(x) + \cdots$$

where

28.18 $$A_k = \frac{1}{(k!)^2} \int_0^\infty e^{-x} f(x) L_k(x)\, dx$$

SPECIAL RESULTS

28.19 $L_n(0) = n!$

28.20 $$\int_0^x L_n(t)\, dt = L_n(x) - \frac{L_{n+1}(x)}{n+1}$$

28.21 $$L_n(x) = (-1)^n \left\{ x^n - \frac{n^2 x^{n-1}}{1!} + \frac{n^2(n-1)^2 x^{n-2}}{2!} - \cdots (-1)^n n! \right\}$$

28.22 $$\int_0^\infty x^p e^{-x} L_n(x)\, dx = \begin{cases} 0 & \text{if } p < n \\ (-1)^n (n!)^2 & \text{if } p = n \end{cases}$$

28.23 $$\sum_{k=0}^{n} \frac{L_k(x) L_k(y)}{(k!)^2} = \frac{L_n(x) L_{n+1}(y) - L_{n+1}(x) L_n(y)}{(n!)^2 (x-y)}$$

28.24 $$\sum_{k=0}^{\infty} \frac{t^k L_k(x)}{(k!)^2} = e^t J_0(2\sqrt{xt})$$

28.25 $$L_n(x) = \int_0^\infty u^n e^{x-u} J_0(2\sqrt{xu})\, du$$

29 ASSOCIATED LAGUERRE POLYNOMIALS

LAGUERRE'S ASSOCIATED DIFFERENTIAL EQUATION

29.1 $$xy'' + (m+1-x)y' + (n-m)y = 0$$

ASSOCIATED LAGUERRE POLYNOMIALS

Solutions of 29.1 for nonnegative integers m and n are given by the associated Laguerre polynomials

29.2 $$L_n^m(x) = \frac{d^m}{dx^m} L_n(x)$$

where $L_n(x)$ are Laguerre polynomials [see page 153].

29.3 $$L_n^0(x) = L_n(x)$$

29.4 $$L_n^m(x) = 0 \qquad \text{if } m > n$$

SPECIAL ASSOCIATED LAGUERRE POLYNOMIALS

29.5 $L_1^1(x) = -1$

29.6 $L_2^1(x) = 2x - 4$

29.7 $L_2^2(x) = 2$

29.8 $L_3^1(x) = -3x^2 + 18x - 18$

29.9 $L_3^2(x) = -6x + 18$

29.10 $L_3^3(x) = -6$

29.11 $L_4^1(x) = 4x^3 - 48x^2 + 144x - 96$

29.12 $L_4^2(x) = 12x^2 - 96x + 144$

29.13 $L_4^3(x) = 24x - 96$

29.14 $L_4^4(x) = 24$

GENERATING FUNCTION FOR $L_n^m(x)$

29.15 $$\frac{(-1)^m t^m}{(1-t)^{m+1}} e^{-xt/(1-t)} = \sum_{n=m}^{\infty} \frac{L_n^m(x)}{n!} t^n$$

RECURRENCE FORMULAS

29.16
$$\frac{n-m+1}{n+1} L_{n+1}^m(x) + (x+m-2n-1) L_n^m(x) + n^2 L_{n-1}^m(x) = 0$$

29.17
$$\frac{d}{dx}\{L_n^m(x)\} = L_n^{m+1}(x)$$

29.18
$$\frac{d}{dx}\{x^m e^{-x} L_n^m(x)\} = (m-n-1)x^{m-1}e^{-x} L_n^{m-1}(x)$$

29.19
$$x\frac{d}{dx}\{L_n^m(x)\} = (x-m) L_n^m(x) + (m-n-1) L_n^{m-1}(x)$$

ORTHOGONALITY

29.20
$$\int_0^\infty x^m e^{-x} L_n^m(x) L_p^m(x)\, dx = 0 \qquad p \neq n$$

29.21
$$\int_0^\infty x^m e^{-x} \{L_n^m(x)\}^2\, dx = \frac{(n!)^3}{(n-m)!}$$

ORTHOGONAL SERIES

29.22
$$f(x) = A_m L_m^m(x) + A_{m+1} L_{m+1}^m(x) + A_{m+2} L_{m+2}^m(x) + \cdots$$

where

29.23
$$A_k = \frac{(k-m)!}{(k!)^3} \int_0^\infty x^m e^{-x} L_k^m(x) f(x)\, dx$$

SPECIAL RESULTS

29.24
$$L_n^m(x) = (-1)^n \frac{n!}{(n-m)!}\left\{x^{n-m} - \frac{n(n-m)}{1!} x^{n-m-1} + \frac{n(n-1)(n-m)(n-m-1)}{2!} x^{n-m-2} + \cdots\right\}$$

29.25
$$\int_0^\infty x^{m+1} e^{-x} \{L_n^m(x)\}^2\, dx = \frac{(2n-m+1)(n!)^3}{(n-m)!}$$

30 CHEBYSHEV POLYNOMIALS

CHEBYSHEV'S DIFFERENTIAL EQUATION

30.1 $$(1-x^2)y'' - xy' + n^2y = 0 \qquad n = 0, 1, 2, \ldots$$

CHEBYSHEV POLYNOMIALS OF THE FIRST KIND

Solutions of 30.1 are given by

30.2 $$T_n(x) = \cos(n\cos^{-1}x) = x^n - \binom{n}{2}x^{n-2}(1-x^2) + \binom{n}{4}x^{n-4}(1-x^2)^2 - \cdots$$

SPECIAL CHEBYSHEV POLYNOMIALS OF THE FIRST KIND

30.3 $T_0(x) = 1$

30.4 $T_1(x) = x$

30.5 $T_2(x) = 2x^2 - 1$

30.6 $T_3(x) = 4x^3 - 3x$

30.7 $T_4(x) = 8x^4 - 8x^2 + 1$

30.8 $T_5(x) = 16x^5 - 20x^3 + 5x$

30.9 $T_6(x) = 32x^6 - 48x^4 + 18x^2 - 1$

30.10 $T_7(x) = 64x^7 - 112x^5 + 56x^3 - 7x$

GENERATING FUNCTION FOR $T_n(x)$

30.11 $$\frac{1-tx}{1-2tx+t^2} = \sum_{n=0}^{\infty} T_n(x)\, t^n$$

SPECIAL VALUES

30.12 $T_n(-x) = (-1)^n T_n(x)$

30.13 $T_n(1) = 1$

30.14 $T_n(-1) = (-1)^n$

30.15 $T_{2n}(0) = (-1)^n$

30.16 $T_{2n+1}(0) = 0$

RECURSION FORMULA FOR $T_n(x)$

30.17 $$T_{n+1}(x) - 2x\,T_n(x) + T_{n-1}(x) = 0$$

ORTHOGONALITY

30.18 $$\int_{-1}^{1} \frac{T_m(x)\,T_n(x)}{\sqrt{1-x^2}}\,dx = 0 \qquad m \neq n$$

30.19 $$\int_{-1}^{1} \frac{\{T_n(x)\}^2}{\sqrt{1-x^2}}\,dx = \begin{cases} \pi & \text{if } n = 0 \\ \pi/2 & \text{if } n = 1, 2, \ldots \end{cases}$$

ORTHOGONAL SERIES

30.20 $$f(x) = \tfrac{1}{2}A_0\,T_0(x) + A_1\,T_1(x) + A_2\,T_2(x) + \cdots$$

where

30.21 $$A_k = \frac{2}{\pi}\int_{-1}^{1} \frac{f(x)\,T_k(x)}{\sqrt{1-x^2}}\,dx$$

CHEBYSHEV POLYNOMIALS OF THE SECOND KIND

30.22 $$\begin{aligned} U_n(x) &= \frac{\sin\{(n+1)\cos^{-1}x\}}{\sin(\cos^{-1}x)} \\ &= \binom{n+1}{1}x^n - \binom{n+1}{3}x^{n-2}(1-x^2) + \binom{n+1}{5}x^{n-4}(1-x^2)^2 - \cdots \end{aligned}$$

SPECIAL CHEBYSHEV POLYNOMIALS OF THE SECOND KIND

30.23 $U_0(x) = 1$

30.24 $U_1(x) = 2x$

30.25 $U_2(x) = 4x^2 - 1$

30.26 $U_3(x) = 8x^3 - 4x$

30.27 $U_4(x) = 16x^4 - 12x^2 + 1$

30.28 $U_5(x) = 32x^5 - 32x^3 + 6x$

30.29 $U_6(x) = 64x^6 - 80x^4 + 24x^2 - 1$

30.30 $U_7(x) = 128x^7 - 192x^5 + 80x^3 - 8x$

GENERATING FUNCTION FOR $U_n(x)$

30.31 $$\frac{1}{1 - 2tx + t^2} = \sum_{n=0}^{\infty} U_n(x)\,t^n$$

SPECIAL VALUES

30.32 $U_n(-x) = (-1)^n U_n(x)$

30.33 $U_n(1) = n+1$

30.34 $U_n(-1) = (-1)^n (n+1)$

30.35 $U_{2n}(0) = (-1)^n$

30.36 $U_{2n+1}(0) = 0$

RECURSION FORMULA FOR $U_n(x)$

30.37
$$U_{n+1}(x) - 2x\,U_n(x) + U_{n-1}(x) = 0$$

ORTHOGONALITY

30.38
$$\int_{-1}^{1} \sqrt{1-x^2}\,U_m(x)\,U_n(x)\,dx = 0 \qquad m \neq n$$

30.39
$$\int_{-1}^{1} \sqrt{1-x^2}\,\{U_n(x)\}^2\,dx = \frac{\pi}{2}$$

ORTHOGONAL SERIES

30.40
$$f(x) = A_0\,U_0(x) + A_1\,U_1(x) + A_2\,U_2(x) + \cdots$$

where

30.41
$$A_k = \frac{2}{\pi}\int_{-1}^{1} \sqrt{1-x^2}\,f(x)\,U_k(x)\,dx$$

RELATIONSHIPS BETWEEN $T_n(x)$ AND $U_n(x)$

30.42
$$T_n(x) = U_n(x) - x\,U_{n-1}(x)$$

30.43
$$(1-x^2)\,U_{n-1}(x) = x\,T_n(x) - T_{n+1}(x)$$

30.44
$$U_n(x) = \frac{1}{\pi}\int_{-1}^{1} \frac{T_{n+1}(v)\,dv}{(v-x)\sqrt{1-v^2}}$$

30.45
$$T_n(x) = \frac{1}{\pi}\int_{-1}^{1} \frac{\sqrt{1-v^2}\,U_{n-1}(v)}{x-v}\,dv$$

GENERAL SOLUTION OF CHEBYSHEV'S DIFFERENTIAL EQUATION

30.46
$$y = \begin{cases} A\,T_n(x) + B\sqrt{1-x^2}\,U_{n-1}(x) & \text{if } n = 1, 2, 3, \ldots \\ A + B\sin^{-1}x & \text{if } n = 0 \end{cases}$$

31 HYPERGEOMETRIC FUNCTIONS

HYPERGEOMETRIC DIFFERENTIAL EQUATION

31.1 $$x(1-x)y'' + \{c - (a+b+1)x\}y' - aby = 0$$

HYPERGEOMETRIC FUNCTIONS

A solution of 31.1 is given by

31.2 $$F(a, b; c; x) = 1 + \frac{a \cdot b}{1 \cdot c}x + \frac{a(a+1)b(b+1)}{1 \cdot 2 \cdot c(c+1)}x^2 + \frac{a(a+1)(a+2)b(b+1)(b+2)}{1 \cdot 2 \cdot 3 \cdot c(c+1)(c+2)}x^3 + \cdots$$

If a, b, c are real, then the series converges for $-1 < x < 1$ provided that $c - (a+b) > -1$.

SPECIAL CASES

31.3 $F(-p, 1; 1; -x) = (1+x)^p$

31.4 $F(1, 1; 2; -x) = [\ln(1+x)]/x$

31.5 $\lim_{n \to \infty} F(1, n; 1; x/n) = e^x$

31.6 $F(\frac{1}{2}, -\frac{1}{2}; \frac{1}{2}; \sin^2 x) = \cos x$

31.7 $F(\frac{1}{2}, 1; 1; \sin^2 x) = \sec x$

31.8 $F(\frac{1}{2}, \frac{1}{2}; \frac{3}{2}; x^2) = (\sin^{-1} x)/x$

31.9 $F(\frac{1}{2}, 1; \frac{3}{2}; -x^2) = (\tan^{-1} x)/x$

31.10 $F(1, p; p; x) = 1/(1-x)$

31.11 $F(n+1, -n; 1; (1-x)/2) = P_n(x)$

31.12 $F(n, -n; \frac{1}{2}; (1-x)/2) = T_n(x)$

GENERAL SOLUTION OF THE HYPERGEOMETRIC EQUATION

If c, $a-b$ and $c-a-b$ are all nonintegers, the general solution valid for $|x| < 1$ is

31.13 $$y = A\,F(a, b; c; x) + Bx^{1-c}F(a-c+1, b-c+1; 2-c; x)$$

MISCELLANEOUS PROPERTIES

31.14 $$F(a, b; c; 1) = \frac{\Gamma(c)\,\Gamma(c-a-b)}{\Gamma(c-a)\,\Gamma(c-b)}$$

31.15 $$\frac{d}{dx}F(a, b; c; x) = \frac{ab}{c}F(a+1, b+1; c+1; x)$$

31.16 $$F(a, b; c; x) = \frac{\Gamma(c)}{\Gamma(b)\,\Gamma(c-b)}\int_0^1 u^{b-1}(1-u)^{c-b-1}(1-ux)^{-a}\,du$$

31.17 $$F(a, b; c; x) = (1-x)^{c-a-b}F(c-a, c-b; c; x)$$

32 LAPLACE TRANSFORMS

DEFINITION OF THE LAPLACE TRANSFORM OF $F(t)$

32.1
$$\mathcal{L}\{F(t)\} = \int_0^\infty e^{-st} F(t)\, dt = f(s)$$

In general $f(s)$ will exist for $s > \alpha$ where α is some constant. $\mathcal{L}$ is called the *Laplace transform operator.*

DEFINITION OF THE INVERSE LAPLACE TRANSFORM OF $f(s)$

If $\mathcal{L}\{F(t)\} = f(s)$, then we say that $F(t) = \mathcal{L}^{-1}\{f(s)\}$ is the *inverse Laplace transform* of $f(s)$. $\mathcal{L}^{-1}$ is called the *inverse Laplace transform operator.*

COMPLEX INVERSION FORMULA

The inverse Laplace transform of $f(s)$ can be found directly by methods of complex variable theory. The result is

32.2
$$F(t) = \frac{1}{2\pi i}\int_{c-i\infty}^{c+i\infty} e^{st} f(s)\, ds = \frac{1}{2\pi i}\lim_{T\to\infty}\int_{c-iT}^{c+iT} e^{st} f(s)\, ds$$

where c is chosen so that all the singular points of $f(s)$ lie to the left of the line $\operatorname{Re}\{s\} = c$ in the complex s plane.

TABLE OF GENERAL PROPERTIES OF LAPLACE TRANSFORMS

	$f(s)$	$F(t)$
32.3	$a\,f_1(s) + b\,f_2(s)$	$a\,F_1(t) + b\,F_2(t)$
32.4	$f(s/a)$	$a\,F(at)$
32.5	$f(s-a)$	$e^{at}\,F(t)$
32.6	$e^{-as}\,f(s)$	$\mathcal{U}(t-a) = \begin{cases} F(t-a) & t > a \\ 0 & t < a \end{cases}$
32.7	$s\,f(s) - F(0)$	$F'(t)$
32.8	$s^2\,f(s) - s\,F(0) - F'(0)$	$F''(t)$
32.9	$s^n\,f(s) - s^{n-1}F(0) - s^{n-2}F'(0) - \cdots - F^{(n-1)}(0)$	$F^{(n)}(t)$
32.10	$f'(s)$	$-t\,F(t)$
32.11	$f''(s)$	$t^2\,F(t)$
32.12	$f^{(n)}(s)$	$(-1)^n t^n\,F(t)$
32.13	$\dfrac{f(s)}{s}$	$\int_0^t F(u)\,du$
32.14	$\dfrac{f(s)}{s^n}$	$\int_0^t \cdots \int_0^t F(u)\,du^n = \int_0^t \dfrac{(t-u)^{n-1}}{(n-1)!} F(u)\,du$
32.15	$f(s)\,g(s)$	$\int_0^t F(u)\,G(t-u)\,du$

	$f(s)$	$F(t)$
32.16	$\int_s^\infty f(u)\,du$	$\frac{F(t)}{t}$
32.17	$\frac{1}{1-e^{-sT}}\int_0^T e^{-su}F(u)\,du$	$F(t) = F(t+T)$
32.18	$\frac{f(\sqrt{s})}{s}$	$\frac{1}{\sqrt{\pi t}}\int_0^\infty e^{-u^2/4t}F(u)\,du$
32.19	$\frac{1}{s}f(1/s)$	$\int_0^\infty J_0(2\sqrt{ut})\,F(u)\,du$
32.20	$\frac{1}{s^{n+1}}f(1/s)$	$t^{n/2}\int_0^\infty u^{-n/2}J_n(2\sqrt{ut})\,F(u)\,du$
32.21	$\frac{f(s+1/s)}{s^2+1}$	$\int_0^t J_0(2\sqrt{u(t-u)})\,F(u)\,du$
32.22	$\frac{1}{2\sqrt{\pi}}\int_0^\infty u^{-3/2}e^{-s^2/4u}f(u)\,du$	$F(t^2)$
32.23	$\frac{f(\ln s)}{s\ln s}$	$\int_0^\infty \frac{t^u F(u)}{\Gamma(u+1)}\,du$
32.24	$\frac{P(s)}{Q(s)}$ $P(s)$ = polynomial of degree less than n, $Q(s) = (s-\alpha_1)(s-\alpha_2)\cdots(s-\alpha_n)$ where $\alpha_1, \alpha_2, \ldots, \alpha_n$ are all distinct.	$\sum_{k=1}^{n}\frac{P(\alpha_k)}{Q'(\alpha_k)}e^{\alpha_k t}$

TABLE OF SPECIAL LAPLACE TRANSFORMS

	$f(s)$	$F(t)$
32.25	$\dfrac{1}{s}$	1
32.26	$\dfrac{1}{s^2}$	t
32.27	$\dfrac{1}{s^n} \qquad n = 1, 2, 3, \ldots$	$\dfrac{t^{n-1}}{(n-1)!}, \quad 0! = 1$
32.28	$\dfrac{1}{s^n} \qquad n > 0$	$\dfrac{t^{n-1}}{\Gamma(n)}$
32.29	$\dfrac{1}{s-a}$	e^{at}
32.30	$\dfrac{1}{(s-a)^n} \qquad n = 1, 2, 3, \ldots$	$\dfrac{t^{n-1}e^{at}}{(n-1)!}, \quad 0! = 1$
32.31	$\dfrac{1}{(s-a)^n} \qquad n > 0$	$\dfrac{t^{n-1}e^{at}}{\Gamma(n)}$
32.32	$\dfrac{1}{s^2+a^2}$	$\dfrac{\sin at}{a}$
32.33	$\dfrac{s}{s^2+a^2}$	$\cos at$
32.34	$\dfrac{1}{(s-b)^2+a^2}$	$\dfrac{e^{bt}\sin at}{a}$
32.35	$\dfrac{s-b}{(s-b)^2+a^2}$	$e^{bt}\cos at$
32.36	$\dfrac{1}{s^2-a^2}$	$\dfrac{\sinh at}{a}$
32.37	$\dfrac{s}{s^2-a^2}$	$\cosh at$
32.38	$\dfrac{1}{(s-b)^2-a^2}$	$\dfrac{e^{bt}\sinh at}{a}$

	$f(s)$	$F(t)$
32.39	$\dfrac{s-b}{(s-b)^2-a^2}$	$e^{bt}\cosh at$
32.40	$\dfrac{1}{(s-a)(s-b)} \quad a \neq b$	$\dfrac{e^{bt}-e^{at}}{b-a}$
32.41	$\dfrac{s}{(s-a)(s-b)} \quad a \neq b$	$\dfrac{be^{bt}-ae^{at}}{b-a}$
32.42	$\dfrac{1}{(s^2+a^2)^2}$	$\dfrac{\sin at - at\cos at}{2a^3}$
32.43	$\dfrac{s}{(s^2+a^2)^2}$	$\dfrac{t\sin at}{2a}$
32.44	$\dfrac{s^2}{(s^2+a^2)^2}$	$\dfrac{\sin at + at\cos at}{2a}$
32.45	$\dfrac{s^3}{(s^2+a^2)^2}$	$\cos at - \frac{1}{2}at\sin at$
32.46	$\dfrac{s^2-a^2}{(s^2+a^2)^2}$	$t\cos at$
32.47	$\dfrac{1}{(s^2-a^2)^2}$	$\dfrac{at\cosh at - \sinh at}{2a^3}$
32.48	$\dfrac{s}{(s^2-a^2)^2}$	$\dfrac{t\sinh at}{2a}$
32.49	$\dfrac{s^2}{(s^2-a^2)^2}$	$\dfrac{\sinh at + at\cosh at}{2a}$
32.50	$\dfrac{s^3}{(s^2-a^2)^2}$	$\cosh at + \frac{1}{2}at\sinh at$
32.51	$\dfrac{s^2+a^2}{(s^2-a^2)^2}$	$t\cosh at$
32.52	$\dfrac{1}{(s^2+a^2)^3}$	$\dfrac{(3-a^2t^2)\sin at - 3at\cos at}{8a^5}$
32.53	$\dfrac{s}{(s^2+a^2)^3}$	$\dfrac{t\sin at - at^2\cos at}{8a^3}$
32.54	$\dfrac{s^2}{(s^2+a^2)^3}$	$\dfrac{(1+a^2t^2)\sin at - at\cos at}{8a^3}$
32.55	$\dfrac{s^3}{(s^2+a^2)^3}$	$\dfrac{3t\sin at + at^2\cos at}{8a}$

	$f(s)$	$F(t)$
32.56	$\frac{s^4}{(s^2+a^2)^3}$	$\frac{(3-a^2t^2)\sin at + 5at\cos at}{8a}$
32.57	$\frac{s^5}{(s^2+a^2)^3}$	$\frac{(8-a^2t^2)\cos at - 7at\sin at}{8}$
32.58	$\frac{3s^2-a^2}{(s^2+a^2)^3}$	$\frac{t^2\sin at}{2a}$
32.59	$\frac{s^3-3a^2s}{(s^2+a^2)^3}$	$\frac{1}{2}t^2\cos at$
32.60	$\frac{s^4-6a^2s^2+a^4}{(s^2+a^2)^4}$	$\frac{1}{6}t^3\cos at$
32.61	$\frac{s^3-a^2s}{(s^2+a^2)^4}$	$\frac{t^3\sin at}{24a}$
32.62	$\frac{1}{(s^2-a^2)^3}$	$\frac{(3+a^2t^2)\sinh at - 3at\cosh at}{8a^5}$
32.63	$\frac{s}{(s^2-a^2)^3}$	$\frac{at^2\cosh at - t\sinh at}{8a^3}$
32.64	$\frac{s^2}{(s^2-a^2)^3}$	$\frac{at\cosh at + (a^2t^2-1)\sinh at}{8a^3}$
32.65	$\frac{s^3}{(s^2-a^2)^3}$	$\frac{3t\sinh at + at^2\cosh at}{8a}$
32.66	$\frac{s^4}{(s^2-a^2)^3}$	$\frac{(3+a^2t^2)\sinh at + 5at\cosh at}{8a}$
32.67	$\frac{s^5}{(s^2-a^2)^3}$	$\frac{(8+a^2t^2)\cosh at + 7at\sinh at}{8}$
32.68	$\frac{3s^2+a^2}{(s^2-a^2)^3}$	$\frac{t^2\sinh at}{2a}$
32.69	$\frac{s^3+3a^2s}{(s^2-a^2)^3}$	$\frac{1}{2}t^2\cosh at$
32.70	$\frac{s^4+6a^2s^2+a^4}{(s^2-a^2)^4}$	$\frac{1}{6}t^3\cosh at$
32.71	$\frac{s^3+a^2s}{(s^2-a^2)^4}$	$\frac{t^3\sinh at}{24a}$
32.72	$\frac{1}{s^3+a^3}$	$\frac{e^{at/2}}{3a^2}\left\{\sqrt{3}\sin\frac{\sqrt{3}\,at}{2} - \cos\frac{\sqrt{3}\,at}{2} + e^{-3at/2}\right\}$

	$f(s)$	$F(t)$
32.73	$\dfrac{s}{s^3+a^3}$	$\dfrac{e^{at/2}}{3a}\left\{\cos\dfrac{\sqrt{3}\,at}{2}+\sqrt{3}\sin\dfrac{\sqrt{3}\,at}{2}-e^{-3at/2}\right\}$
32.74	$\dfrac{s^2}{s^3+a^3}$	$\dfrac{1}{3}\left(e^{-at}+2e^{at/2}\cos\dfrac{\sqrt{3}\,at}{2}\right)$
32.75	$\dfrac{1}{s^3-a^3}$	$\dfrac{e^{-at/2}}{3a^2}\left\{e^{3at/2}-\cos\dfrac{\sqrt{3}\,at}{2}-\sqrt{3}\sin\dfrac{\sqrt{3}\,at}{2}\right\}$
32.76	$\dfrac{s}{s^3-a^3}$	$\dfrac{e^{-at/2}}{3a}\left\{\sqrt{3}\sin\dfrac{\sqrt{3}\,at}{2}-\cos\dfrac{\sqrt{3}\,at}{2}+e^{3at/2}\right\}$
32.77	$\dfrac{s^2}{s^3-a^3}$	$\dfrac{1}{3}\left(e^{at}+2e^{-at/2}\cos\dfrac{\sqrt{3}\,at}{2}\right)$
32.78	$\dfrac{1}{s^4+4a^4}$	$\dfrac{1}{4a^3}(\sin at\cosh at-\cos at\sinh at)$
32.79	$\dfrac{s}{s^4+4a^4}$	$\dfrac{\sin at\sinh at}{2a^2}$
32.80	$\dfrac{s^2}{s^4+4a^4}$	$\dfrac{1}{2a}(\sin at\cosh at+\cos at\sinh at)$
32.81	$\dfrac{s^3}{s^4+4a^4}$	$\cos at\cosh at$
32.82	$\dfrac{1}{s^4-a^4}$	$\dfrac{1}{2a^3}(\sinh at-\sin at)$
32.83	$\dfrac{s}{s^4-a^4}$	$\dfrac{1}{2a^2}(\cosh at-\cos at)$
32.84	$\dfrac{s^2}{s^4-a^4}$	$\dfrac{1}{2a}(\sinh at+\sin at)$
32.85	$\dfrac{s^3}{s^4-a^4}$	$\frac{1}{2}(\cosh at+\cos at)$
32.86	$\dfrac{1}{\sqrt{s+a}+\sqrt{s+b}}$	$\dfrac{e^{-bt}-e^{-at}}{2(b-a)\sqrt{\pi t^3}}$
32.87	$\dfrac{1}{s\sqrt{s+a}}$	$\dfrac{\operatorname{erf}\sqrt{at}}{\sqrt{a}}$
32.88	$\dfrac{1}{\sqrt{s}\,(s-a)}$	$\dfrac{e^{at}\operatorname{erf}\sqrt{at}}{\sqrt{a}}$
32.89	$\dfrac{1}{\sqrt{s-a}+b}$	$e^{at}\left\{\dfrac{1}{\sqrt{\pi t}}-b\,e^{b^2t}\operatorname{erfc}(b\sqrt{t})\right\}$

	$f(s)$	$F(t)$
32.90	$\dfrac{1}{\sqrt{s^2+a^2}}$	$J_0(at)$
32.91	$\dfrac{1}{\sqrt{s^2-a^2}}$	$I_0(at)$
32.92	$\dfrac{(\sqrt{s^2+a^2}-s)^n}{\sqrt{s^2+a^2}} \quad n>-1$	$a^n J_n(at)$
32.93	$\dfrac{(s-\sqrt{s^2-a^2})^n}{\sqrt{s^2-a^2}} \quad n>-1$	$a^n I_n(at)$
32.94	$\dfrac{e^{b(s-\sqrt{s^2+a^2})}}{\sqrt{s^2+a^2}}$	$J_0(a\sqrt{t(t+2b)})$
32.95	$\dfrac{e^{-b\sqrt{s^2+a^2}}}{\sqrt{s^2+a^2}}$	$\begin{cases} J_0(a\sqrt{t^2-b^2}) & t>b \\ 0 & t<b \end{cases}$
32.96	$\dfrac{1}{(s^2+a^2)^{3/2}}$	$\dfrac{tJ_1(at)}{a}$
32.97	$\dfrac{s}{(s^2+a^2)^{3/2}}$	$tJ_0(at)$
32.98	$\dfrac{s^2}{(s^2+a^2)^{3/2}}$	$J_0(at) - atJ_1(at)$
32.99	$\dfrac{1}{(s^2-a^2)^{3/2}}$	$\dfrac{tI_1(at)}{a}$
32.100	$\dfrac{s}{(s^2-a^2)^{3/2}}$	$tI_0(at)$
32.101	$\dfrac{s^2}{(s^2-a^2)^{3/2}}$	$I_0(at) + atI_1(at)$
32.102	$\dfrac{1}{s(e^s-1)} = \dfrac{e^{-s}}{s(1-e^{-s})}$ See also entry 32.165.	$F(t)=n, \; n \leqq t < n+1, \; n=0,1,2,\ldots$
32.103	$\dfrac{1}{s(e^s-r)} = \dfrac{e^{-s}}{s(1-re^{-s})}$	$F(t) = \sum_{k=1}^{[t]} r^k$ where $[t]$ = greatest integer $\leqq t$
32.104	$\dfrac{e^s-1}{s(e^s-r)} = \dfrac{1-e^{-s}}{s(1-re^{-s})}$ See also entry 32.167.	$F(t)=r^n, \; n \leqq t < n+1, \; n=0,1,2,\ldots$
32.105	$\dfrac{e^{-a/s}}{\sqrt{s}}$	$\dfrac{\cos 2\sqrt{at}}{\sqrt{\pi t}}$

	$f(s)$	$F(t)$
32.106	$\dfrac{e^{-a/s}}{s^{3/2}}$	$\dfrac{\sin 2\sqrt{at}}{\sqrt{\pi a}}$
32.107	$\dfrac{e^{-a/s}}{s^{n+1}} \quad n > -1$	$\left(\dfrac{t}{a}\right)^{n/2} J_n(2\sqrt{at})$
32.108	$\dfrac{e^{-a\sqrt{s}}}{\sqrt{s}}$	$\dfrac{e^{-a^2/4t}}{\sqrt{\pi t}}$
32.109	$e^{-a\sqrt{s}}$	$\dfrac{a}{2\sqrt{\pi t^3}} e^{-a^2/4t}$
32.110	$\dfrac{1 - e^{-a\sqrt{s}}}{s}$	$\operatorname{erf}(a/2\sqrt{t})$
32.111	$\dfrac{e^{-a\sqrt{s}}}{s}$	$\operatorname{erfc}(a/2\sqrt{t})$
32.112	$\dfrac{e^{-a\sqrt{s}}}{\sqrt{s}(\sqrt{s}+b)}$	$e^{b(bt+a)} \operatorname{erfc}\left(b\sqrt{t} + \dfrac{a}{2\sqrt{t}}\right)$
32.113	$\dfrac{e^{-a/\sqrt{s}}}{s^{n+1}} \quad n > -1$	$\dfrac{1}{\sqrt{\pi t}\, a^{2n+1}} \displaystyle\int_0^\infty u^n e^{-u^2/4a^2t} J_{2n}(2\sqrt{u})\, du$
32.114	$\ln\left(\dfrac{s+a}{s+b}\right)$	$\dfrac{e^{-bt} - e^{-at}}{t}$
32.115	$\dfrac{\ln[(s^2+a^2)/a^2]}{2s}$	$Ci(at)$
32.116	$\dfrac{\ln[(s+a)/a]}{s}$	$Ei(at)$
32.117	$-\dfrac{(\gamma + \ln s)}{s}$ γ = Euler's constant = .5772156...	$\ln t$
32.118	$\ln\left(\dfrac{s^2+a^2}{s^2+b^2}\right)$	$\dfrac{2(\cos at - \cos bt)}{t}$
32.119	$\dfrac{\pi^2}{6s} + \dfrac{(\gamma + \ln s)^2}{s}$ γ = Euler's constant = .5772156...	$\ln^2 t$
32.120	$\dfrac{\ln s}{s}$	$-(\ln t + \gamma)$ γ = Euler's constant = .5772156...
32.121	$\dfrac{\ln^2 s}{s}$	$(\ln t + \gamma)^2 - \frac{1}{6}\pi^2$ γ = Euler's constant = .5772156...

	$f(s)$	$F(t)$
32.122	$\dfrac{\Gamma'(n+1) - \Gamma(n+1)\ln s}{s^{n+1}} \quad n > -1$	$t^n \ln t$
32.123	$\tan^{-1}(a/s)$	$\dfrac{\sin at}{t}$
32.124	$\dfrac{\tan^{-1}(a/s)}{s}$	$Si(at)$
32.125	$\dfrac{e^{a/s}}{\sqrt{s}}\operatorname{erfc}(\sqrt{a/s})$	$\dfrac{e^{-2\sqrt{at}}}{\sqrt{\pi t}}$
32.126	$e^{s^2/4a^2}\operatorname{erfc}(s/2a)$	$\dfrac{2a}{\sqrt{\pi}}e^{-a^2t^2}$
32.127	$\dfrac{e^{s^2/4a^2}\operatorname{erfc}(s/2a)}{s}$	$\operatorname{erf}(at)$
32.128	$\dfrac{e^{as}\operatorname{erfc}\sqrt{as}}{\sqrt{s}}$	$\dfrac{1}{\sqrt{\pi(t+a)}}$
32.129	$e^{as}\,Ei(as)$	$\dfrac{1}{t+a}$
32.130	$\dfrac{1}{a}\left[\cos as\left\{\dfrac{\pi}{2} - Si(as)\right\} - \sin as\, Ci(as)\right]$	$\dfrac{1}{t^2+a^2}$
32.131	$\sin as\left\{\dfrac{\pi}{2} - Si(as)\right\} + \cos as\, Ci(as)$	$\dfrac{t}{t^2+a^2}$
32.132	$\dfrac{\cos as\left\{\dfrac{\pi}{2} - Si(as)\right\} - \sin as\, Ci(as)}{s}$	$\tan^{-1}(t/a)$
32.133	$\dfrac{\sin as\left\{\dfrac{\pi}{2} - Si(as)\right\} + \cos as\, Ci(as)}{s}$	$\dfrac{1}{2}\ln\left(\dfrac{t^2+a^2}{a^2}\right)$
32.134	$\left[\dfrac{\pi}{2} - Si(as)\right]^2 + Ci^2(as)$	$\dfrac{1}{t}\ln\left(\dfrac{t^2+a^2}{a^2}\right)$
32.135	0	$\mathcal{N}(t)$ = null function
32.136	1	$\delta(t)$ = delta function
32.137	e^{-as}	$\delta(t-a)$
32.138	$\dfrac{e^{-as}}{s}$ See also entry 32.163.	$\mathcal{U}(t-a)$

	$f(s)$	$F(t)$
32.139	$\dfrac{\sinh sx}{s \sinh sa}$	$\dfrac{x}{a} + \dfrac{2}{\pi} \sum_{n=1}^{\infty} \dfrac{(-1)^n}{n} \sin \dfrac{n\pi x}{a} \cos \dfrac{n\pi t}{a}$
32.140	$\dfrac{\sinh sx}{s \cosh sa}$	$\dfrac{4}{\pi} \sum_{n=1}^{\infty} \dfrac{(-1)^n}{2n-1} \sin \dfrac{(2n-1)\pi x}{2a} \sin \dfrac{(2n-1)\pi t}{2a}$
32.141	$\dfrac{\cosh sx}{s \sinh as}$	$\dfrac{t}{a} + \dfrac{2}{\pi} \sum_{n=1}^{\infty} \dfrac{(-1)^n}{n} \cos \dfrac{n\pi x}{a} \sin \dfrac{n\pi t}{a}$
32.142	$\dfrac{\cosh sx}{s \cosh sa}$	$1 + \dfrac{4}{\pi} \sum_{n=1}^{\infty} \dfrac{(-1)^n}{2n-1} \cos \dfrac{(2n-1)\pi x}{2a} \cos \dfrac{(2n-1)\pi t}{2a}$
32.143	$\dfrac{\sinh sx}{s^2 \sinh sa}$	$\dfrac{xt}{a} + \dfrac{2a}{\pi^2} \sum_{n=1}^{\infty} \dfrac{(-1)^n}{n^2} \sin \dfrac{n\pi x}{a} \sin \dfrac{n\pi t}{a}$
32.144	$\dfrac{\sinh sx}{s^2 \cosh sa}$	$x + \dfrac{8a}{\pi^2} \sum_{n=1}^{\infty} \dfrac{(-1)^n}{(2n-1)^2} \sin \dfrac{(2n-1)\pi x}{2a} \cos \dfrac{(2n-1)\pi t}{2a}$
32.145	$\dfrac{\cosh sx}{s^2 \sinh sa}$	$\dfrac{t^2}{2a} + \dfrac{2a}{\pi^2} \sum_{n=1}^{\infty} \dfrac{(-1)^n}{n^2} \cos \dfrac{n\pi x}{a} \left(1 - \cos \dfrac{n\pi t}{a}\right)$
32.146	$\dfrac{\cosh sx}{s^2 \cosh sa}$	$t + \dfrac{8a}{\pi^2} \sum_{n=1}^{\infty} \dfrac{(-1)^n}{(2n-1)^2} \cos \dfrac{(2n-1)\pi x}{2a} \sin \dfrac{(2n-1)\pi t}{2a}$
32.147	$\dfrac{\cosh sx}{s^3 \cosh sa}$	$\tfrac{1}{2}(t^2 + x^2 - a^2) - \dfrac{16a^2}{\pi^3} \sum_{n=1}^{\infty} \dfrac{(-1)^n}{(2n-1)^3} \cos \dfrac{(2n-1)\pi x}{2a} \cos \dfrac{(2n-1)\pi t}{2a}$
32.148	$\dfrac{\sinh x\sqrt{s}}{\sinh a\sqrt{s}}$	$\dfrac{2\pi}{a^2} \sum_{n=1}^{\infty} (-1)^n n\, e^{-n^2\pi^2 t/a^2} \sin \dfrac{n\pi x}{a}$
32.149	$\dfrac{\cosh x\sqrt{s}}{\cosh a\sqrt{s}}$	$\dfrac{\pi}{a^2} \sum_{n=1}^{\infty} (-1)^{n-1} (2n-1)\, e^{-(2n-1)^2\pi^2 t/4a^2} \cos \dfrac{(2n-1)\pi x}{2a}$
32.150	$\dfrac{\sinh x\sqrt{s}}{\sqrt{s} \cosh a\sqrt{s}}$	$\dfrac{2}{a} \sum_{n=1}^{\infty} (-1)^{n-1} e^{-(2n-1)^2\pi^2 t/4a^2} \sin \dfrac{(2n-1)\pi x}{2a}$
32.151	$\dfrac{\cosh x\sqrt{s}}{\sqrt{s} \sinh a\sqrt{s}}$	$\dfrac{1}{a} + \dfrac{2}{a} \sum_{n=1}^{\infty} (-1)^n e^{-n^2\pi^2 t/a^2} \cos \dfrac{n\pi x}{a}$
32.152	$\dfrac{\sinh x\sqrt{s}}{s \sinh a\sqrt{s}}$	$\dfrac{x}{a} + \dfrac{2}{\pi} \sum_{n=1}^{\infty} \dfrac{(-1)^n}{n} e^{-n^2\pi^2 t/a^2} \sin \dfrac{n\pi x}{a}$
32.153	$\dfrac{\cosh x\sqrt{s}}{s \cosh a\sqrt{s}}$	$1 + \dfrac{4}{\pi} \sum_{n=1}^{\infty} \dfrac{(-1)^n}{2n-1} e^{-(2n-1)^2\pi^2 t/4a^2} \cos \dfrac{(2n-1)\pi x}{2a}$
32.154	$\dfrac{\sinh x\sqrt{s}}{s^2 \sinh a\sqrt{s}}$	$\dfrac{xt}{a} + \dfrac{2a^2}{\pi^3} \sum_{n=1}^{\infty} \dfrac{(-1)^n}{n^3} (1 - e^{-n^2\pi^2 t/a^2}) \sin \dfrac{n\pi x}{a}$
32.155	$\dfrac{\cosh x\sqrt{s}}{s^2 \cosh a\sqrt{s}}$	$\tfrac{1}{2}(x^2 - a^2) + t - \dfrac{16a^2}{\pi^3} \sum_{n=1}^{\infty} \dfrac{(-1)^n}{(2n-1)^3} e^{-(2n-1)^2\pi^2 t/4a^2} \cos \dfrac{(2n-1)\pi x}{2a}$

	$f(s)$	$F(t)$
32.156	$\dfrac{J_0(ix\sqrt{s})}{s\,J_0(ia\sqrt{s})}$	$1 - 2\displaystyle\sum_{n=1}^{\infty} \frac{e^{-\lambda_n^2 t/a^2} J_0(\lambda_n x/a)}{\lambda_n J_1(\lambda_n)}$ where $\lambda_1, \lambda_2, \ldots$ are the positive roots of $J_0(\lambda) = 0$
32.157	$\dfrac{J_0(ix\sqrt{s})}{s^2 J_0(ia\sqrt{s})}$	$\frac{1}{4}(x^2 - a^2) + t + 2a^2 \displaystyle\sum_{n=1}^{\infty} \frac{e^{-\lambda_n^2 t/a^2} J_0(\lambda_n x/a)}{\lambda_n^3 J_1(\lambda_n)}$ where $\lambda_1, \lambda_2, \ldots$ are the positive roots of $J_0(\lambda) = 0$
32.158	$\dfrac{1}{as^2}\tanh\left(\dfrac{as}{2}\right)$	**Triangular wave function** 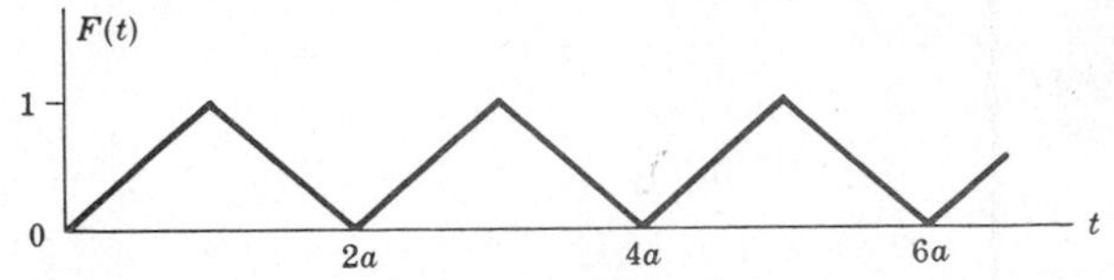 Fig. 32-1
32.159	$\dfrac{1}{s}\tanh\left(\dfrac{as}{2}\right)$	**Square wave function** 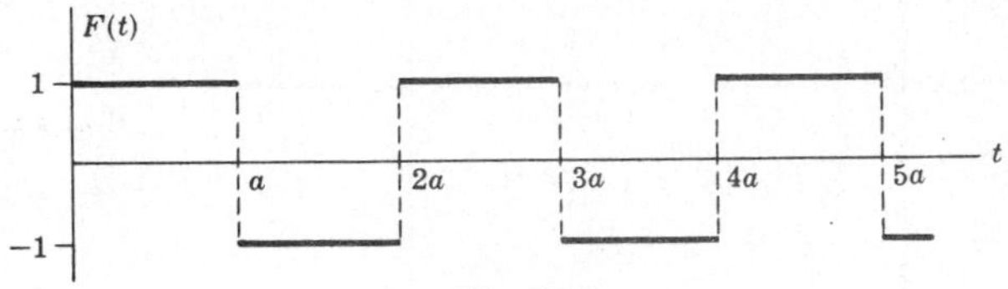 Fig. 32-2
32.160	$\dfrac{\pi a}{a^2 s^2 + \pi^2}\coth\left(\dfrac{as}{2}\right)$	**Rectified sine wave function** 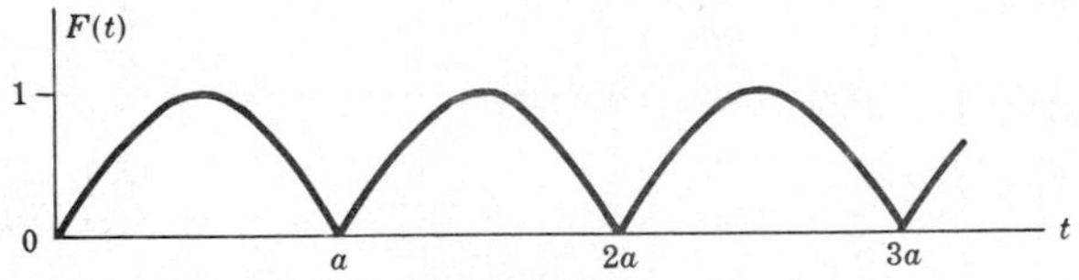 Fig. 32-3
32.161	$\dfrac{\pi a}{(a^2 s^2 + \pi^2)(1 - e^{-as})}$	**Half rectified sine wave function** 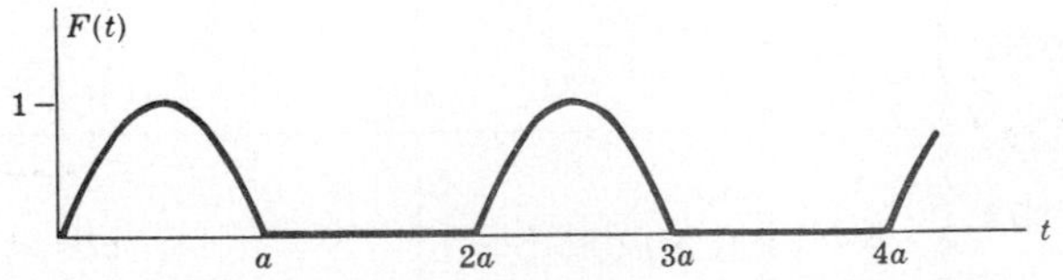 Fig. 32-4
32.162	$\dfrac{1}{as^2} - \dfrac{e^{-as}}{s(1 - e^{-as})}$	**Saw tooth wave function** 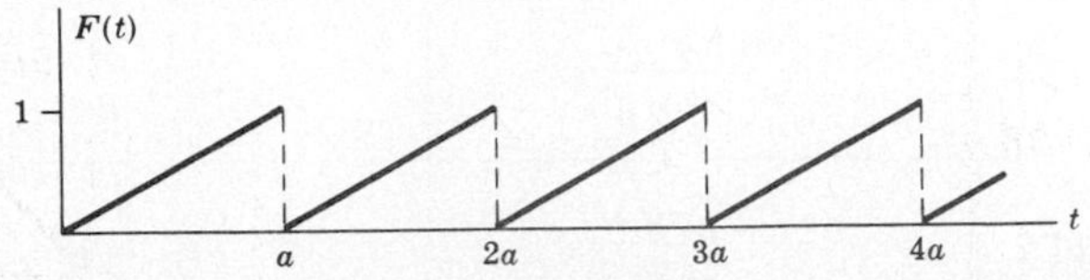 Fig. 32-5

	$f(s)$	$F(t)$
32.163	$\dfrac{e^{-as}}{s}$ See also entry 32.138.	**Heaviside's unit function** $\mathcal{U}(t-a)$ $F(t)$ 1 0 a t Fig. 32-6
32.164	$\dfrac{e^{-as}(1-e^{-\epsilon s})}{s}$	**Pulse function** $F(t)$ 1 0 a $a+\epsilon$ t Fig. 32-7
32.165	$\dfrac{1}{s(1-e^{-as})}$ See also entry 32.102.	**Step function** $F(t)$ 3 2 1 0 a $2a$ $3a$ $4a$ t Fig. 32-8
32.166	$\dfrac{e^{-s}+e^{-2s}}{s(1-e^{-s})^2}$	$F(t) = n^2, \ n \leqq t < n+1, \ n = 0, 1, 2, \ldots$ $F(t)$ 4 3 2 1 0 1 2 3 t Fig. 32-9
32.167	$\dfrac{1-e^{-s}}{s(1-re^{-s})}$ See also entry 32.104.	$F(t) = r^n, \ n \leqq t < n+1, \ n = 0, 1, 2, \ldots$ $F(t)$ r 1 0 1 2 3 t Fig. 32-10
32.168	$\dfrac{\pi a(1+e^{-as})}{a^2s^2+\pi^2}$	$F(t) = \begin{cases} \sin(\pi t/a) & 0 \leqq t \leqq a \\ 0 & t > a \end{cases}$ $F(t)$ 1 0 a t Fig. 32-11

33 FOURIER TRANSFORMS

FOURIER'S INTEGRAL THEOREM

33.1
$$f(x) = \int_0^\infty \{A(\alpha)\cos\alpha x + B(\alpha)\sin\alpha x\}\, d\alpha$$

where

33.2
$$\begin{cases} A(\alpha) = \dfrac{1}{\pi}\displaystyle\int_{-\infty}^{\infty} f(x)\cos\alpha x\, dx \\ B(\alpha) = \dfrac{1}{\pi}\displaystyle\int_{-\infty}^{\infty} f(x)\sin\alpha x\, dx \end{cases}$$

Sufficient conditions under which this theorem holds are:

(i) $f(x)$ and $f'(x)$ are piecewise continuous in every finite interval $-L < x < L$;

(ii) $\int_{-\infty}^{\infty} |f(x)|\, dx$ converges;

(iii) $f(x)$ is replaced by $\frac{1}{2}\{f(x+0) + f(x-0)\}$ if x is a point of discontinuity.

EQUIVALENT FORMS OF FOURIER'S INTEGRAL THEOREM

33.3
$$f(x) = \frac{1}{2\pi}\int_{\alpha=-\infty}^{\infty}\int_{u=-\infty}^{\infty} f(u)\cos\alpha\,(x-u)\, du\, d\alpha$$

33.4
$$f(x) = \frac{1}{2\pi}\int_{-\infty}^{\infty} e^{i\alpha x}\, d\alpha \int_{-\infty}^{\infty} f(u)\, e^{-i\alpha u}\, du$$
$$= \frac{1}{2\pi}\int_{-\infty}^{\infty}\int_{-\infty}^{\infty} f(u)\, e^{i\alpha(x-u)}\, du\, d\alpha$$

33.5
$$f(x) = \frac{2}{\pi}\int_0^\infty \sin\alpha x\, d\alpha \int_0^\infty f(u)\sin\alpha u\, du$$

where $f(x)$ is an *odd function* $[f(-x) = -f(x)]$.

33.6
$$f(x) = \frac{2}{\pi}\int_0^\infty \cos\alpha x\, d\alpha \int_0^\infty f(u)\cos\alpha u\, du$$

where $f(x)$ is an *even function* $[f(-x) = f(x)]$.

FOURIER TRANSFORMS

The Fourier transform of $f(x)$ is defined as

33.7
$$\mathcal{F}\{f(x)\} = F(\alpha) = \int_{-\infty}^{\infty} f(x)\, e^{-i\alpha x}\, dx$$

Then from 33.7 the inverse Fourier transform of $F(\alpha)$ is

33.8
$$\mathcal{F}^{-1}\{F(\alpha)\} = f(x) = \frac{1}{2\pi}\int_{-\infty}^{\infty} F(\alpha)\, e^{i\alpha x}\, d\alpha$$

We call $f(x)$ and $F(\alpha)$ *Fourier transform pairs.*

CONVOLUTION THEOREM FOR FOURIER TRANSFORMS

If $F(\alpha) = \mathcal{F}\{f(x)\}$ and $G(\alpha) = \mathcal{F}\{g(x)\}$, then

33.9
$$\frac{1}{2\pi}\int_{-\infty}^{\infty} F(\alpha)\, G(\alpha)\, e^{i\alpha x}\, d\alpha = \int_{-\infty}^{\infty} f(u)\, g(x-u)\, du = f*g$$

where $f*g$ is called the *convolution* of f and g. Thus

33.10
$$\mathcal{F}\{f*g\} = \mathcal{F}\{f\}\,\mathcal{F}\{g\}$$

PARSEVAL'S IDENTITY

If $F(\alpha) = \mathcal{F}\{f(x)\}$, then

33.11
$$\int_{-\infty}^{\infty} |f(x)|^2\, dx = \frac{1}{2\pi}\int_{-\infty}^{\infty} |F(\alpha)|^2\, d\alpha$$

More generally if $F(\alpha) = \mathcal{F}\{f(x)\}$ and $G(\alpha) = \mathcal{F}\{g(x)\}$, then

33.12
$$\int_{-\infty}^{\infty} f(x)\, \overline{g(x)}\, dx = \frac{1}{2\pi}\int_{-\infty}^{\infty} F(\alpha)\, \overline{G(\alpha)}\, d\alpha$$

where the bar denotes complex conjugate.

FOURIER SINE TRANSFORMS

The Fourier sine transform of $f(x)$ is defined as

33.13
$$F_S(\alpha) = \mathcal{F}_S\{f(x)\} = \int_0^{\infty} f(x)\sin \alpha x\, dx$$

Then from 33.13 the inverse Fourier sine transform of $F_S(\alpha)$ is

33.14
$$f(x) = \mathcal{F}_S^{-1}\{F_S(\alpha)\} = \frac{2}{\pi}\int_0^{\infty} F_S(\alpha)\sin \alpha x\, d\alpha$$

FOURIER COSINE TRANSFORMS

The Fourier cosine transform of $f(x)$ is defined as

33.15 $$F_C(\alpha) = \mathcal{F}_C\{f(x)\} = \int_0^\infty f(x)\cos\alpha x\,dx$$

Then from 33.15 the inverse Fourier cosine transform of $F_C(\alpha)$ is

33.16 $$f(x) = \mathcal{F}_C^{-1}\{F_C(\alpha)\} = \frac{2}{\pi}\int_0^\infty F_C(\alpha)\cos\alpha x\,d\alpha$$

SPECIAL FOURIER TRANSFORM PAIRS

	$f(x)$	$F(\alpha)$
33.17	$\begin{cases} 1 & \lvert x\rvert < b \\ 0 & \lvert x\rvert > b \end{cases}$	$\dfrac{2\sin b\alpha}{\alpha}$
33.18	$\dfrac{1}{x^2+b^2}$	$\dfrac{\pi e^{-b\alpha}}{b}$
33.19	$\dfrac{x}{x^2+b^2}$	$-\dfrac{\pi i\alpha}{b}e^{-b\alpha}$
33.20	$f^{(n)}(x)$	$i^n\alpha^n F(\alpha)$
33.21	$x^n f(x)$	$i^n \dfrac{d^nF}{d\alpha^n}$
33.22	$f(bx)e^{itx}$	$\dfrac{1}{b}F\left(\dfrac{\alpha-t}{b}\right)$

SPECIAL FOURIER SINE TRANSFORMS

	$f(x)$	$F_C(\alpha)$
33.23	$\begin{cases} 1 & 0 < x < b \\ 0 & x > b \end{cases}$	$\dfrac{1 - \cos b\alpha}{\alpha}$
33.24	x^{-1}	$\dfrac{\pi}{2}$
33.25	$\dfrac{x}{x^2 + b^2}$	$\dfrac{\pi}{2} e^{-b\alpha}$
33.26	e^{-bx}	$\dfrac{\alpha}{\alpha^2 + b^2}$
33.27	$x^{n-1} e^{-bx}$	$\dfrac{\Gamma(n) \sin (n \tan^{-1} \alpha/b)}{(\alpha^2 + b^2)^{n/2}}$
33.28	xe^{-bx^2}	$\dfrac{\sqrt{\pi}}{4b^{3/2}} \alpha e^{-\alpha^2/4b}$
33.29	$x^{-1/2}$	$\sqrt{\dfrac{\pi}{2\alpha}}$
33.30	x^{-n}	$\dfrac{\pi\alpha^{n-1} \csc (n\pi/2)}{2\,\Gamma(n)} \quad 0 < n < 2$
33.31	$\dfrac{\sin bx}{x}$	$\dfrac{1}{2} \ln \left(\dfrac{\alpha + b}{\alpha - b}\right)$
33.32	$\dfrac{\sin bx}{x^2}$	$\begin{cases} \pi\alpha/2 & \alpha < b \\ \pi b/2 & \alpha > b \end{cases}$
33.33	$\dfrac{\cos bx}{x}$	$\begin{cases} 0 & \alpha < b \\ \pi/4 & \alpha = b \\ \pi/2 & \alpha > b \end{cases}$
33.34	$\tan^{-1} (x/b)$	$\dfrac{\pi}{2\alpha} e^{-b\alpha}$
33.35	$\csc bx$	$\dfrac{\pi}{2b} \tanh \dfrac{\pi\alpha}{2b}$
33.36	$\dfrac{1}{e^{2x} - 1}$	$\dfrac{\pi}{4} \coth \left(\dfrac{\pi\alpha}{2}\right) - \dfrac{1}{2\alpha}$

SPECIAL FOURIER COSINE TRANSFORMS

	$f(x)$	$F_C(\alpha)$
33.37	$\begin{cases} 1 & 0 < x < b \\ 0 & x > b \end{cases}$	$\dfrac{\sin b\alpha}{\alpha}$
33.38	$\dfrac{1}{x^2 + b^2}$	$\dfrac{\pi e^{-b\alpha}}{2b}$
33.39	e^{-bx}	$\dfrac{b}{\alpha^2 + b^2}$
33.40	$x^{n-1} e^{-bx}$	$\dfrac{\Gamma(n) \cos (n \tan^{-1} \alpha/b)}{(\alpha^2 + b^2)^{n/2}}$
33.41	e^{-bx^2}	$\dfrac{1}{2}\sqrt{\dfrac{\pi}{b}} e^{-\alpha^2/4b}$
33.42	$x^{-1/2}$	$\sqrt{\dfrac{\pi}{2\alpha}}$
33.43	x^{-n}	$\dfrac{\pi \alpha^{n-1} \sec (n\pi/2)}{2\,\Gamma(n)}, \quad 0 < n < 1$
33.44	$\ln\left(\dfrac{x^2 + b^2}{x^2 + c^2}\right)$	$\dfrac{e^{-c\alpha} - e^{-b\alpha}}{\pi \alpha}$
33.45	$\dfrac{\sin bx}{x}$	$\begin{cases} \pi/2 & \alpha < b \\ \pi/4 & \alpha = b \\ 0 & \alpha > b \end{cases}$
33.46	$\sin bx^2$	$\sqrt{\dfrac{\pi}{8b}}\left(\cos \dfrac{\alpha^2}{4b} - \sin \dfrac{\alpha^2}{4b}\right)$
33.47	$\cos bx^2$	$\sqrt{\dfrac{\pi}{8b}}\left(\cos \dfrac{\alpha^2}{4b} + \sin \dfrac{\alpha^2}{4b}\right)$
33.48	$\operatorname{sech} bx$	$\dfrac{\pi}{2b} \operatorname{sech} \dfrac{\pi\alpha}{2b}$
33.49	$\dfrac{\cosh (\sqrt{\pi}\, x/2)}{\cosh (\sqrt{\pi}\, x)}$	$\sqrt{\dfrac{\pi}{2}} \dfrac{\cosh (\sqrt{\pi}\, \alpha/2)}{\cosh (\sqrt{\pi}\, \alpha)}$
33.50	$\dfrac{e^{-b\sqrt{x}}}{\sqrt{x}}$	$\sqrt{\dfrac{\pi}{2\alpha}} \{\cos (2b\sqrt{\alpha}) - \sin (2b\sqrt{\alpha})\}$

34 ELLIPTIC FUNCTIONS

INCOMPLETE ELLIPTIC INTEGRAL OF THE FIRST KIND

34.1 $$u = F(k, \phi) = \int_0^\phi \frac{d\theta}{\sqrt{1-k^2\sin^2\theta}} = \int_0^x \frac{dv}{\sqrt{(1-v^2)(1-k^2v^2)}}$$

where $\phi = \operatorname{am} u$ is called the *amplitude* of u and $x = \sin\phi$, and where here and below $0 < k < 1$.

COMPLETE ELLIPTIC INTEGRAL OF THE FIRST KIND

34.2 $$K = F(k, \pi/2) = \int_0^{\pi/2} \frac{d\theta}{\sqrt{1-k^2\sin^2\theta}} = \int_0^1 \frac{dv}{\sqrt{(1-v^2)(1-k^2v^2)}}$$

$$= \frac{\pi}{2}\left\{1 + \left(\frac{1}{2}\right)^2 k^2 + \left(\frac{1\cdot 3}{2\cdot 4}\right)^2 k^4 + \left(\frac{1\cdot 3\cdot 5}{2\cdot 4\cdot 6}\right)^2 k^6 + \cdots\right\}$$

INCOMPLETE ELLIPTIC INTEGRAL OF THE SECOND KIND

34.3 $$E(k, \phi) = \int_0^\phi \sqrt{1-k^2\sin^2\theta}\,d\theta = \int_0^x \frac{\sqrt{1-k^2v^2}}{\sqrt{1-v^2}}\,dv$$

COMPLETE ELLIPTIC INTEGRAL OF THE SECOND KIND

34.4 $$E = E(k, \pi/2) = \int_0^{\pi/2} \sqrt{1-k^2\sin^2\theta}\,d\theta = \int_0^1 \frac{\sqrt{1-k^2v^2}}{\sqrt{1-v^2}}\,dv$$

$$= \frac{\pi}{2}\left\{1 - \left(\frac{1}{2}\right)^2 k^2 - \left(\frac{1\cdot 3}{2\cdot 4}\right)^2 \frac{k^4}{3} - \left(\frac{1\cdot 3\cdot 5}{2\cdot 4\cdot 6}\right)^2 \frac{k^6}{5} - \cdots\right\}$$

INCOMPLETE ELLIPTIC INTEGRAL OF THE THIRD KIND

34.5 $$\Pi(k, n, \phi) = \int_0^\phi \frac{d\theta}{(1+n\sin^2\theta)\sqrt{1-k^2\sin^2\theta}} = \int_0^x \frac{dv}{(1+nv^2)\sqrt{(1-v^2)(1-k^2v^2)}}$$

COMPLETE ELLIPTIC INTEGRAL OF THE THIRD KIND

34.6 $$\Pi(k, n, \pi/2) = \int_0^{\pi/2} \frac{d\theta}{(1 + n\sin^2\theta)\sqrt{1 - k^2\sin^2\theta}} = \int_0^1 \frac{dv}{(1 + nv^2)\sqrt{(1 - v^2)(1 - k^2v^2)}}$$

LANDEN'S TRANSFORMATION

34.7 $$\tan\phi = \frac{\sin 2\phi_1}{k + \cos 2\phi_1} \quad \text{or} \quad k\sin\phi = \sin(2\phi_1 - \phi)$$

This yields

34.8 $$F(k, \phi) = \int_0^{\phi} \frac{d\theta}{\sqrt{1 - k^2\sin^2\theta}} = \frac{2}{1 + k}\int_0^{\phi_1} \frac{d\theta_1}{\sqrt{1 - k_1^2\sin^2\theta_1}}$$

where $k_1 = 2\sqrt{k}/(1 + k)$. By successive applications, sequences $k_1, k_2, k_3, \ldots$ and $\phi_1, \phi_2, \phi_3, \ldots$ are obtained such that $k < k_1 < k_2 < k_3 < \cdots < 1$ where $\lim_{n\to\infty} k_n = 1$. It follows that

34.9 $$F(k, \Phi) = \sqrt{\frac{k_1k_2k_3\ldots}{k}}\int_0^{\Phi} \frac{d\theta}{\sqrt{1 - \sin^2\theta}} = \sqrt{\frac{k_1k_2k_3\ldots}{k}}\ln\tan\left(\frac{\pi}{4} + \frac{\Phi}{2}\right)$$

where

34.10 $$k_1 = \frac{2\sqrt{k}}{1 + k}, \quad k_2 = \frac{2\sqrt{k_1}}{1 + k_1}, \quad \ldots \quad \text{and} \quad \Phi = \lim_{n\to\infty} \phi_n$$

The result is used in the approximate evaluation of $F(k, \phi)$.

JACOBI'S ELLIPTIC FUNCTIONS

From 34.1 we define the following elliptic functions.

34.11 $$x = \sin(\text{am}\, u) = \text{sn}\, u$$

34.12 $$\sqrt{1 - x^2} = \cos(\text{am}\, u) = \text{cn}\, u$$

34.13 $$\sqrt{1 - k^2x^2} = \sqrt{1 - k^2\,\text{sn}^2 u} = \text{dn}\, u$$

We can also define the inverse functions $\text{sn}^{-1} x$, $\text{cn}^{-1} x$, $\text{dn}^{-1} x$ and the following

34.14 $$\text{ns}\, u = \frac{1}{\text{sn}\, u}$$

34.15 $$\text{nc}\, u = \frac{1}{\text{cn}\, u}$$

34.16 $$\text{nd}\, u = \frac{1}{\text{dn}\, u}$$

34.17 $$\text{sc}\, u = \frac{\text{sn}\, u}{\text{cn}\, u}$$

34.18 $$\text{sd}\, u = \frac{\text{sn}\, u}{\text{dn}\, u}$$

34.19 $$\text{cd}\, u = \frac{\text{cn}\, u}{\text{dn}\, u}$$

34.20 $$\text{cs}\, u = \frac{\text{cn}\, u}{\text{sn}\, u}$$

34.21 $$\text{dc}\, u = \frac{\text{dn}\, u}{\text{cn}\, u}$$

34.22 $$\text{ds}\, u = \frac{\text{dn}\, u}{\text{sn}\, u}$$

ADDITION FORMULAS

34.23 $$\text{sn}(u + v) = \frac{\text{sn}\, u\,\text{cn}\, v\,\text{dn}\, v + \text{cn}\, u\,\text{sn}\, v\,\text{dn}\, u}{1 - k^2\,\text{sn}^2 u\,\text{sn}^2 v}$$

34.24 $$\text{cn}(u + v) = \frac{\text{cn}\, u\,\text{cn}\, v - \text{sn}\, u\,\text{sn}\, v\,\text{dn}\, u\,\text{dn}\, v}{1 - k^2\,\text{sn}^2 u\,\text{sn}^2 v}$$

34.25 $$\text{dn}(u + v) = \frac{\text{dn}\, u\,\text{dn}\, v - k^2\,\text{sn}\, u\,\text{sn}\, v\,\text{cn}\, u\,\text{cn}\, v}{1 - k^2\,\text{sn}^2 u\,\text{sn}^2 v}$$

DERIVATIVES

34.26 $\dfrac{d}{du}\operatorname{sn} u = \operatorname{cn} u \operatorname{dn} u$

34.27 $\dfrac{d}{du}\operatorname{cn} u = -\operatorname{sn} u \operatorname{dn} u$

34.28 $\dfrac{d}{du}\operatorname{dn} u = -k^2 \operatorname{sn} u \operatorname{cn} u$

34.29 $\dfrac{d}{du}\operatorname{sc} u = \operatorname{dc} u \operatorname{nc} u$

SERIES EXPANSIONS

34.30 $$\operatorname{sn} u = u - (1+k^2)\frac{u^3}{3!} + (1+14k^2+k^4)\frac{u^5}{5!} - (1+135k^2+135k^4+k^6)\frac{u^7}{7!} + \cdots$$

34.31 $$\operatorname{cn} u = 1 - \frac{u^2}{2!} + (1+4k^2)\frac{u^4}{4!} - (1+44k^2+16k^4)\frac{u^6}{6!} + \cdots$$

34.32 $$\operatorname{dn} u = 1 - k^2\frac{u^2}{2!} + k^2(4+k^2)\frac{u^4}{4!} - k^2(16+44k^2+k^4)\frac{u^6}{6!} + \cdots$$

CATALAN'S CONSTANT

34.33 $$\frac{1}{2}\int_0^1 K\,dk = \frac{1}{2}\int_{k=0}^{1}\int_{\theta=0}^{\pi/2}\frac{d\theta\,dk}{\sqrt{1-k^2\sin^2\theta}} = \frac{1}{1^2} - \frac{1}{3^2} + \frac{1}{5^2} - \cdots = .915965594\ldots$$

PERIODS OF ELLIPTIC FUNCTIONS

Let

34.34 $$K = \int_0^{\pi/2}\frac{d\theta}{\sqrt{1-k^2\sin^2\theta}}, \quad K' = \int_0^{\pi/2}\frac{d\theta}{\sqrt{1-k'^2\sin^2\theta}} \quad \text{where } k' = \sqrt{1-k^2}$$

Then

34.35 $\operatorname{sn} u$ has periods $4K$ and $2iK'$

34.36 $\operatorname{cn} u$ has periods $4K$ and $2K + 2iK'$

34.37 $\operatorname{dn} u$ has periods $2K$ and $4iK'$

IDENTITIES INVOLVING ELLIPTIC FUNCTIONS

34.38 $\operatorname{sn}^2 u + \operatorname{cn}^2 u = 1$

34.39 $\operatorname{dn}^2 u + k^2 \operatorname{sn}^2 u = 1$

34.40 $\operatorname{dn}^2 u - k^2 \operatorname{cn}^2 u = k'^2$ where $k' = \sqrt{1-k^2}$

34.41 $\operatorname{sn}^2 u = \dfrac{1-\operatorname{cn} 2u}{1+\operatorname{dn} 2u}$

34.42 $\operatorname{cn}^2 u = \dfrac{\operatorname{dn} 2u + \operatorname{cn} 2u}{1+\operatorname{dn} 2u}$

34.43 $\operatorname{dn}^2 u = \dfrac{1-k^2+\operatorname{dn} 2u + k^2 \operatorname{cn} u}{1+\operatorname{dn} 2u}$

34.44 $\sqrt{\dfrac{1-\operatorname{cn} 2u}{1+\operatorname{cn} 2u}} = \dfrac{\operatorname{sn} u \operatorname{dn} u}{\operatorname{cn} u}$

34.45 $\sqrt{\dfrac{1-\operatorname{dn} 2u}{1+\operatorname{dn} 2u}} = \dfrac{k \operatorname{sn} u \operatorname{cn} u}{\operatorname{dn} u}$

SPECIAL VALUES

34.46 $\operatorname{sn} 0 = 0$ **34.47** $\operatorname{cn} 0 = 1$ **34.48** $\operatorname{dn} 0 = 1$ **34.49** $\operatorname{sc} 0 = 0$ **34.50** $\operatorname{am} 0 = 0$

INTEGRALS

34.51 $$\int \operatorname{sn} u\, du = \frac{1}{k} \ln(\operatorname{dn} u - k \operatorname{cn} u)$$

34.52 $$\int \operatorname{cn} u\, du = \frac{1}{k} \cos^{-1}(\operatorname{dn} u)$$

34.53 $$\int \operatorname{dn} u\, du = \sin^{-1}(\operatorname{sn} u)$$

34.54 $$\int \operatorname{sc} u\, du = \frac{1}{\sqrt{1-k^2}} \ln(\operatorname{dc} u + \sqrt{1-k^2} \operatorname{nc} u)$$

34.55 $$\int \operatorname{cs} u\, du = \ln(\operatorname{ns} u - \operatorname{ds} u)$$

34.56 $$\int \operatorname{cd} u\, du = \frac{1}{k} \ln(\operatorname{nd} u + k \operatorname{sd} u)$$

34.57 $$\int \operatorname{dc} u\, du = \ln(\operatorname{nc} u + \operatorname{sc} u)$$

34.58 $$\int \operatorname{sd} u\, du = \frac{-1}{k\sqrt{1-k^2}} \sin^{-1}(k \operatorname{cd} u)$$

34.59 $$\int \operatorname{ds} u\, du = \ln(\operatorname{ns} u - \operatorname{cs} u)$$

34.60 $$\int \operatorname{ns} u\, du = \ln(\operatorname{ds} u - \operatorname{cs} u)$$

34.61 $$\int \operatorname{nc} u\, du = \frac{1}{\sqrt{1-k^2}} \ln\left(\operatorname{dc} u + \frac{\operatorname{sc} u}{\sqrt{1-k^2}}\right)$$

34.62 $$\int \operatorname{nd} u\, du = \frac{1}{\sqrt{1-k^2}} \cos^{-1}(\operatorname{cd} u)$$

LEGENDRE'S RELATION

34.63 $$EK' + E'K - KK' = \pi/2$$

where

34.64 $$E = \int_0^{\pi/2} \sqrt{1-k^2 \sin^2\theta}\, d\theta \qquad K = \int_0^{\pi/2} \frac{d\theta}{\sqrt{1-k^2\sin^2\theta}}$$

34.65 $$E' = \int_0^{\pi/2} \sqrt{1-k'^2 \sin^2\theta} \qquad K' = \int_0^{\pi/2} \frac{d\theta}{\sqrt{1-k'^2\sin^2\theta}}$$

35 MISCELLANEOUS SPECIAL FUNCTIONS

ERROR FUNCTION $\operatorname{erf}(x) = \dfrac{2}{\sqrt{\pi}}\displaystyle\int_0^x e^{-u^2}\,du$

35.1 $$\operatorname{erf}(x) = \frac{2}{\sqrt{\pi}}\left(x - \frac{x^3}{3\cdot 1!} + \frac{x^5}{5\cdot 2!} - \frac{x^7}{7\cdot 3!} + \cdots\right)$$

35.2 $$\operatorname{erf}(x) \sim 1 - \frac{e^{-x^2}}{\sqrt{\pi}\,x}\left(1 - \frac{1}{2x^2} + \frac{1\cdot 3}{(2x^2)^2} - \frac{1\cdot 3\cdot 5}{(2x^2)^3} + \cdots\right)$$

35.3 $$\operatorname{erf}(-x) = -\operatorname{erf}(x), \quad \operatorname{erf}(0) = 0, \quad \operatorname{erf}(\infty) = 1$$

COMPLEMENTARY ERROR FUNCTION $\operatorname{erfc}(x) = 1 - \operatorname{erf}(x) = \dfrac{2}{\sqrt{\pi}}\displaystyle\int_x^\infty e^{-u^2}\,du$

35.4 $$\operatorname{erfc}(x) = 1 - \frac{2}{\sqrt{\pi}}\left(x - \frac{x^3}{3\cdot 1!} + \frac{x^5}{5\cdot 2!} - \frac{x^7}{7\cdot 3!} + \cdots\right)$$

35.5 $$\operatorname{erfc}(x) \sim \frac{e^{-x^2}}{\sqrt{\pi}\,x}\left(1 - \frac{1}{2x^2} + \frac{1\cdot 3}{(2x^2)^2} - \frac{1\cdot 3\cdot 5}{(2x^2)^3} + \cdots\right)$$

35.6 $$\operatorname{erfc}(0) = 1, \quad \operatorname{erfc}(\infty) = 0$$

EXPONENTIAL INTEGRAL $Ei(x) = \displaystyle\int_x^\infty \frac{e^{-u}}{u}\,du$

35.7 $$Ei(x) = -\gamma - \ln x + \int_0^x \frac{1 - e^{-u}}{u}\,du$$

35.8 $$Ei(x) = -\gamma - \ln x + \left(\frac{x}{1\cdot 1!} - \frac{x^2}{2\cdot 2!} + \frac{x^3}{3\cdot 3!} - \cdots\right)$$

35.9 $$Ei(x) \sim \frac{e^{-x}}{x}\left(1 - \frac{1!}{x} + \frac{2!}{x^2} - \frac{3!}{x^3} + \cdots\right)$$

35.10 $$Ei(\infty) = 0$$

SINE INTEGRAL $Si(x) = \displaystyle\int_0^x \frac{\sin u}{u}\,du$

35.11 $$Si(x) = \frac{x}{1\cdot 1!} - \frac{x^3}{3\cdot 3!} + \frac{x^5}{5\cdot 5!} - \frac{x^7}{7\cdot 7!} + \cdots$$

35.12 $$Si(x) \sim \frac{\pi}{2} - \frac{\sin x}{x}\left(\frac{1}{x} - \frac{3!}{x^3} + \frac{5!}{x^5} - \cdots\right) - \frac{\cos x}{x}\left(1 - \frac{2!}{x^2} + \frac{4!}{x^4} - \cdots\right)$$

35.13 $$Si(-x) = -Si(x), \quad Si(0) = 0, \quad Si(\infty) = \pi/2$$

COSINE INTEGRAL $Ci(x) = \int_x^\infty \frac{\cos u}{u}\,du$

35.14 $$Ci(x) = -\gamma - \ln x + \int_0^x \frac{1-\cos u}{u}\,du$$

35.15 $$Ci(x) = -\gamma - \ln x + \frac{x^2}{2\cdot 2!} - \frac{x^4}{4\cdot 4!} + \frac{x^6}{6\cdot 6!} - \frac{x^8}{8\cdot 8!} + \cdots$$

35.16 $$Ci(x) \sim \frac{\cos x}{x}\left(\frac{1}{x} - \frac{3!}{x^3} + \frac{5!}{x^5} - \cdots\right) - \frac{\sin x}{x}\left(1 - \frac{2!}{x^2} + \frac{4!}{x^4} - \cdots\right)$$

35.17 $$Ci(\infty) = 0$$

FRESNEL SINE INTEGRAL $S(x) = \sqrt{\frac{2}{\pi}}\int_0^x \sin u^2\,du$

35.18 $$S(x) = \sqrt{\frac{2}{\pi}}\left(\frac{x^3}{3\cdot 1!} - \frac{x^7}{7\cdot 3!} + \frac{x^{11}}{11\cdot 5!} - \frac{x^{15}}{15\cdot 7!} + \cdots\right)$$

35.19 $$S(x) \sim \frac{1}{2} - \frac{1}{\sqrt{2\pi}}\left\{(\cos x^2)\left(\frac{1}{x} - \frac{1\cdot 3}{2^2x^5} + \frac{1\cdot 3\cdot 5\cdot 7}{2^4x^9} - \cdots\right) + (\sin x^2)\left(\frac{1}{2x^3} - \frac{1\cdot 3\cdot 5}{2^3x^7} + \cdots\right)\right\}$$

35.20 $$S(-x) = -S(x), \quad S(0) = 0, \quad S(\infty) = \frac{1}{2}$$

FRESNEL COSINE INTEGRAL $C(x) = \sqrt{\frac{2}{\pi}}\int_0^x \cos u^2\,du$

35.21 $$C(x) = \sqrt{\frac{2}{\pi}}\left(\frac{x}{1!} - \frac{x^5}{5\cdot 2!} + \frac{x^9}{9\cdot 4!} - \frac{x^{13}}{13\cdot 6!} + \cdots\right)$$

35.22 $$C(x) \sim \frac{1}{2} + \frac{1}{\sqrt{2\pi}}\left\{(\sin x^2)\left(\frac{1}{x} - \frac{1\cdot 3}{2^2x^5} + \frac{1\cdot 3\cdot 5\cdot 7}{2^4x^9} - \cdots\right) - (\cos x^2)\left(\frac{1}{2x^3} - \frac{1\cdot 3\cdot 5}{2^3x^7} + \cdots\right)\right\}$$

35.23 $$C(-x) = -C(x), \quad C(0) = 0, \quad C(\infty) = \frac{1}{2}$$

RIEMANN ZETA FUNCTION $\zeta(x) = \frac{1}{1^x} + \frac{1}{2^x} + \frac{1}{3^x} + \cdots$

35.24 $$\zeta(x) = \frac{1}{\Gamma(x)}\int_0^\infty \frac{u^{x-1}}{e^u - 1}\,du, \qquad x > 1$$

35.25 $$\zeta(1-x) = 2^{1-x}\pi^{-x}\,\Gamma(x)\cos(\pi x/2)\,\zeta(x) \qquad \text{[extension to other values]}$$

35.26 $$\zeta(2k) = \frac{2^{2k-1}\pi^{2k}B_k}{(2k)!} \qquad k = 1, 2, 3, \ldots$$

36 INEQUALITIES

TRIANGLE INEQUALITY

36.1 $$|a_1| - |a_2| \leqq |a_1 + a_2| \leqq |a_1| + |a_2|$$

36.2 $$|a_1 + a_2 + \cdots + a_n| \leqq |a_1| + |a_2| + \cdots + |a_n|$$

CAUCHY-SCHWARZ INEQUALITY

36.3 $$|a_1b_1 + a_2b_2 + \cdots + a_nb_n|^2 \leqq (|a_1|^2 + |a_2|^2 + \cdots + |a_n|^2)(|b_1|^2 + |b_2|^2 + \cdots + |b_n|^2)$$

The equality holds if and only if $a_1/b_1 = a_2/b_2 = \cdots = a_n/b_n$.

INEQUALITIES INVOLVING ARITHMETIC, GEOMETRIC AND HARMONIC MEANS

If A, G and H are the arithmetic, geometric and harmonic means of the positive numbers $a_1, a_2, \ldots, a_n$, then

36.4 $$H \leqq G \leqq A$$

where

36.5 $$A = \frac{a_1 + a_2 + \cdots + a_n}{n}$$

36.6 $$G = \sqrt[n]{a_1 a_2 \ldots a_n}$$

36.7 $$\frac{1}{H} = \frac{1}{n}\left(\frac{1}{a_1} + \frac{1}{a_2} + \cdots + \frac{1}{a_n}\right)$$

The equality holds if and only if $a_1 = a_2 = \cdots = a_n$.

HOLDER'S INEQUALITY

36.8 $$|a_1b_1 + a_2b_2 + \cdots + a_nb_n| \leqq (|a_1|^p + |a_2|^p + \cdots + |a_n|^p)^{1/p}(|b_1|^q + |b_2|^q + \cdots + |b_n|^q)^{1/q}$$

where

36.9 $$\frac{1}{p} + \frac{1}{q} = 1 \qquad p > 1,\ q > 1$$

The equality holds if and only if $|a_1|^{p-1}/|b_1| = |a_2|^{p-1}/|b_2| = \cdots = |a_n|^{p-1}/|b_n|$. For $p = q = 2$ it reduces to 36.3.

CHEBYSHEV'S INEQUALITY

If $a_1 \geqq a_2 \geqq \cdots \geqq a_n$ and $b_1 \geqq b_2 \geqq \cdots \geqq b_n$, then

36.10 $$\left(\frac{a_1 + a_2 + \cdots + a_n}{n}\right)\left(\frac{b_1 + b_2 + \cdots + b_n}{n}\right) \leqq \frac{a_1b_1 + a_2b_2 + \cdots + a_nb_n}{n}$$

or

36.11 $$(a_1 + a_2 + \cdots + a_n)(b_1 + b_2 + \cdots + b_n) \leqq n(a_1b_1 + a_2b_2 + \cdots + a_nb_n)$$

MINKOWSKI'S INEQUALITY

If $a_1, a_2, \ldots, a_n, b_1, b_2, \ldots b_n$ are all positive and $p > 1$, then

36.12 $$\{(a_1 + b_1)^p + (a_2 + b_2)^p + \cdots + (a_n + b_n)^p\}^{1/p} \leqq (a_1^p + a_2^p + \cdots + a_n^p)^{1/p} + (b_1^p + b_2^p + \cdots + b_n^p)^{1/p}$$

The equality holds if and only if $a_1/b_1 = a_2/b_2 = \cdots = a_n/b_n$.

CAUCHY-SCHWARZ INEQUALITY FOR INTEGRALS

36.13 $$\left|\int_a^b f(x)\, g(x)\, dx\right|^2 \leqq \left\{\int_a^b |f(x)|^2\, dx\right\}\left\{\int_a^b |g(x)|^2\, dx\right\}$$

The equality holds if and only if $f(x)/g(x)$ is a constant.

HOLDER'S INEQUALITY FOR INTEGRALS

36.14 $$\int_a^b |f(x)\, g(x)|\, dx \leqq \left\{\int_a^b |f(x)|^p\, dx\right\}^{1/p}\left\{\int_a^b |g(x)|^q\, dx\right\}^{1/q}$$

where $1/p + 1/q = 1$, $p > 1$, $q > 1$. If $p = q = 2$, this reduces to 36.13.

The equality holds if and only if $|f(x)|^{p-1}/|g(x)|$ is a constant.

MINKOWSKI'S INEQUALITY FOR INTEGRALS

If $p > 1$,

36.15 $$\left\{\int_a^b |f(x) + g(x)|^p\, dx\right\}^{1/p} \leqq \left\{\int_a^b |f(x)|^p\, dx\right\}^{1/p} + \left\{\int_a^b |g(x)|^p\, dx\right\}^{1/p}$$

The equality holds if and only if $f(x)/g(x)$ is a constant.

37 PARTIAL FRACTION EXPANSIONS

37.1 $\cot x = \frac{1}{x} + 2x\left\{\frac{1}{x^2-\pi^2} + \frac{1}{x^2-4\pi^2} + \frac{1}{x^2-9\pi^2} + \cdots\right\}$

37.2 $\csc x = \frac{1}{x} - 2x\left\{\frac{1}{x^2-\pi^2} - \frac{1}{x^2-4\pi^2} + \frac{1}{x^2-9\pi^2} - \cdots\right\}$

37.3 $\sec x = 4\pi\left\{\frac{1}{\pi^2-4x^2} - \frac{3}{9\pi^2-4x^2} + \frac{5}{25\pi^2-4x^2} - \cdots\right\}$

37.4 $\tan x = 8x\left\{\frac{1}{\pi^2-4x^2} + \frac{1}{9\pi^2-4x^2} + \frac{1}{25\pi^2-4x^2} + \cdots\right\}$

37.5 $\sec^2 x = 4\left\{\frac{1}{(\pi-2x)^2} + \frac{1}{(\pi+2x)^2} + \frac{1}{(3\pi-2x)^2} + \frac{1}{(3\pi+2x)^2} + \cdots\right\}$

37.6 $\csc^2 x = \frac{1}{x^2} + \frac{1}{(x-\pi)^2} + \frac{1}{(x+\pi)^2} + \frac{1}{(x-2\pi)^2} + \frac{1}{(x+2\pi)^2} + \cdots$

37.7 $\coth x = \frac{1}{x} + 2x\left\{\frac{1}{x^2+\pi^2} + \frac{1}{x^2+4\pi^2} + \frac{1}{x^2+9\pi^2} + \cdots\right\}$

37.8 $\operatorname{csch} x = \frac{1}{x} - 2x\left\{\frac{1}{x^2+\pi^2} - \frac{1}{x^2+4\pi^2} + \frac{1}{x^2+9\pi^2} - \cdots\right\}$

37.9 $\operatorname{sech} x = 4\pi\left\{\frac{1}{\pi^2+4x^2} - \frac{3}{9\pi^2+4x^2} + \frac{5}{25\pi^2+4x^2} - \cdots\right\}$

37.10 $\tanh x = 8x\left\{\frac{1}{\pi^2+4x^2} + \frac{1}{9\pi^2+4x^2} + \frac{1}{25\pi^2+4x^2} + \cdots\right\}$

38 INFINITE PRODUCTS

38.1 $\sin x = x\left(1-\frac{x^2}{\pi^2}\right)\left(1-\frac{x^2}{4\pi^2}\right)\left(1-\frac{x^2}{9\pi^2}\right)\cdots$

38.2 $\cos x = \left(1-\frac{4x^2}{\pi^2}\right)\left(1-\frac{4x^2}{9\pi^2}\right)\left(1-\frac{4x^2}{25\pi^2}\right)\cdots$

38.3 $\sinh x = x\left(1+\frac{x^2}{\pi^2}\right)\left(1+\frac{x^2}{4\pi^2}\right)\left(1+\frac{x^2}{9\pi^2}\right)\cdots$

38.4 $\cosh x = \left(1+\frac{4x^2}{\pi^2}\right)\left(1+\frac{4x^2}{9\pi^2}\right)\left(1+\frac{4x^2}{25\pi^2}\right)\cdots$

38.5 $\frac{1}{\Gamma(x)} = xe^{\gamma x}\left\{\left(1+\frac{x}{1}\right)e^{-x}\right\}\left\{\left(1+\frac{x}{2}\right)e^{-x/2}\right\}\left\{\left(1+\frac{x}{3}\right)e^{-x/3}\right\}\cdots$

See also 16.12, page 102.

38.6 $J_0(x) = \left(1-\frac{x^2}{\lambda_1^2}\right)\left(1-\frac{x^2}{\lambda_2^2}\right)\left(1-\frac{x^2}{\lambda_3^2}\right)\cdots$

where $\lambda_1, \lambda_2, \lambda_3, \ldots$ are the positive roots of $J_0(x) = 0$.

38.7 $J_1(x) = x\left(1-\frac{x^2}{\lambda_1^2}\right)\left(1-\frac{x^2}{\lambda_2^2}\right)\left(1-\frac{x^2}{\lambda_3^2}\right)\cdots$

where $\lambda_1, \lambda_2, \lambda_3, \ldots$ are the positive roots of $J_1(x) = 0$.

38.8 $\frac{\sin x}{x} = \cos\frac{x}{2}\cos\frac{x}{4}\cos\frac{x}{8}\cos\frac{x}{16}\cdots$

38.9 $\frac{\pi}{2} = \frac{2}{1}\cdot\frac{2}{3}\cdot\frac{4}{3}\cdot\frac{4}{5}\cdot\frac{6}{5}\cdot\frac{6}{7}\cdot\cdots$

This is called *Wallis' product.*

39 PROBABILITY DISTRIBUTIONS

BINOMIAL DISTRIBUTION

39.1 $$\Phi(x) = \sum_{t \leqq x} \binom{n}{t} p^t q^{n-t} \qquad p > 0,\; q > 0,\; p + q = 1$$

POISSON DISTRIBUTION

39.2 $$\Phi(x) = \sum_{t \leqq x} \frac{\lambda^t e^{-\lambda}}{t!} \qquad \lambda > 0$$

HYPERGEOMETRIC DISTRIBUTION

39.3 $$\Phi(x) = \sum_{t \leqq x} \frac{\binom{r}{t}\binom{s}{n-t}}{\binom{r+s}{n}}$$

NORMAL DISTRIBUTION

39.4 $$\Phi(x) = \frac{1}{\sqrt{2\pi}} \int_{-\infty}^{x} e^{-t^2}\, dt$$

STUDENT'S t DISTRIBUTION

39.5 $$\Phi(x) = \frac{1}{\sqrt{n\pi}} \frac{\Gamma\left(\frac{n+1}{2}\right)}{\Gamma(n/2)} \int_{-\infty}^{x} \left(1 + \frac{t^2}{n}\right)^{-(n+1)/2} dt$$

CHI SQUARE DISTRIBUTION

39.6 $$\Phi(x) = \frac{1}{2^{n/2}\,\Gamma(n/2)} \int_{0}^{x} t^{(n-2)/2} e^{-t/2}\, dt$$

F DISTRIBUTION

39.7 $$\Phi(x) = \frac{\Gamma\left(\frac{n_1+n_2}{2}\right) n_1^{n_1/2} n_2^{n_2/2}}{\Gamma(n_1/2)\,\Gamma(n_2/2)} \int_{0}^{x} t^{n_1/2}(n_2 + n_1 t)^{-(n_1+n_2)/2}\, dt$$

40 SPECIAL MOMENTS OF INERTIA

The table below shows the moments of inertia of various rigid bodies of mass M. In all cases it is assumed the body has uniform [i.e. constant] density.

TYPE OF RIGID BODY	MOMENT OF INERTIA
40.1 Thin rod of length a	
(a) about axis perpendicular to the rod through the center of mass,	$\frac{1}{12}Ma^2$
(b) about axis perpendicular to the rod through one end.	$\frac{1}{3}Ma^2$
40.2 Rectangular parallelepiped with sides a, b, c	
(a) about axis parallel to c and through center of face ab,	$\frac{1}{12}M(a^2+b^2)$
(b) about axis through center of face bc and parallel to c.	$\frac{1}{12}M(4a^2+b^2)$
40.3 Thin rectangular plate with sides a, b	
(a) about axis perpendicular to the plate through center,	$\frac{1}{12}M(a^2+b^2)$
(b) about axis parallel to side b through center.	$\frac{1}{12}Ma^2$
40.4 Circular cylinder of radius a and height h	
(a) about axis of cylinder,	$\frac{1}{2}Ma^2$
(b) about axis through center of mass and perpendicular to cylindrical axis,	$\frac{1}{12}M(h^2+3a^2)$
(c) about axis coinciding with diameter at one end.	$\frac{1}{12}M(4h^2+3a^2)$
40.5 Hollow circular cylinder of outer radius a, inner radius b and height h	
(a) about axis of cylinder,	$\frac{1}{2}M(a^2+b^2)$
(b) about axis through center of mass and perpendicular to cylindrical axis,	$\frac{1}{12}M(3a^2+3b^2+h^2)$
(c) about axis coinciding with diameter at one end.	$\frac{1}{12}M(3a^2+3b^2+4h^2)$

40.6 Circular plate of radius a	
(a) about axis perpendicular to plate through center,	$\frac{1}{2}Ma^2$
(b) about axis coinciding with a diameter.	$\frac{1}{4}Ma^2$
40.7 Hollow circular plate or ring with outer radius a and inner radius b	
(a) about axis perpendicular to plane of plate through center,	$\frac{1}{2}M(a^2+b^2)$
(b) about axis coinciding with a diameter.	$\frac{1}{4}M(a^2+b^2)$
40.8 Thin circular ring of radius a	
(a) about axis perpendicular to plane of ring through center,	Ma^2
(b) about axis coinciding with diameter.	$\frac{1}{2}Ma^2$
40.9 Sphere of radius a	
(a) about axis coinciding with a diameter,	$\frac{2}{5}Ma^2$
(b) about axis tangent to the surface.	$\frac{7}{5}Ma^2$
40.10 Hollow sphere of outer radius a and inner radius b	
(a) about axis coinciding with a diameter,	$\frac{2}{5}M(a^5-b^5)/(a^3-b^3)$
(b) about axis tangent to the surface.	$\frac{2}{5}M(a^5-b^5)/(a^3-b^3) + Ma^2$
40.11 Hollow spherical shell of radius a	
(a) about axis coinciding with a diameter,	Ma^2
(b) about axis tangent to the surface.	$2Ma^2$
40.12 Ellipsoid with semi-axes a, b, c	
(a) about axis coinciding with semi-axis c,	$\frac{1}{5}M(a^2+b^2)$
(b) about axis tangent to surface, parallel to semi-axis c and at distance a from center.	$\frac{1}{5}M(6a^2+b^2)$
40.13 Circular cone of radius a and height h	
(a) about axis of cone,	$\frac{3}{10}Ma^2$
(b) about axis through vertex and perpendicular to axis,	$\frac{3}{20}M(a^2+4h^2)$
(c) about axis through center of mass and perpendicular to axis.	$\frac{3}{80}M(4a^2+h^2)$
40.14 Torus with outer radius a and inner radius b	
(a) about axis through center of mass and perpendicular to plane of torus,	$\frac{1}{4}M(7a^2-6ab+3b^2)$
(b) about axis through center of mass and in the plane of the torus.	$\frac{1}{4}M(9a^2-10ab+5b^2)$

41 CONVERSION FACTORS

Length

1 kilometer (km)	=	1000 meters (m)
1 meter (m)	=	100 centimeters (cm)
1 centimeter (cm)	=	10^{-2} m
1 millimeter (mm)	=	10^{-3} m
1 micron (μ)	=	10^{-6} m
1 millimicron (mμ)	=	10^{-9} m
1 angstrom (A)	=	10^{-10} m
1 inch (in.)	=	2.540 cm
1 foot (ft)	=	30.48 cm
1 mile (mi)	=	1.609 km
1 mil	=	10^{-3} in.
1 centimeter	=	0.3937 in.
1 meter	=	39.37 in.
1 kilometer	=	0.6214 mile

Area

1 square meter (m^2) = 10.76 ft^2
1 square foot (ft^2) = 929 cm^2
1 square mile (mi^2) = 640 acres
1 acre = 43,560 ft^2

Volume

1 liter (l) = 1000 cm^3 = 1.057 quart (qt) = 61.02 in^3 = 0.03532 ft^3
1 cubic meter (m^3) = 1000 l = 35.32 ft^3
1 cubic foot (ft^3) = 7.481 U.S. gal = 0.02832 m^3 = 28.32 l
1 U.S. gallon (gal) = 231 in^3 = 3.785 l; 1 British gallon = 1.201 U.S. gallon = 277.4 in^3

Mass

1 kilogram (kg) = 2.2046 pounds (lb) = 0.06852 slug; 1 lb = 453.6 gm = 0.03108 slug
1 slug = 32.174 lb = 14.59 kg

Speed

1 km/hr = 0.2778 m/sec = 0.6214 mi/hr = 0.9113 ft/sec
1 mi/hr = 1.467 ft/sec = 1.609 km/hr = 0.4470 m/sec

Density

1 gm/cm^3 = 10^3 kg/m^3 = 62.43 lb/ft^3 = 1.940 slug/ft^3
1 lb/ft^3 = 0.01602 gm/cm^3; 1 slug/ft^3 = 0.5154 gm/cm^3

Force

1 newton (nt) = 10^5 dynes = 0.1020 kgwt = 0.2248 lbwt
1 pound weight (lbwt) = 4.448 nt = 0.4536 kgwt = 32.17 poundals
1 kilogram weight (kgwt) = 2.205 lbwt = 9.807 nt
1 U.S. short ton = 2000 lbwt; 1 long ton = 2240 lbwt; 1 metric ton = 2205 lbwt

Energy

1 joule = 1 nt m = 10^7 ergs = 0.7376 ft lbwt = 0.2389 cal = 9.481×10^{-4} Btu
1 ft lbwt = 1.356 joules = 0.3239 cal = 1.285×10^{-3} Btu
1 calorie (cal) = 4.186 joules = 3.087 ft lbwt = 3.968×10^{-3} Btu
1 Btu (British thermal unit) = 778 ft lbwt = 1055 joules = 0.293 watt hr
1 kilowatt hour (kw hr) = 3.60×10^6 joules = 860.0 kcal = 3413 Btu
1 electron volt (ev) = 1.602×10^{-19} joule

Power

1 watt = 1 joule/sec = 10^7 ergs/sec = 0.2389 cal/sec
1 horsepower (hp) = 550 ft lbwt/sec = 33,000 ft lbwt/min = 745.7 watts
1 kilowatt (kw) = 1.341 hp = 737.6 ft lbwt/sec = 0.9483 Btu/sec

Pressure

1 nt/m^2 = 10 dynes/cm^2 = 9.869×10^{-6} atmosphere = 2.089×10^{-2} lbwt/ft^2
1 lbwt/in^2 = 6895 nt/m^2 = 5.171 cm mercury = 27.68 in. water
1 atmosphere (atm) = 1.013×10^5 nt/m^2 = 1.013×10^6 dynes/cm^2 = 14.70 lbwt/in^2
= 76 cm mercury = 406.8 in. water

Part II

TABLES

SAMPLE PROBLEMS

ILLUSTRATING USE OF THE TABLES

COMMON LOGARITHMS

1. Find log 2.36.

We must find the number p such that $10^p = 2.36 = N$. Since $10^0 = 1$ and $10^1 = 10$, p lies between 0 and 1 and can be found from the tables of common logarithms on page 202.

Thus to find log 2.36 we glance down the *left* column headed N until we come to the first two digits, 23. Then we proceed *right* to the column headed 6. We find the entry 3729. Thus $\log 2.36 = 0.3729$, i.e. $2.36 = 10^{0.3729}$.

2. Find (*a*) log 23.6, (*b*) log 236, (*c*) log 2360.

From Problem 1, $2.36 = 10^{0.3729}$. Then multiplying successively by 10 we have

$$23.6 = 10^{1.3729}, \quad 236 = 10^{2.3729}, \quad 2360 = 10^{3.3729}$$

Thus

(*a*) $\log 23.6 = 1.3729$

(*b*) $\log 236 = 2.3729$

(*c*) $\log 2360 = 3.3729$.

The number .3729 obtained from the table is called the *mantissa* of the logarithm. The number before the decimal point is called the *characteristic*. Thus in (*b*) the characteristic is 2.

The following rule is easily demonstrated.

Rule 1. For a number greater than 1, the characteristic is one less than the number of digits before the decimal point. For example since 2360 has four digits before the decimal point, the characteristic is $4 - 1 = 3$.

3. Find (*a*) log .236, (*b*) log .0236, (*c*) log .00236.

From Problem 1, $2.36 = 10^{0.3729}$. Then dividing successively by 10 we have

$$.236 = 10^{0.3729-1} = 10^{9.3729-10} = 10^{-.6271}$$

$$.0236 = 10^{0.3729-2} = 10^{8.3729-10} = 10^{-1.6271}$$

$$.00236 = 10^{0.3729-3} = 10^{7.3729-10} = 10^{-2.6271}$$

Then

(*a*) $\log .236 = 9.3729 - 10 = -.6271$

(*b*) $\log .0236 = 8.3729 - 10 = -1.6271$

(*c*) $\log .00236 = 7.3729 - 10 = -2.6271$.

The number .3729 is the mantissa of the logarithm. The number apart from the mantissa [for example $9 - 10$, $8 - 10$ or $7 - 10$] is the characteristic.

The following rule is easily demonstrated.

Rule 2. For a positive number less than 1, the characteristic is negative and numerically one more than the number of zeros immediately following the decimal point. For example since .00236 has two zeros immediately following the decimal point, the characteristic is -3 or $7 - 10$.

4. Verify each of the following logarithms.

(*a*) $\log 87.2$. Mantissa $= .9405$, characteristic $= 1$; then $\log 87.2 = 1.9405$.

(*b*) $\log 395{,}000 = 5.5966$.

(*c*) $\log .0482$. Mantissa $= .6830$, characteristic $= 8 - 10$; then $\log_{10} .0482 = 8.6830 - 10$.

(*d*) $\log .000827 = 6.9175 - 10$.

5. Find $\log 4.638$.

Since the number has four digits, we must use interpolation to find the mantissa. The mantissa of $\log 4638$ is .8 of the way between the mantissas of $\log 4630$ and $\log 4640$.

$$\begin{aligned} \text{Mantissa of } \log 4640 &= .6665 \\ \text{Mantissa of } \log 4630 &= \underline{.6656} \\ \text{Tabular difference} &= .0009 \end{aligned} \qquad \begin{aligned} \text{Mantissa of } \log 4.638 &= .6656 + (.8)(.0009) \\ &= .6663 \text{ to four digits} \\ \text{Then } \log 4.638 &= 0.6663 \end{aligned}$$

If desired the proportional parts table on page 202 can be used to give the mantissa directly $(6656 + 7)$.

6. Verify each of the following logarithms.

(*a*) $\log 183.2 = 2.2630$ $\quad (2625 + 5)$

(*b*) $\log 87{,}640 = 4.9427$ $\quad (9425 + 2)$

(*c*) $\log .2548 = 9.4062 - 10$ $\quad (4048 + 14)$

(*d*) $\log .009848 = 7.9933 - 10$ $\quad (9930 + 3)$

COMMON ANTILOGARITHMS

7. Find (*a*) antilog 1.7530, (*b*) antilog $(7.7530 - 10)$.

(*a*) We must find the value of $10^{1.7530}$. Since the mantissa is .7530 we glance down the *left* column headed p in the table on page 205 until we come to the first two digits 75. Then we proceed *right* to the column headed 3. We find the entry 5662. Since the characteristic is 1, there are two digits before the decimal point. Then the required number is 56.62.

(*b*) As in part (*a*) we find the entry 5662 corresponding to the mantissa .7530. Then since the characteristic is $7 - 10$, the number must have two zeros immediately following the decimal point. Thus the required number is .005662.

8. Find antilog $(9.3842 - 10)$.

The mantissa .3842 lies between .3840 and .3850 and we must use interpolation. From the table on page 204 we have

$$\begin{aligned} \text{Number corresponding to } .3850 &= 2427 \\ \text{Number corresponding to } .3840 &= \underline{2421} \\ \text{Tabular difference} &= \quad 6 \end{aligned} \qquad \begin{aligned} \text{Given mantissa} &= .3842 \\ \text{Next smaller mantissa} &= \underline{.3840} \\ \text{Difference} &= .0002 \end{aligned}$$

Then $2421 + \frac{2}{10}(2427 - 2421) = 2422$ to four digits, and the required number is 0.2422.

The proportional parts table on page 204 can also be used.

9. Verify each of the following antilogarithm.

(*a*) antilog $2.6715 = 469.3$

(*b*) antilog $9.6089 - 10 = .4063$

(*c*) antilog $4.2023 = 15{,}930$

COMPUTATIONS USING LOGARITHMS

10. $P = \dfrac{(784.6)(.0431)}{28.23}$. $\log P = \log 784.6 + \log .0431 - \log 28.23$.

$$\begin{aligned}
\log 784.6 &= 2.8947 \\
(+)\ \log .0431 &= \underline{8.6345-10} \\
&\ 11.5292-10 \\
(-)\ \log 28.23 &= \underline{1.4507} \\
\log P &= 10.0785-10 = .0785. \quad \text{Then } P = 1.198.
\end{aligned}$$

Note the exponential significance of the computation, i.e.

$$\frac{(784.6)(.0431)}{28.23} = \frac{(10^{2.8947})(10^{8.6345-10})}{10^{1.4507}} = 10^{2.8947+8.6345-10-1.4507} = 10^{.0785} = 1.198$$

11. $P = (5.395)^8$. $\log P = 8 \log 5.395 = 8(0.7320) = 5.8560$, and $P = 717{,}800$.

12. $P = \sqrt{387.2} = (387.2)^{1/2}$. $\log P = \frac{1}{2}\log 387.2 = \frac{1}{2}(2.5879) = 1.2940$ and $P = 19.68$.

13. $P = \sqrt[5]{.08317} = (.08317)^{1/5}$. $\log P = \frac{1}{5}\log .08317 = \frac{1}{5}(8.9200-10) = \frac{1}{5}(48.9200-50) = 9.7840-10$ and $P = .6081$.

14. $P = \dfrac{\sqrt{.003654}\,(18.37)^3}{(8.724)^4\sqrt[4]{743.8}}$. $\log P = \frac{1}{2}\log .003654 + 3\log 18.37 - (4\log 8.724 + \frac{1}{4}\log 743.8)$

Numerator N

$$\begin{aligned}
\tfrac{1}{2}\log .003654 &= \tfrac{1}{2}(7.5628-10) \\
&= \tfrac{1}{2}(17.5628-20) = 8.7814-10 \\
3\log 18.37 &= 3(1.2641) = \underline{3.7923} \\
\text{Add:} \quad \log N &= 12.5737-10
\end{aligned}$$

Denominator D

$$\begin{aligned}
4\log 8.724 &= 4(0.9407) = 3.7628 \\
\tfrac{1}{4}\log 743.6 &= \tfrac{1}{4}(2.8714) = \underline{0.7178} \\
\text{Add:} \quad \log D &= 4.4806
\end{aligned}$$

$$\begin{aligned}
\log N &= 12.5737-10 \\
(-)\ \log D &= \underline{4.4806} \\
\log P &= 8.0931-10. \quad \text{Then } P = .01239
\end{aligned}$$

NATURAL OR NAPIERIAN LOGARITHMS

15. Find (*a*) ln 7.236, (*b*) ln 836.2, (*c*) ln .002548.

(*a*) Use the table on page 225.

$$\begin{aligned}
\ln 7.240 &= 1.97962 \\
\ln 7.230 &= \underline{1.97824} \\
\text{Tabular difference} &= .00138
\end{aligned}$$

Then $\ln 7.236 = 1.97824 + \frac{6}{10}(.00138) = 1.97907$

In terms of exponentials this means that $e^{1.97907} = 7.236$.

(*b*) As in part (*a*) we find

$$\ln 8.362 = 2.12346 + \tfrac{2}{10}(2.12465 - 2.12346) = 2.12370$$

Then

$$\ln 836.2 = \ln(8.362 \times 10^2) = \log 8.362 + 2\ln 10 = 2.12370 + 4.60517 = 6.72887$$

In terms of exponentials this means that $e^{6.72887} = 836.2$.

(*c*) As in part (*a*) we find

$$\ln 2.548 = 0.93216 + \tfrac{8}{10}(0.93609 - 0.93216) = 0.93530$$

Then

$$\ln .002548 = \ln(2.548 \times 10^{-3}) = \ln 2.548 - 3\ln 10 = 0.93530 - 6.90776 = -5.97246$$

In terms of exponentials this means that $e^{-5.97246} = .002548$.

TRIGONOMETRIC FUNCTIONS (DEGREES AND MINUTES)

16. Find (*a*) $\sin 74°23'$, (*b*) $\cos 35°42'$, (*c*) $\tan 82°56'$.

(*a*) Refer to the table on page 206.

$$\begin{aligned} \sin 74°30' &= .9636 \\ \sin 74°20' &= \underline{.9628} \\ \text{Tabular difference} &= .0008 \end{aligned}$$

Then $\sin 74°23' = .9628 + \frac{3}{10}(.0008) = .9630$

(*b*) Refer to the table on page 207.

$$\begin{aligned} \cos 35°40' &= .8124 \\ \cos 35°50' &= \underline{.8107} \\ \text{Tabular difference} &= .0017 \end{aligned}$$

Then $\cos 35°42' = .8124 - \frac{2}{10}(.0017) = .8121$

or $\cos 35°42' = .8107 + \frac{8}{10}(.0017) = .8121$

(*c*) Refer to the table on page 208.

$$\begin{aligned} \tan 82°60' = \tan 83°0' &= 8.1443 \\ \tan 82°50' &= \underline{7.9530} \\ \text{Tabular difference} &= .1913 \end{aligned}$$

Then $\tan 82°56' = 7.9530 + \frac{6}{10}(.1913) = 8.0678$

17. Find (*a*) $\cot 45°16'$, (*b*) $\sec 73°48'$, (*c*) $\csc 28°33'$.

(*a*) Refer to the table on page 209.

$$\begin{aligned} \cot 45°10' &= .9942 \\ \cot 45°20' &= \underline{.9884} \\ \text{Tabular difference} &= .0058 \end{aligned}$$

Then $\cot 45°16' = .9942 - \frac{6}{10}(.0058) = .9907$

or $\cot 45°16' = .9884 + \frac{4}{10}(.0058) = .9907$

(*b*) Refer to the table on page 210.

$$\begin{aligned} \sec 73°50' &= 3.592 \\ \sec 73°40' &= \underline{3.556} \\ \text{Tabular difference} &= .036 \end{aligned}$$

Then $\sec 73°48' = 3.556 + \frac{8}{10}(.036) = 3.585$

(*c*) Refer to the table on page 211.

$$\begin{aligned} \csc 28°30' &= 2.096 \\ \csc 28°40' &= \underline{2.085} \\ \text{Tabular difference} &= .011 \end{aligned}$$

Then $\csc 28°33' = 2.096 - \frac{3}{10}(.011) = 2.093$

or $\csc 28°33' = 2.085 + \frac{7}{10}(.011) = 2.093$

INVERSE TRIGONOMETRIC FUNCTIONS (DEGREES AND MINUTES)

18. Find (*a*) $\sin^{-1}(.2143)$, (*b*) $\cos^{-1}(.5412)$, (*c*) $\tan^{-1}(1.1536)$.

(*a*) Refer to the table on page 206.

$$\begin{aligned} \sin 12°30' &= .2164 \\ \sin 12°20' &= \underline{.2136} \\ \text{Tabular difference} &= .0028 \end{aligned}$$

Since .2143 is $\dfrac{.2143 - .2136}{.0028} = \frac{1}{4}$ of the way between .2136 and .2164, the required angle is $12°20' + \frac{1}{4}(10') = 12°22.5'$.

(*b*) Refer to the table on page 207.

$$\begin{aligned} \cos 57°10' &= .5422 \\ \cos 57°20' &= \underline{.5398} \\ \text{Tabular difference} &= .0024 \end{aligned}$$

Then $\cos^{-1}(.5412) = 57°20' - \dfrac{.5412 - .5398}{.0024}(10') = 57°14.2'$

or $\cos^{-1}(.5412) = 57°10' + \dfrac{.5422 - .5412}{.0024}(10') = 57°14.2'$

(*c*) Refer to the table on page 208.

$$\begin{aligned} \tan 49°10' &= 1.1571 \\ \tan 49°0' &= \underline{1.1504} \\ \text{Tabular difference} &= .0067 \end{aligned}$$

Then $\tan^{-1}(1.1536) = 49°0' + \dfrac{1.1536 - 1.1504}{.0067}(10') = 49°4.8'$

Other inverse trigonometric functions can be obtained similarly.

TRIGONOMETRIC AND INVERSE TRIGONOMETRIC FUNCTIONS (RADIANS)

19. Find (*a*) $\sin(.627)$, (*b*) $\cos(1.056)$, (*c*) $\tan(.153)$.

(*a*) Refer to the table on page 213.

$$\begin{aligned} \sin(.630) &= .58914 \\ \sin(.620) &= \underline{.58104} \\ \text{Tabular difference} &= .00810 \end{aligned}$$

Then $\sin(.627) = .58104 + \frac{7}{10}(.00810) = .58671$

(*b*) Refer to the table on page 214.

$$\begin{aligned} \cos(1.050) &= .49757 \\ \cos(1.060) &= \underline{.48887} \\ \text{Tabular difference} &= .00870 \end{aligned}$$

Then $\cos(1.056) = .49757 - \frac{6}{10}(.00870) = .49235$

or $\cos(1.056) = .48887 + \frac{4}{10}(.00870) = .49235$

(*c*) Refer to the table on page 212.

$$\begin{aligned} \tan(.160) &= .16138 \\ \tan(.150) &= \underline{.15114} \\ \text{Tabular difference} &= .01024 \end{aligned}$$

Then $\tan(.153) = .15114 + \frac{3}{10}(.01024) = .15421$

Similarly other trigonometric functions are obtained.

20. Find $\sin^{-1}(.512)$ in radians.

Refer to the table on page 213.

$$\begin{aligned} \sin(.540) &= .51414 \\ \sin(.530) &= .50553 \\ \text{Tabular difference} &= .00861 \end{aligned}$$

Then
$$\sin^{-1}(.512) = .530 + \frac{.512 - .50553}{.00861}(.01) = .5375 \text{ radians}$$

Similarly the other inverse trigonometric functions are obtained.

COMMON LOGARITHMS OF TRIGONOMETRIC FUNCTIONS

21. Find (*a*) $\log \sin 63°17'$, (*b*) $\log \cos 48°44'$.

(*a*) Refer to the table on page 217.

$$\begin{aligned} \log \sin 63°20' &= 9.9512 - 10 \\ \log \sin 63°10' &= 9.9505 - 10 \\ \text{Tabular difference} &= .0007 \end{aligned}$$

Then
$$\log \sin 63°17' = 9.9505 - 10 + \tfrac{7}{10}(.0007) = 9.9510 - 10$$

(*b*) Refer to the table on page 219.

$$\begin{aligned} \log \cos 48°40' &= 9.8198 - 10 \\ \log \cos 48°50' &= 9.8184 - 10 \\ \text{Tabular difference} &= .0014 \end{aligned}$$

Then
$$\log \cos 48°44' = 9.8198 - 10 - \tfrac{4}{10}(.0014) = 9.8192 - 10$$

or
$$\log \cos 48°44' = 9.8184 - 10 + \tfrac{6}{10}(.0014) = 9.8192 - 10$$

Similarly we can find logarithms of other trigonometric functions. Note that $\log \sec x = -\log \cos x$, $\log \cot x = -\log \tan x$, $\log \csc x = -\log \sin x$.

22. If $\log \tan x = 9.6845 - 10$, find x.

Refer to the table on page 220.

$$\begin{aligned} \log \tan 25°50' &= 9.6850 - 10 \\ \log \tan 25°40' &= 9.6817 - 10 \\ \text{Tabular difference} &= .0033 \end{aligned}$$

Then
$$x = 25°40' + \frac{9.6845 - 9.6817}{.0033}(10') = 25°48.5'$$

CONVERSION OF DEGREES, MINUTES AND SECONDS TO RADIANS

23. Find $75°\,28'\,47''$ in radians.

Refer to the table on page 223.

$$\begin{aligned} 70° &= 1.221730 \text{ radians} \\ 5° &= .087267 \\ 20' &= .005818 \\ 8' &= .002327 \\ 40'' &= .000194 \\ 7'' &= .000034 \end{aligned}$$

Adding,
$$75°\,28'\,47'' = 1.317370 \text{ radians}$$

CONVERSION OF RADIANS TO DEGREES, MINUTES AND SECONDS

24. Find 2.547 radians in degrees, minutes and seconds.

Refer to the table on page 222.

$$\begin{aligned}
2 \text{ radians} &= 114^\circ\ 35'\ 29.6'' \\
.5 \phantom{\text{ radians}} &= 28^\circ\ 38'\ 52.4'' \\
.04 \phantom{\text{ radians}} &= 2^\circ\ 17'\ 30.6'' \\
.007 \phantom{\text{ radians}} &= 0^\circ\ 24'\ 3.9''
\end{aligned}$$

Adding, $\quad 2.547 \text{ radians} = 144^\circ\ 114'\ 116.5'' = 145^\circ\ 55'\ 56.5''$

CONVERSION OF RADIANS TO FRACTIONS OF A DEGREE

25. Find 1.382 radians in terms of degrees.

Refer to the table on page 222.

$$\begin{aligned}
1 \text{ radian} &= 57.2958^\circ \\
.3 \phantom{\text{ radian}} &= 17.1887^\circ \\
.08 \phantom{\text{ radian}} &= 4.5837^\circ \\
.002 \phantom{\text{ radian}} &= .1146^\circ
\end{aligned}$$

Adding, $\quad 1.382 \text{ radians} = 79.1828^\circ$

EXPONENTIAL AND HYPERBOLIC FUNCTIONS

26. Find (*a*) $e^{5.24}$, (*b*) $e^{-.158}$.

(*a*) Refer to the table on page 226.

$$\begin{aligned}
e^{5.30} &= 200.34 \\
e^{5.20} &= 181.27 \\
\text{Tabular difference} &= 19.07
\end{aligned}$$

Then $\quad e^{5.24} = 181.27 + \frac{4}{10}(19.07) = 188.90$

(*b*) Refer to the table on page 227.

$$\begin{aligned}
e^{-.150} &= .86071 \\
e^{-.160} &= .85214 \\
\text{Tabular difference} &= .00857
\end{aligned}$$

Then $\quad e^{-.158} = .86071 - \frac{8}{10}(.00857) = .85385$

or $\quad e^{-.158} = .85214 + \frac{2}{10}(.00857) = .85385$

27. Find (*a*) sinh (4.846), (*b*) sech (.163).

(*a*) Refer to the table on page 229.

$$\begin{aligned}
\sinh(4.850) &= 63.866 \\
\sinh(4.840) &= 63.231 \\
\text{Tabular difference} &= .635
\end{aligned}$$

Then $\quad \sinh(4.846) = 63.231 + \frac{6}{10}(.635) = 63.612$

(*b*) Refer to the table on page 230.

$$\begin{aligned}
\cosh(.170) &= 1.0145 \\
\cosh(.160) &= 1.0128 \\
\text{Tabular difference} &= .0017
\end{aligned}$$

Then $\quad \cosh(.163) = 1.0128 + \frac{3}{10}(.0017) = 1.0133$

and so $\quad \operatorname{sech}(.163) = \dfrac{1}{\cosh(.163)} = \dfrac{1}{1.0133} = .98687$

28. Find $\tanh^{-1}(.71423)$.

Refer to the table on page 232.

$$\begin{aligned} \tanh(.900) &= .71630 \\ \tanh(.890) &= .71139 \\ \text{Tabular difference} &= .00491 \end{aligned}$$

Then $$\tanh^{-1}(.71423) = .890 + \frac{.71423 - .71139}{.00491}(10) = .8958$$

INTEREST AND ANNUITIES

29. A man deposits $2800 in a bank which pays 5% compounded quarterly. What will the deposit amount to in 8 years?

There are $n = 8 \cdot 4 = 32$ payment periods at interest rate $r = .05/4 = .0125$ per period. Then the amount is

$$A = \$2800(1 + .0125)^{32} = \$2800(1.4881) = \$4166.68$$

using the table on page 240.

30. A man expects to receive $12,000 in 10 years. How much is that money worth now, considering interest at 6% compounded semi-annually?

We are asked for the present value P which will amount to $A = \$12{,}000$ in 10 years. Since there are $n = 10 \cdot 2 = 20$ payment periods at interest rate $r = .06/2 = .03$ per period, the present value is

$$P = \$12{,}000(1 + .03)^{-20} = \$12{,}000(.55368) = \$6644.16$$

using the table on page 241.

31. An investor has an annuity in which a payment of $500 is made at the end of each year. If interest is 4% compounded annually, what is the amount of the annuity after 20 years?

Here $r = .04$, $n = 20$ and the amount is [see table on page 242],

$$\$500\left[\frac{(1 + .04)^{20} - 1}{.04}\right] = \$500(29.7781) = \$14{,}889.05$$

32. What is the present value of an annuity of $120 at the end of each 3 months for 12 years at 6% compounded quarterly?

Here $n = 4 \cdot 12 = 48$ payment periods, $r = .06/4 = .015$ and the present value is

$$\$120\left[\frac{1 - (1.015)^{-48}}{.015}\right] = \$120(34.0426) = \$4085.11$$

using the table on page 243.

TABLE 1

FOUR PLACE COMMON LOGARITHMS

$\log_{10} N$ or $\log N$

N	0	1	2	3	4	5	6	7	8	9	Proportional Parts								
											1	2	3	4	5	6	7	8	9
10	0000	0043	0086	0128	0170	0212	0253	0294	0334	0374	4	8	12	17	21	25	29	33	37
11	0414	0453	0492	0531	0569	0607	0645	0682	0719	0755	4	8	11	15	19	23	26	30	34
12	0792	0828	0864	0899	0934	0969	1004	1038	1072	1106	3	7	10	14	17	21	24	28	31
13	1139	1173	1206	1239	1271	1303	1335	1367	1399	1430	3	6	10	13	16	19	23	26	29
14	1461	1492	1523	1553	1584	1614	1644	1673	1703	1732	3	6	9	12	15	18	21	24	27
15	1761	1790	1818	1847	1875	1903	1931	1959	1987	2014	3	6	8	11	14	17	20	22	25
16	2041	2068	2095	2122	2148	2175	2201	2227	2253	2279	3	5	8	11	13	16	18	21	24
17	2304	2330	2355	2380	2405	2430	2455	2480	2504	2529	2	5	7	10	12	15	17	20	22
18	2553	2577	2601	2625	2648	2672	2695	2718	2742	2765	2	5	7	9	12	14	16	19	21
19	2788	2810	2833	2856	2878	2900	2923	2945	2967	2989	2	4	7	9	11	13	16	18	20
20	3010	3032	3054	3075	3096	3118	3139	3160	3181	3201	2	4	6	8	11	13	15	17	19
21	3222	3243	3263	3284	3304	3324	3345	3365	3385	3404	2	4	6	8	10	12	14	16	18
22	3424	3444	3464	3483	3502	3522	3541	3560	3579	3598	2	4	6	8	10	12	14	15	17
23	3617	3636	3655	3674	3692	3711	3729	3747	3766	3784	2	4	6	7	9	11	13	15	17
24	3802	3820	3838	3856	3874	3892	3909	3927	3945	3962	2	4	5	7	9	11	12	14	16
25	3979	3997	4014	4031	4048	4065	4082	4099	4116	4133	2	3	5	7	9	10	12	14	15
26	4150	4166	4183	4200	4216	4232	4249	4265	4281	4298	2	3	5	7	8	10	11	13	15
27	4314	4330	4346	4362	4378	4393	4409	4425	4440	4456	2	3	5	6	8	9	11	13	14
28	4472	4487	4502	4518	4533	4548	4564	4579	4594	4609	2	3	5	6	8	9	11	12	14
29	4624	4639	4654	4669	4683	4698	4713	4728	4742	4757	1	3	4	6	7	9	10	12	13
30	4771	4786	4800	4814	4829	4843	4857	4871	4886	4900	1	3	4	6	7	9	10	11	13
31	4914	4928	4942	4955	4969	4983	4997	5011	5024	5038	1	3	4	6	7	8	10	11	12
32	5051	5065	5079	5092	5105	5119	5132	5145	5159	5172	1	3	4	5	7	8	9	11	12
33	5185	5198	5211	5224	5237	5250	5263	5276	5289	5302	1	3	4	5	6	8	9	10	12
34	5315	5328	5340	5353	5366	5378	5391	5403	5416	5428	1	3	4	5	6	8	9	10	11
35	5441	5453	5465	5478	5490	5502	5514	5527	5539	5551	1	2	4	5	6	7	9	10	11
36	5563	5575	5587	5599	5611	5623	5635	5647	5658	5670	1	2	4	5	6	7	8	10	11
37	5682	5694	5705	5717	5729	5740	5752	5763	5775	5786	1	2	3	5	6	7	8	9	10
38	5798	5809	5821	5832	5843	5855	5866	5877	5888	5899	1	2	3	5	6	7	8	9	10
39	5911	5922	5933	5944	5955	5966	5977	5988	5999	6010	1	2	3	4	5	7	8	9	10
40	6021	6031	6042	6053	6064	6075	6085	6096	6107	6117	1	2	3	4	5	6	8	9	10
41	6128	6138	6149	6160	6170	6180	6191	6201	6212	6222	1	2	3	4	5	6	7	8	9
42	6232	6243	6253	6263	6274	6284	6294	6304	6314	6325	1	2	3	4	5	6	7	8	9
43	6335	6345	6355	6365	6375	6385	6395	6405	6415	6425	1	2	3	4	5	6	7	8	9
44	6435	6444	6454	6464	6474	6484	6493	6503	6513	6522	1	2	3	4	5	6	7	8	9
45	6532	6542	6551	6561	6571	6580	6590	6599	6609	6618	1	2	3	4	5	6	7	8	9
46	6628	6637	6646	6656	6665	6675	6684	6693	6702	6712	1	2	3	4	5	6	7	7	8
47	6721	6730	6739	6749	6758	6767	6776	6785	6794	6803	1	2	3	4	5	5	6	7	8
48	6812	6821	6830	6839	6848	6857	6866	6875	6884	6893	1	2	3	4	4	5	6	7	8
49	6902	6911	6920	6928	6937	6946	6955	6964	6972	6981	1	2	3	4	4	5	6	7	8
50	6990	6998	7007	7016	7024	7033	7042	7050	7059	7067	1	2	3	3	4	5	6	7	8
51	7076	7084	7093	7101	7110	7118	7126	7135	7143	7152	1	2	3	3	4	5	6	7	8
52	7160	7168	7177	7185	7193	7202	7210	7218	7226	7235	1	2	2	3	4	5	6	7	7
53	7243	7251	7259	7267	7275	7284	7292	7300	7308	7316	1	2	2	3	4	5	6	6	7
54	7324	7332	7340	7348	7356	7364	7372	7380	7388	7396	1	2	2	3	4	5	6	6	7
N	0	1	2	3	4	5	6	7	8	9	1	2	3	4	5	6	7	8	9

Table 1 (continued)

FOUR PLACE COMMON LOGARITHMS

$\log_{10} N$ or $\log N$

N	0	1	2	3	4	5	6	7	8	9	Proportional Parts 1	2	3	4	5	6	7	8	9
55	7404	7412	7419	7427	7435	7443	7451	7459	7466	7474	1	2	2	3	4	5	5	6	7
56	7482	7490	7497	7505	7513	7520	7528	7536	7543	7551	1	2	2	3	4	5	5	6	7
57	7559	7566	7574	7582	7589	7597	7604	7612	7619	7627	1	2	2	3	4	5	5	6	7
58	7634	7642	7649	7657	7664	7672	7679	7686	7694	7701	1	1	2	3	4	4	5	6	7
59	7709	7716	7723	7731	7738	7745	7752	7760	7767	7774	1	1	2	3	4	4	5	6	7
60	7782	7789	7796	7803	7810	7818	7825	7832	7839	7846	1	1	2	3	4	4	5	6	6
61	7853	7860	7868	7875	7882	7889	7896	7903	7910	7917	1	1	2	3	4	4	5	6	6
62	7924	7931	7938	7945	7952	7959	7966	7973	7980	7987	1	1	2	3	3	4	5	6	6
63	7993	8000	8007	8014	8021	8028	8035	8041	8048	8055	1	1	2	3	3	4	5	5	6
64	8062	8069	8075	8082	8089	8096	8102	8109	8116	8122	1	1	2	3	3	4	5	5	6
65	8129	8136	8142	8149	8156	8162	8169	8176	8182	8189	1	1	2	3	3	4	5	5	6
66	8195	8202	8209	8215	8222	8228	8235	8241	8248	8254	1	1	2	3	3	4	5	5	6
67	8261	8267	8274	8280	8287	8293	8299	8306	8312	8319	1	1	2	3	3	4	5	5	6
68	8325	8331	8338	8344	8351	8357	8363	8370	8376	8382	1	1	2	3	3	4	4	5	6
69	8388	8395	8401	8407	8414	8420	8426	8432	8439	8445	1	1	2	2	3	4	4	5	6
70	8451	8457	8463	8470	8476	8482	8488	8494	8500	8506	1	1	2	2	3	4	4	5	6
71	8513	8519	8525	8531	8537	8543	8549	8555	8561	8567	1	1	2	2	3	4	4	5	5
72	8573	8579	8585	8591	8597	8603	8609	8615	8621	8627	1	1	2	2	3	4	4	5	5
73	8633	8639	8645	8651	8657	8663	8669	8675	8681	8686	1	1	2	2	3	4	4	5	5
74	8692	8698	8704	8710	8716	8722	8727	8733	8739	8745	1	1	2	2	3	4	4	5	5
75	8751	8756	8762	8768	8774	8779	8785	8791	8797	8802	1	1	2	2	3	3	4	5	5
76	8808	8814	8820	8825	8831	8837	8842	8848	8854	8859	1	1	2	2	3	3	4	5	5
77	8865	8871	8876	8882	8887	8893	8899	8904	8910	8915	1	1	2	2	3	3	4	4	5
78	8921	8927	8932	8938	8943	8949	8954	8960	8965	8971	1	1	2	2	3	3	4	4	5
79	8976	8982	8987	8993	8998	9004	9009	9015	9020	9025	1	1	2	2	3	3	4	4	5
80	9031	9036	9042	9047	9053	9058	9063	9069	9074	9079	1	1	2	2	3	3	4	4	5
81	9085	9090	9096	9101	9106	9112	9117	9122	9128	9133	1	1	2	2	3	3	4	4	5
82	9138	9143	9149	9154	9159	9165	9170	9175	9180	9186	1	1	2	2	3	3	4	4	5
83	9191	9196	9201	9206	9212	9217	9222	9227	9232	9238	1	1	2	2	3	3	4	4	5
84	9243	9248	9253	9258	9263	9269	9274	9279	9284	9289	1	1	2	2	3	3	4	4	5
85	9294	9299	9304	9309	9315	9320	9325	9330	9335	9340	1	1	2	2	3	3	4	4	5
86	9345	9350	9355	9360	9365	9370	9375	9380	9385	9390	1	1	2	2	3	3	4	4	5
87	9395	9400	9405	9410	9415	9420	9425	9430	9435	9440	0	1	1	2	2	3	3	4	4
88	9445	9450	9455	9460	9465	9469	9474	9479	9484	9489	0	1	1	2	2	3	3	4	4
89	9494	9499	9504	9509	9513	9518	9523	9528	9533	9538	0	1	1	2	2	3	3	4	4
90	9542	9547	9552	9557	9562	9566	9571	9576	9581	9586	0	1	1	2	2	3	3	4	4
91	9590	9595	9600	9605	9609	9614	9619	9624	9628	9633	0	1	1	2	2	3	3	4	4
92	9638	9643	9647	9652	9657	9661	9666	9671	9675	9680	0	1	1	2	2	3	3	4	4
93	9685	9689	9694	9699	9703	9708	9713	9717	9722	9727	0	1	1	2	2	3	3	4	4
94	9731	9736	9741	9745	9750	9754	9759	9763	9768	9773	0	1	1	2	2	3	3	4	4
95	9777	9782	9786	9791	9795	9800	9805	9809	9814	9818	0	1	1	2	2	3	3	4	4
96	9823	9827	9832	9836	9841	9845	9850	9854	9859	9863	0	1	1	2	2	3	3	4	4
97	9868	9872	9877	9881	9886	9890	9894	9899	9903	9908	0	1	1	2	2	3	3	4	4
98	9912	9917	9921	9926	9930	9934	9939	9943	9948	9952	0	1	1	2	2	3	3	4	4
99	9956	9961	9965	9969	9974	9978	9983	9987	9991	9996	0	1	1	2	2	3	3	3	4
N	0	1	2	3	4	5	6	7	8	9	1	2	3	4	5	6	7	8	9

TABLE 2

FOUR PLACE COMMON ANTILOGARITHMS

10^p or antilog p

p	0	1	2	3	4	5	6	7	8	9	Proportional Parts								
											1	2	3	4	5	6	7	8	9
.00	1000	1002	1005	1007	1009	1012	1014	1016	1019	1021	0	0	1	1	1	1	2	2	2
.01	1023	1026	1028	1030	1033	1035	1038	1040	1042	1045	0	0	1	1	1	1	2	2	2
.02	1047	1050	1052	1054	1057	1059	1062	1064	1067	1069	0	0	1	1	1	1	2	2	2
.03	1072	1074	1076	1079	1081	1084	1086	1089	1091	1094	0	0	1	1	1	1	2	2	2
.04	1096	1099	1102	1104	1107	1109	1112	1114	1117	1119	0	1	1	1	1	2	2	2	2
.05	1122	1125	1127	1130	1132	1135	1138	1140	1143	1146	0	1	1	1	1	2	2	2	2
.06	1148	1151	1153	1156	1159	1161	1164	1167	1169	1172	0	1	1	1	1	2	2	2	2
.07	1175	1178	1180	1183	1186	1189	1191	1194	1197	1199	0	1	1	1	1	2	2	2	2
.08	1202	1205	1208	1211	1213	1216	1219	1222	1225	1227	0	1	1	1	1	2	2	2	3
.09	1230	1233	1236	1239	1242	1245	1247	1250	1253	1256	0	1	1	1	1	2	2	2	3
.10	1259	1262	1265	1268	1271	1274	1276	1279	1282	1285	0	1	1	1	1	2	2	2	3
.11	1288	1291	1294	1297	1300	1303	1306	1309	1312	1315	0	1	1	1	2	2	2	2	3
.12	1318	1321	1324	1327	1330	1334	1337	1340	1343	1346	0	1	1	1	2	2	2	2	3
.13	1349	1352	1355	1358	1361	1365	1368	1371	1374	1377	0	1	1	1	2	2	2	3	3
.14	1380	1384	1387	1390	1393	1396	1400	1403	1406	1409	0	1	1	1	2	2	2	3	3
.15	1413	1416	1419	1422	1426	1429	1432	1435	1439	1442	0	1	1	1	2	2	2	3	3
.16	1445	1449	1452	1455	1459	1462	1466	1469	1472	1476	0	1	1	1	2	2	2	3	3
.17	1479	1483	1486	1489	1493	1496	1500	1503	1507	1510	0	1	1	1	2	2	2	3	3
.18	1514	1517	1521	1524	1528	1531	1535	1538	1542	1545	0	1	1	1	2	2	2	3	3
.19	1549	1552	1556	1560	1563	1567	1570	1574	1578	1581	0	1	1	1	2	2	3	3	3
.20	1585	1589	1592	1596	1600	1603	1607	1611	1614	1618	0	1	1	1	2	2	3	3	3
.21	1622	1626	1629	1633	1637	1641	1644	1648	1652	1656	0	1	1	2	2	2	3	3	3
.22	1660	1663	1667	1671	1675	1679	1683	1687	1690	1694	0	1	1	2	2	2	3	3	3
.23	1698	1702	1706	1710	1714	1718	1722	1726	1730	1734	0	1	1	2	2	2	3	3	4
.24	1738	1742	1746	1750	1754	1758	1762	1766	1770	1774	0	1	1	2	2	2	3	3	4
.25	1778	1782	1786	1791	1795	1799	1803	1807	1811	1816	0	1	1	2	2	2	3	3	4
.26	1820	1824	1828	1832	1837	1841	1845	1849	1854	1858	0	1	1	2	2	3	3	3	4
.27	1862	1866	1871	1875	1879	1884	1888	1892	1897	1901	0	1	1	2	2	3	3	3	4
.28	1905	1910	1914	1919	1923	1928	1932	1936	1941	1945	0	1	1	2	2	3	3	4	4
.29	1950	1954	1959	1963	1968	1972	1977	1982	1986	1991	0	1	1	2	2	3	3	4	4
.30	1995	2000	2004	2009	2014	2018	2023	2028	2032	2037	0	1	1	2	2	3	3	4	4
.31	2042	2046	2051	2056	2061	2065	2070	2075	2080	2084	0	1	1	2	2	3	3	4	4
.32	2089	2094	2099	2104	2109	2113	2118	2123	2128	2133	0	1	1	2	2	3	3	4	4
.33	2138	2143	2148	2153	2158	2163	2168	2173	2178	2183	0	1	1	2	2	3	3	4	4
.34	2188	2193	2198	2203	2208	2213	2218	2223	2228	2234	1	1	2	2	3	3	4	4	5
.35	2239	2244	2249	2254	2259	2265	2270	2275	2280	2286	1	1	2	2	3	3	4	4	5
.36	2291	2296	2301	2307	2312	2317	2323	2328	2333	2339	1	1	2	2	3	3	4	4	5
.37	2344	2350	2355	2360	2366	2371	2377	2382	2388	2393	1	1	2	2	3	3	4	4	5
.38	2399	2404	2410	2415	2421	2427	2432	2438	2443	2449	1	1	2	2	3	3	4	4	5
.39	2455	2460	2466	2472	2477	2483	2489	2495	2500	2506	1	1	2	2	3	3	4	5	5
.40	2512	2518	2523	2529	2535	2541	2547	2553	2559	2564	1	1	2	2	3	4	4	5	5
.41	2570	2576	2582	2588	2594	2600	2606	2612	2618	2624	1	1	2	2	3	4	4	5	5
.42	2630	2636	2642	2649	2655	2661	2667	2673	2679	2685	1	1	2	2	3	4	4	5	6
.43	2692	2698	2704	2710	2716	2723	2729	2735	2742	2748	1	1	2	3	3	4	4	5	6
.44	2754	2761	2767	2773	2780	2786	2793	2799	2805	2812	1	1	2	3	3	4	4	5	6
.45	2818	2825	2831	2838	2844	2851	2858	2864	2871	2877	1	1	2	3	3	4	5	5	6
.46	2884	2891	2897	2904	2911	2917	2924	2931	2938	2944	1	1	2	3	3	4	5	5	6
.47	2951	2958	2965	2972	2979	2985	2992	2999	3006	3013	1	1	2	3	3	4	5	5	6
.48	3020	3027	3034	3041	3048	3055	3062	3069	3076	3083	1	1	2	3	4	4	5	6	6
.49	3090	3097	3105	3112	3119	3126	3133	3141	3148	3155	1	1	2	3	4	4	5	6	6
p	0	1	2	3	4	5	6	7	8	9	1	2	3	4	5	6	7	8	9

Table 2
(continued)

FOUR PLACE COMMON ANTILOGARITHMS

10^p or antilog p

p	0	1	2	3	4	5	6	7	8	9	Proportional Parts 1	2	3	4	5	6	7	8	9
.50	3162	3170	3177	3184	3192	3199	3206	3214	3221	3228	1	1	2	3	4	4	5	6	7
.51	3236	3243	3251	3258	3266	3273	3281	3289	3296	3304	1	2	2	3	4	5	5	6	7
.52	3311	3319	3327	3334	3342	3350	3357	3365	3373	3381	1	2	2	3	4	5	5	6	7
.53	3388	3396	3404	3412	3420	3428	3436	3443	3451	3459	1	2	2	3	4	5	5	6	7
.54	3467	3475	3483	3491	3499	3508	3516	3524	3532	3540	1	2	2	3	4	5	6	6	7
.55	3548	3556	3565	3573	3581	3589	3597	3606	3614	3622	1	2	2	3	4	5	6	7	7
.56	3631	3639	3648	3656	3664	3673	3681	3690	3698	3707	1	2	3	3	4	5	6	7	8
.57	3715	3724	3733	3741	3750	3758	3767	3776	3784	3793	1	2	3	3	4	5	6	7	8
.58	3802	3811	3819	3828	3837	3846	3855	3864	3873	3882	1	2	3	4	4	5	6	7	8
.59	3890	3899	3908	3917	3926	3936	3945	3954	3963	3972	1	2	3	4	5	5	6	7	8
.60	3981	3990	3999	4009	4018	4027	4036	4046	4055	4064	1	2	3	4	5	6	6	7	8
.61	4074	4083	4093	4102	4111	4121	4130	4140	4150	4159	1	2	3	4	5	6	7	8	9
.62	4169	4178	4188	4198	4207	4217	4227	4236	4246	4256	1	2	3	4	5	6	7	8	9
.63	4266	4276	4285	4295	4305	4315	4325	4335	4345	4355	1	2	3	4	5	6	7	8	9
.64	4365	4375	4385	4395	4406	4416	4426	4436	4446	4457	1	2	3	4	5	6	7	8	9
.65	4467	4477	4487	4498	4508	4519	4529	4539	4550	4560	1	2	3	4	5	6	7	8	9
.66	4571	4581	4592	4603	4613	4624	4634	4645	4656	4667	1	2	3	4	5	6	7	9	10
.67	4677	4688	4699	4710	4721	4732	4742	4753	4764	4775	1	2	3	4	5	7	8	9	10
.68	4786	4797	4808	4819	4831	4842	4853	4864	4875	4887	1	2	3	4	6	7	8	9	10
.69	4898	4909	4920	4932	4943	4955	4966	4977	4989	5000	1	2	3	5	6	7	8	9	10
.70	5012	5023	5035	5047	5058	5070	5082	5093	5105	5117	1	2	4	5	6	7	8	9	11
.71	5129	5140	5152	5164	5176	5188	5200	5212	5224	5236	1	2	4	5	6	7	8	10	11
.72	5248	5260	5272	5284	5297	5309	5321	5333	5346	5358	1	2	4	5	6	7	9	10	11
.73	5370	5383	5395	5408	5420	5433	5445	5458	5470	5483	1	3	4	5	6	8	9	10	11
.74	5495	5508	5521	5534	5546	5559	5572	5585	5598	5610	1	3	4	5	6	8	9	10	12
.75	5623	5636	5649	5662	5675	5689	5702	5715	5728	5741	1	3	4	5	7	8	9	10	12
.76	5754	5768	5781	5794	5808	5821	5834	5848	5861	5875	1	3	4	5	7	8	9	11	12
.77	5888	5902	5916	5929	5943	5957	5970	5984	5998	6012	1	3	4	5	7	8	10	11	12
.78	6026	6039	6053	6067	6081	6095	6109	6124	6138	6152	1	3	4	6	7	8	10	11	13
.79	6166	6180	6194	6209	6223	6237	6252	6266	6281	6295	1	3	4	6	7	9	10	11	13
.80	6310	6324	6339	6353	6368	6383	6397	6412	6427	6442	1	3	4	6	7	9	10	12	13
.81	6457	6471	6486	6501	6516	6531	6546	6561	6577	6592	2	3	5	6	8	9	11	12	14
.82	6607	6622	6637	6653	6668	6683	6699	6714	6730	6745	2	3	5	6	8	9	11	12	14
.83	6761	6776	6792	6808	6823	6839	6855	6871	6887	6902	2	3	5	6	8	9	11	13	14
.84	6918	6934	6950	6966	6982	6998	7015	7031	7047	7063	2	3	5	6	8	10	11	13	15
.85	7079	7096	7112	7129	7145	7161	7178	7194	7211	7228	2	3	5	7	8	10	12	13	15
.86	7244	7261	7278	7295	7311	7328	7345	7362	7379	7396	2	3	5	7	8	10	12	13	15
.87	7413	7430	7447	7464	7482	7499	7516	7534	7551	7568	2	3	5	7	9	10	12	14	16
.88	7586	7603	7621	7638	7656	7674	7691	7709	7727	7745	2	4	5	7	9	11	12	14	16
.89	7762	7780	7798	7816	7834	7852	7870	7889	7907	7925	2	4	5	7	9	11	13	14	16
.90	7943	7962	7980	7998	8017	8035	8054	8072	8091	8110	2	4	6	7	9	11	13	15	17
.91	8128	8147	8166	8185	8204	8222	8241	8260	8279	8299	2	4	6	8	9	11	13	15	17
.92	8318	8337	8356	8375	8395	8414	8433	8453	8472	8492	2	4	6	8	10	12	14	15	17
.93	8511	8531	8551	8570	8590	8610	8630	8650	8670	8690	2	4	6	8	10	12	14	16	18
.94	8710	8730	8750	8770	8790	8810	8831	8851	8872	8892	2	4	6	8	10	12	14	16	18
.95	8913	8933	8954	8974	8995	9016	9036	9057	9078	9099	2	4	6	8	10	12	15	17	19
.96	9120	9141	9162	9183	9204	9226	9247	9268	9290	9311	2	4	6	8	11	13	15	17	19
.97	9333	9354	9376	9397	9419	9441	9462	9484	9506	9528	2	4	7	9	11	13	15	17	20
.98	9550	9572	9594	9616	9638	9661	9683	9705	9727	9750	2	4	7	9	11	13	16	18	20
.99	9772	9795	9817	9840	9863	9886	9908	9931	9954	9977	2	5	7	9	11	14	16	18	20
p	0	1	2	3	4	5	6	7	8	9	1	2	3	4	5	6	7	8	9

TABLE 3

Sin x (x in degrees and minutes)

x	0′	10′	20′	30′	40′	50′
0°	.0000	.0029	.0058	.0087	.0116	.0145
1	.0175	.0204	.0233	.0262	.0291	.0320
2	.0349	.0378	.0407	.0436	.0465	.0494
3	.0523	.0552	.0581	.0610	.0640	.0669
4	.0698	.0727	.0756	.0785	.0814	.0843
5°	.0872	.0901	.0929	.0958	.0987	.1016
6	.1045	.1074	.1103	.1132	.1161	.1190
7	.1219	.1248	.1276	.1305	.1334	.1363
8	.1392	.1421	.1449	.1478	.1507	.1536
9	.1564	.1593	.1622	.1650	.1679	.1708
10°	.1736	.1765	.1794	.1822	.1851	.1880
11	.1908	.1937	.1965	.1994	.2022	.2051
12	.2079	.2108	.2136	.2164	.2193	.2221
13	.2250	.2278	.2306	.2334	.2363	.2391
14	.2419	.2447	.2476	.2504	.2532	.2560
15°	.2588	.2616	.2644	.2672	.2700	.2728
16	.2756	.2784	.2812	.2840	.2868	.2896
17	.2924	.2952	.2979	.3007	.3035	.3062
18	.3090	.3118	.3145	.3173	.3201	.3228
19	.3256	.3283	.3311	.3338	.3365	.3393
20°	.3420	.3448	.3475	.3502	.3529	.3557
21	.3584	.3611	.3638	.3665	.3692	.3719
22	.3746	.3773	.3800	.3827	.3854	.3881
23	.3907	.3934	.3961	.3987	.4014	.4041
24	.4067	.4094	.4120	.4147	.4173	.4200
25°	.4226	.4253	.4279	.4305	.4331	.4358
26	.4384	.4410	.4436	.4462	.4488	.4514
27	.4540	.4566	.4592	.4617	.4643	.4669
28	.4695	.4720	.4746	.4772	.4797	.4823
29	.4848	.4874	.4899	.4924	.4950	.4975
30°	.5000	.5025	.5050	.5075	.5100	.5125
31	.5150	.5175	.5200	.5225	.5250	.5275
32	.5299	.5324	.5348	.5373	.5398	.5422
33	.5446	.5471	.5495	.5519	.5544	.5568
34	.5592	.5616	.5640	.5664	.5688	.5712
35°	.5736	.5760	.5783	.5807	.5831	.5854
36	.5878	.5901	.5925	.5948	.5972	.5995
37	.6018	.6041	.6065	.6088	.6111	.6134
38	.6157	.6180	.6202	.6225	.6248	.6271
39	.6293	.6316	.6338	.6361	.6383	.6406
40°	.6428	.6450	.6472	.6494	.6517	.6539
41	.6561	.6583	.6604	.6626	.6648	.6670
42	.6691	.6713	.6734	.6756	.6777	.6799
43	.6820	.6841	.6862	.6884	.6905	.6926
44	.6947	.6967	.6988	.7009	.7030	.7050
45°	.7071	.7092	.7112	.7133	.7153	.7173

x	0′	10′	20′	30′	40′	50′
45°	.7071	.7092	.7112	.7133	.7153	.7173
46	.7193	.7214	.7234	.7254	.7274	.7294
47	.7314	.7333	.7353	.7373	.7392	.7412
48	.7431	.7451	.7470	.7490	.7509	.7528
49	.7547	.7566	.7585	.7604	.7623	.7642
50°	.7660	.7679	.7698	.7716	.7735	.7753
51	.7771	.7790	.7808	.7826	.7844	.7862
52	.7880	.7898	.7916	.7934	.7951	.7969
53	.7986	.8004	.8021	.8039	.8056	.8073
54	.8090	.8107	.8124	.8141	.8158	.8175
55°	.8192	.8208	.8225	.8241	.8258	.8274
56	.8290	.8307	.8323	.8339	.8355	.8371
57	.8387	.8403	.8418	.8434	.8450	.8465
58	.8480	.8496	.8511	.8526	.8542	.8557
59	.8572	.8587	.8601	.8616	.8631	.8646
60°	.8660	.8675	.8689	.8704	.8718	.8732
61	.8746	.8760	.8774	.8788	.8802	.8816
62	.8829	.8843	.8857	.8870	.8884	.8897
63	.8910	.8923	.8936	.8949	.8962	.8975
64	.8988	.9001	.9013	.9026	.9038	.9051
65°	.9063	.9075	.9088	.9100	.9112	.9124
66	.9135	.9147	.9159	.9171	.9182	.9194
67	.9205	.9216	.9228	.9239	.9250	.9261
68	.9272	.9283	.9293	.9304	.9315	.9325
69	.9336	.9346	.9356	.9367	.9377	.9387
70°	.9397	.9407	.9417	.9426	.9436	.9446
71	.9455	.9465	.9474	.9483	.9492	.9502
72	.9511	.9520	.9528	.9537	.9546	.9555
73	.9563	.9572	.9580	.9588	.9596	.9605
74	.9613	.9621	.9628	.9636	.9644	.9652
75°	.9659	.9667	.9674	.9681	.9689	.9696
76	.9703	.9710	.9717	.9724	.9730	.9737
77	.9744	.9750	.9757	.9763	.9769	.9775
78	.9781	.9787	.9793	.9799	.9805	.9811
79	.9816	.9822	.9827	.9833	.9838	.9843
80°	.9848	.9853	.9858	.9863	.9868	.9872
81	.9877	.9881	.9886	.9890	.9894	.9899
82	.9903	.9907	.9911	.9914	.9918	.9922
83	.9925	.9929	.9932	.9936	.9939	.9942
84	.9945	.9948	.9951	.9954	.9957	.9959
85°	.9962	.9964	.9967	.9969	.9971	.9974
86	.9976	.9978	.9980	.9981	.9983	.9985
87	.9986	.9988	.9989	.9990	.9992	.9993
88	.9994	.9995	.9996	.9997	.9997	.9998
89	.9998	.9999	.9999	1.0000	1.0000	1.0000
90°	1.0000					

TABLE

4

Cos x (x in degrees and minutes)

x	0′	10′	20′	30′	40′	50′
0°	1.0000	1.0000	1.0000	1.0000	.9999	.9999
1	.9998	.9998	.9997	.9997	.9996	.9995
2	.9994	.9993	.9992	.9990	.9989	.9988
3	.9986	.9985	.9983	.9981	.9980	.9978
4	.9976	.9974	.9971	.9969	.9967	.9964
5°	.9962	.9959	.9957	.9954	.9951	.9948
6	.9945	.9942	.9939	.9936	.9932	.9929
7	.9925	.9922	.9918	.9914	.9911	.9907
8	.9903	.9899	.9894	.9890	.9886	.9881
9	.9877	.9872	.9868	.9863	.9858	.9853
10°	.9848	.9843	.9838	.9833	.9827	.9822
11	.9816	.9811	.9805	.9799	.9793	.9787
12	.9781	.9775	.9769	.9763	.9757	.9750
13	.9744	.9737	.9730	.9724	.9717	.9710
14	.9703	.9696	.9689	.9681	.9674	.9667
15°	.9659	.9652	.9644	.9636	.9628	.9621
16	.9613	.9605	.9596	.9588	.9580	.9572
17	.9563	.9555	.9546	.9537	.9528	.9520
18	.9511	.9502	.9492	.9483	.9474	.9465
19	.9455	.9446	.9436	.9426	.9417	.9407
20°	.9397	.9387	.9377	.9367	.9356	.9346
21	.9336	.9325	.9315	.9304	.9293	.9283
22	.9272	.9261	.9250	.9239	.9228	.9216
23	.9205	.9194	.9182	.9171	.9159	.9147
24	.9135	.9124	.9112	.9100	.9088	.9075
25°	.9063	.9051	.9038	.9026	.9013	.9001
26	.8988	.8975	.8962	.8949	.8936	.8923
27	.8910	.8897	.8884	.8870	.8857	.8843
28	.8829	.8816	.8802	.8788	.8774	.8760
29	.8746	.8732	.8718	.8704	.8689	.8675
30°	.8660	.8646	.8631	.8616	.8601	.8587
31	.8572	.8557	.8542	.8526	.8511	.8496
32	.8480	.8465	.8450	.8434	.8418	.8403
33	.8387	.8371	.8355	.8339	.8323	.8307
34	.8290	.8274	.8258	.8241	.8225	.8208
35°	.8192	.8175	.8158	.8141	.8124	.8107
36	.8090	.8073	.8056	.8039	.8021	.8004
37	.7986	.7969	.7951	.7934	.7916	.7898
38	.7880	.7862	.7844	.7826	.7808	.7790
39	.7771	.7753	.7735	.7716	.7698	.7679
40°	.7660	.7642	.7623	.7604	.7585	.7566
41	.7547	.7528	.7509	.7490	.7470	.7451
42	.7431	.7412	.7392	.7373	.7353	.7333
43	.7314	.7294	.7274	.7254	.7234	.7214
44	.7193	.7173	.7153	.7133	.7112	.7092
45°	.7071	.7050	.7030	.7009	.6988	.6967

x	0′	10′	20′	30′	40′	50′
45°	.7071	.7050	.7030	.7009	.6988	.6967
46	.6947	.6926	.6905	.6884	.6862	.6841
47	.6820	.6799	.6777	.6756	.6734	.6713
48	.6691	.6670	.6648	.6626	.6604	.6583
49	.6561	.6539	.6517	.6494	.6472	.6450
50°	.6428	.6406	.6383	.6361	.6338	.6316
51	.6293	.6271	.6248	.6225	.6202	.6180
52	.6157	.6134	.6111	.6088	.6065	.6041
53	.6018	.5995	.5972	.5948	.5925	.5901
54	.5878	.5854	.5831	.5807	.5783	.5760
55°	.5736	.5712	.5688	.5664	.5640	.5616
56	.5592	.5568	.5544	.5519	.5495	.5471
57	.5446	.5422	.5398	.5373	.5348	.5324
58	.5299	.5275	.5250	.5225	.5200	.5175
59	.5150	.5125	.5100	.5075	.5050	.5025
60°	.5000	.4975	.4950	.4924	.4899	.4874
61	.4848	.4823	.4797	.4772	.4746	.4720
62	.4695	.4669	.4643	.4617	.4592	.4566
63	.4540	.4514	.4488	.4462	.4436	.4410
64	.4384	.4358	.4331	.4305	.4279	.4253
65°	.4226	.4200	.4173	.4147	.4120	.4094
66	.4067	.4041	.4014	.3987	.3961	.3934
67	.3907	.3881	.3854	.3827	.3800	.3773
68	.3746	.3719	.3692	.3665	.3638	.3611
69	.3584	.3557	.3529	.3502	.3475	.3448
70°	.3420	.3393	.3365	.3338	.3311	.3283
71	.3256	.3228	.3201	.3173	.3145	.3118
72	.3090	.3062	.3035	.3007	.2979	.2952
73	.2924	.2896	.2868	.2840	.2812	.2784
74	.2756	.2728	.2700	.2672	.2644	.2616
75°	.2588	.2560	.2532	.2504	.2476	.2447
76	.2419	.2391	.2363	.2334	.2306	.2278
77	.2250	.2221	.2193	.2164	.2136	.2108
78	.2079	.2051	.2022	.1994	.1965	.1937
79	.1908	.1880	.1851	.1822	.1794	.1765
80°	.1736	.1708	.1679	.1650	.1622	.1593
81	.1564	.1536	.1507	.1478	.1449	.1421
82	.1392	.1363	.1334	.1305	.1276	.1248
83	.1219	.1190	.1161	.1132	.1103	.1074
84	.1045	.1016	.0987	.0958	.0929	.0901
85°	.0872	.0843	.0814	.0785	.0756	.0727
86	.0698	.0669	.0640	.0610	.0581	.0552
87	.0523	.0494	.0465	.0436	.0407	.0378
88	.0349	.0320	.0291	.0262	.0233	.0204
89	.0175	.0145	.0116	.0087	.0058	.0029
90°	.0000					

TABLE 5

Tan x (x in degrees and minutes)

x	0′	10′	20′	30′	40′	50′
0°	.0000	.0029	.0058	.0087	.0116	.0145
1	.0175	.0204	.0233	.0262	.0291	.0320
2	.0349	.0378	.0407	.0437	.0466	.0495
3	.0524	.0553	.0582	.0612	.0641	.0670
4	.0699	.0729	.0758	.0787	.0816	.0846
5°	.0875	.0904	.0934	.0963	.0992	.1022
6	.1051	.1080	.1110	.1139	.1169	.1198
7	.1228	.1257	.1287	.1317	.1346	.1376
8	.1405	.1435	.1465	.1495	.1524	.1554
9	.1584	.1614	.1644	.1673	.1703	.1733
10°	.1763	.1793	.1823	.1853	.1883	.1914
11	.1944	.1974	.2004	.2035	.2065	.2095
12	.2126	.2156	.2186	.2217	.2247	.2278
13	.2309	.2339	.2370	.2401	.2432	.2462
14	.2493	.2524	.2555	.2586	.2617	.2648
15°	.2679	.2711	.2742	.2773	.2805	.2836
16	.2867	.2899	.2931	.2962	.2994	.3026
17	.3057	.3089	.3121	.3153	.3185	.3217
18	.3249	.3281	.3314	.3346	.3378	.3411
19	.3443	.3476	.3508	.3541	.3574	.3607
20°	.3640	.3673	.3706	.3739	.3772	.3805
21	.3839	.3872	.3906	.3939	.3973	.4006
22	.4040	.4074	.4108	.4142	.4176	.4210
23	.4245	.4279	.4314	.4348	.4383	.4417
24	.4452	.4487	.4522	.4557	.4592	.4628
25°	.4663	.4699	.4734	.4770	.4806	.4841
26	.4877	.4913	.4950	.4986	.5022	.5059
27	.5095	.5132	.5169	.5206	.5243	.5280
28	.5317	.5354	.5392	.5430	.5467	.5505
29	.5543	.5581	.5619	.5658	.5696	.5735
30°	.5774	.5812	.5851	.5890	.5930	.5969
31	.6009	.6048	.6088	.6128	.6168	.6208
32	.6249	.6289	.6330	.6371	.6412	.6453
33	.6494	.6536	.6577	.6619	.6661	.6703
34	.6745	.6787	.6830	.6873	.6916	.6959
35°	.7002	.7046	.7089	.7133	.7177	.7221
36	.7265	.7310	.7355	.7400	.7445	.7490
37	.7536	.7581	.7627	.7673	.7720	.7766
38	.7813	.7860	.7907	.7954	.8002	.8050
39	.8098	.8146	.8195	.8243	.8292	.8342
40°	.8391	.8441	.8491	.8541	.8591	.8642
41	.8693	.8744	.8796	.8847	.8899	.8952
42	.9004	.9057	.9110	.9163	.9217	.9271
43	.9325	.9380	.9435	.9490	.9545	.9601
44	.9657	.9713	.9770	.9827	.9884	.9942
45°	1.0000	1.0058	1.0117	1.0176	1.0235	1.0295

x	0′	10′	20′	30′	40′	50′
45°	1.0000	1.0058	1.0117	1.0176	1.0235	1.0295
46	1.0355	1.0416	1.0477	1.0538	1.0599	1.0661
47	1.0724	1.0786	1.0850	1.0913	1.0977	1.1041
48	1.1106	1.1171	1.1237	1.1303	1.1369	1.1436
49	1.1504	1.1571	1.1640	1.1708	1.1778	1.1847
50°	1.1918	1.1988	1.2059	1.2131	1.2203	1.2276
51	1.2349	1.2423	1.2497	1.2572	1.2647	1.2723
52	1.2799	1.2876	1.2954	1.3032	1.3111	1.3190
53	1.3270	1.3351	1.3432	1.3514	1.3597	1.3680
54	1.3764	1.3848	1.3934	1.4019	1.4106	1.4193
55°	1.4281	1.4370	1.4460	1.4550	1.4641	1.4733
56	1.4826	1.4919	1.5013	1.5108	1.5204	1.5301
57	1.5399	1.5497	1.5597	1.5697	1.5798	1.5900
58	1.6003	1.6107	1.6212	1.6319	1.6426	1.6534
59	1.6643	1.6753	1.6864	1.6977	1.7090	1.7205
60°	1.7321	1.7437	1.7556	1.7675	1.7796	1.7917
61	1.8040	1.8165	1.8291	1.8418	1.8546	1.8676
62	1.8807	1.8940	1.9074	1.9210	1.9347	1.9486
63	1.9626	1.9768	1.9912	2.0057	2.0204	2.0353
64	2.0503	2.0655	2.0809	2.0965	2.1123	2.1283
65°	2.1445	2.1609	2.1775	2.1943	2.2113	2.2286
66	2.2460	2.2637	2.2817	2.2998	2.3183	2.3369
67	2.3559	2.3750	2.3945	2.4142	2.4342	2.4545
68	2.4751	2.4960	2.5172	2.5386	2.5605	2.5826
69	2.6051	2.6279	2.6511	2.6746	2.6985	2.7228
70°	2.7475	2.7725	2.7980	2.8239	2.8502	2.8770
71	2.9042	2.9319	2.9600	2.9887	3.0178	3.0475
72	3.0777	3.1084	3.1397	3.1716	3.2041	3.2371
73	3.2709	3.3052	3.3402	3.3759	3.4124	3.4495
74	3.4874	3.5261	3.5656	3.6059	3.6470	3.6891
75°	3.7321	3.7760	3.8208	3.8667	3.9136	3.9617
76	4.0108	4.0611	4.1126	4.1653	4.2193	4.2747
77	4.3315	4.3897	4.4494	4.5107	4.5736	4.6382
78	4.7046	4.7729	4.8430	4.9152	4.9894	5.0658
79	5.1446	5.2257	5.3093	5.3955	5.4845	5.5764
80°	5.6713	5.7694	5.8708	5.9758	6.0844	6.1970
81	6.3138	6.4348	6.5606	6.6912	6.8269	6.9682
82	7.1154	7.2687	7.4287	7.5958	7.7704	7.9530
83	8.1443	8.3450	8.5555	8.7769	9.0098	9.2553
84	9.5144	9.7882	10.078	10.385	10.712	11.059
85°	11.430	11.826	12.251	12.706	13.197	13.727
86	14.301	14.924	15.605	16.350	17.169	18.075
87	19.081	20.206	21.470	22.904	24.542	26.432
88	28.636	31.242	34.368	38.188	42.964	49.104
89	57.290	68.750	85.940	114.59	171.89	343.77
90°	∞					

TABLE

6

Cot x (x in degrees and minutes)

x	0′	10′	20′	30′	40′	50′
0°	∞	343.77	171.89	114.59	85.940	68.750
1	57.290	49.104	42.964	38.188	34.368	31.242
2	28.636	26.432	24.542	22.904	21.470	20.206
3	19.081	18.075	17.169	16.350	15.605	14.924
4	14.301	13.727	13.197	12.706	12.251	11.826
5°	11.430	11.059	10.712	10.385	10.078	9.7882
6	9.5144	9.2553	9.0098	8.7769	8.5555	8.3450
7	8.1443	7.9530	7.7704	7.5958	7.4287	7.2687
8	7.1154	6.9682	6.8269	6.6912	6.5606	6.4348
9	6.3138	6.1970	6.0844	5.9758	5.8708	5.7694
10°	5.6713	5.5764	5.4845	5.3955	5.3093	5.2257
11	5.1446	5.0658	4.9894	4.9152	4.8430	4.7729
12	4.7046	4.6382	4.5736	4.5107	4.4494	4.3897
13	4.3315	4.2747	4.2193	4.1653	4.1126	4.0611
14	4.0108	3.9617	3.9136	3.8667	3.8208	3.7760
15°	3.7321	3.6891	3.6470	3.6059	3.5656	3.5261
16	3.4874	3.4495	3.4124	3.3759	3.3402	3.3052
17	3.2709	3.2371	3.2041	3.1716	3.1397	3.1084
18	3.0777	3.0475	3.0178	2.9887	2.9600	2.9319
19	2.9042	2.8770	2.8502	2.8239	2.7980	2.7725
20°	2.7475	2.7228	2.6985	2.6746	2.6511	2.6279
21	2.6051	2.5826	2.5605	2.5386	2.5172	2.4960
22	2.4751	2.4545	2.4342	2.4142	2.3945	2.3750
23	2.3559	2.3369	3.3183	2.2998	2.2817	2.2637
24	2.2460	2.2286	2.2113	2.1943	2.1775	2.1609
25°	2.1445	2.1283	2.1123	2.0965	2.0809	2.0655
26	2.0503	2.0353	2.0204	2.0057	1.9912	1.9768
27	1.9626	1.9486	1.9347	1.9210	1.9074	1.8940
28	1.8807	1.8676	1.8546	1.8418	1.8291	1.8165
29	1.8040	1.7917	1.7796	1.7675	1.7556	1.7437
30°	1.7321	1.7205	1.7090	1.6977	1.6864	1.6753
31	1.6643	1.6534	1.6426	1.6319	1.6212	1.6107
32	1.6003	1.5900	1.5798	1.5697	1.5597	1.5497
33	1.5399	1.5301	1.5204	1.5108	1.5013	1.4919
34	1.4826	1.4733	1.4641	1.4550	1.4460	1.4370
35°	1.4281	1.4193	1.4106	1.4019	1.3934	1.3848
36	1.3764	1.3680	1.3597	1.3514	1.3432	1.3351
37	1.3270	1.3190	1.3111	1.3032	1.2954	1.2876
38	1.2799	1.2723	1.2647	1.2572	1.2497	1.2423
39	1.2349	1.2276	1.2203	1.2131	1.2059	1.1988
40°	1.1918	1.1847	1.1778	1.1708	1.1640	1.1571
41	1.1504	1.1436	1.1369	1.1303	1.1237	1.1171
42	1.1106	1.1041	1.0977	1.0913	1.0850	1.0786
43	1.0724	1.0661	1.0599	1.0538	1.0477	1.0416
44	1.0355	1.0295	1.0235	1.0176	1.0117	1.0058
45°	1.0000	.9942	.9884	.9827	.9770	.9713

x	0′	10′	20′	30′	40′	50′
45°	1.000	.9942	.9884	.9827	.9770	.9713
46	.9657	.9601	.9545	.9490	.9435	.9380
47	.9325	.9271	.9217	.9163	.9110	.9057
48	.9004	.8952	.8899	.8847	.8796	.8744
49	.8693	.8642	.8591	.8541	.8491	.8441
50°	.8391	.8342	.8292	.8243	.8195	.8146
51	.8098	.8050	.8002	.7954	.7907	.7860
52	.7813	.7766	.7720	.7673	.7627	.7581
53	.7536	.7490	.7445	.7400	.7355	.7310
54	.7265	.7221	.7177	.7133	.7089	.7046
55°	.7002	.6959	.6916	.6873	.6830	.6787
56	.6745	.6703	.6661	.6619	.6577	.6536
57	.6494	.6453	.6412	.6371	.6330	.6289
58	.6249	.6208	.6168	.6128	.6088	.6048
59	.6009	.5969	.5930	.5890	.5851	.5812
60°	.5774	.5735	.5696	.5658	.5619	.5581
61	.5543	.5505	.5467	.5430	.5392	.5354
62	.5317	.5280	.5243	.5206	.5169	.5132
63	.5095	.5059	.5022	.4986	.4950	.4913
64	.4877	.4841	.4806	.4770	.4734	.4699
65°	.4663	.4628	.4592	.4557	.4522	.4487
66	.4452	.4417	.4383	.4348	.4314	.4279
67	.4245	.4210	.4176	.4142	.4108	.4074
68	.4040	.4006	.3973	.3939	.3906	.3872
69	.3839	.3805	.3772	.3739	.3706	.3673
70°	.3640	.3607	.3574	.3541	.3508	.3476
71	.3443	.3411	.3378	.3346	.3314	.3281
72	.3249	.3217	.3185	.3153	.3121	.3089
73	.3057	.3026	.2994	.2962	.2931	.2899
74	.2867	.2836	.2805	.2773	.2742	.2711
75°	.2679	.2648	.2617	.2586	.2555	.2524
76	.2493	.2462	.2432	.2401	.2370	.2339
77	.2309	.2278	.2247	.2217	.2186	.2156
78	.2126	.2095	.2065	.2035	.2004	.1974
79	.1944	.1914	.1883	.1853	.1823	.1793
80°	.1763	.1733	.1703	.1673	.1644	.1614
81	.1584	.1554	.1524	.1495	.1465	.1435
82	.1405	.1376	.1346	.1317	.1287	.1257
83	.1228	.1198	.1169	.1139	.1110	.1080
84	.1051	.1022	.0992	.0963	.0934	.0904
85°	.0875	.0846	.0816	.0787	.0758	.0729
86	.0699	.0670	.0641	.0612	.0582	.0553
87	.0524	.0495	.0466	.0437	.0407	.0378
88	.0349	.0320	.0291	.0262	.0233	.0204
89	.0175	.0145	.0116	.0087	.0058	.0029
90°	.0000					

TABLE 7

Sec x (x in degrees and minutes)

x	0′	10′	20′	30′	40′	50′
0°	1.000	1.000	1.000	1.000	1.000	1.000
1	1.000	1.000	1.000	1.000	1.000	1.001
2	1.001	1.001	1.001	1.001	1.001	1.001
3	1.001	1.002	1.002	1.002	1.002	1.002
4	1.002	1.003	1.003	1.003	1.003	1.004
5°	1.004	1.004	1.004	1.005	1.005	1.005
6	1.006	1.006	1.006	1.006	1.007	1.007
7	1.008	1.008	1.008	1.009	1.009	1.009
8	1.010	1.010	1.011	1.011	1.012	1.012
9	1.012	1.013	1.013	1.014	1.014	1.015
10°	1.015	1.016	1.016	1.017	1.018	1.018
11	1.019	1.019	1.020	1.020	1.021	1.022
12	1.022	1.023	1.024	1.024	1.025	1.026
13	1.026	1.027	1.028	1.028	1.029	1.030
14	1.031	1.031	1.032	1.033	1.034	1.034
15°	1.035	1.036	1.037	1.038	1.039	1.039
16	1.040	1.041	1.042	1.043	1.044	1.045
17	1.046	1.047	1.048	1.048	1.049	1.050
18	1.051	1.052	1.053	1.054	1.056	1.057
19	1.058	1.059	1.060	1.061	1.062	1.063
20°	1.064	1.065	1.066	1.068	1.069	1.070
21	1.071	1.072	1.074	1.075	1.076	1.077
22	1.079	1.080	1.081	1.082	1.084	1.085
23	1.086	1.088	1.089	1.090	1.092	1.093
24	1.095	1.096	1.097	1.099	1.100	1.102
25°	1.103	1.105	1.106	1.108	1.109	1.111
26	1.113	1.114	1.116	1.117	1.119	1.121
27	1.122	1.124	1.126	1.127	1.129	1.131
28	1.133	1.134	1.136	1.138	1.140	1.142
29	1.143	1.145	1.147	1.149	1.151	1.153
30°	1.155	1.157	1.159	1.161	1.163	1.165
31	1.167	1.169	1.171	1.173	1.175	1.177
32	1.179	1.181	1.184	1.186	1.188	1.190
33	1.192	1.195	1.197	1.199	1.202	1.204
34	1.206	1.209	1.211	1.213	1.216	1.218
35°	1.221	1.223	1.226	1.228	1.231	1.233
36	1.236	1.239	1.241	1.244	1.247	1.249
37	1.252	1.255	1.258	1.260	1.263	1.266
38	1.269	1.272	1.275	1.278	1.281	1.284
39	1.287	1.290	1.293	1.296	1.299	1.302
40°	1.305	1.309	1.312	1.315	1.318	1.322
41	1.325	1.328	1.332	1.335	1.339	1.342
42	1.346	1.349	1.353	1.356	1.360	1.364
43	1.367	1.371	1.375	1.379	1.382	1.386
44	1.390	1.394	1.398	1.402	1.406	1.410
45°	1.414	1.418	1.423	1.427	1.431	1.435

x	0′	10′	20′	30′	40′	50′
45°	1.414	1.418	1.423	1.427	1.431	1.435
46	1.440	1.444	1.448	1.453	1.457	1.462
47	1.466	1.471	1.476	1.480	1.485	1.490
48	1.494	1.499	1.504	1.509	1.514	1.519
49	1.524	1.529	1.535	1.540	1.545	1.550
50°	1.556	1.561	1.567	1.572	1.578	1.583
51	1.589	1.595	1.601	1.606	1.612	1.618
52	1.624	1.630	1.636	1.643	1.649	1.655
53	1.662	1.668	1.675	1.681	1.688	1.695
54	1.701	1.708	1.715	1.722	1.729	1.736
55°	1.743	1.751	1.758	1.766	1.773	1.781
56	1.788	1.796	1.804	1.812	1.820	1.828
57	1.836	1.844	1.853	1.861	1.870	1.878
58	1.887	1.896	1.905	1.914	1.923	1.932
59	1.942	1.951	1.961	1.970	1.980	1.990
60°	2.000	2.010	2.020	2.031	2.041	2.052
61	2.063	2.074	2.085	2.096	2.107	2.118
62	2.130	2.142	2.154	2.166	2.178	2.190
63	2.203	2.215	2.228	2.241	2.254	2.268
64	2.281	2.295	2.309	2.323	2.337	2.352
65°	2.366	2.381	2.396	2.411	2.427	2.443
66	2.459	2.475	2.491	2.508	2.525	2.542
67	2.559	2.577	2.595	2.613	2.632	2.650
68	2.669	2.689	2.709	2.729	2.749	2.769
69	2.790	2.812	2.833	2.855	2.878	2.901
70°	2.924	2.947	2.971	2.996	3.021	3.046
71	3.072	3.098	3.124	3.152	3.179	3.207
72	3.236	3.265	3.295	3.326	3.357	3.388
73	3.420	3.453	3.487	3.521	3.556	3.592
74	3.628	3.665	3.703	3.742	3.782	3.822
75°	3.864	3.906	3.950	3.994	4.039	4.086
76	4.134	4.182	4.232	4.284	4.336	4.390
77	4.445	4.502	4.560	4.620	4.682	4.745
78	4.810	4.876	4.945	5.016	5.089	5.164
79	5.241	5.320	5.403	5.487	5.575	5.665
80°	5.759	5.855	5.955	6.059	6.166	6.277
81	6.392	6.512	6.636	6.765	6.900	7.040
82	7.185	7.337	7.496	7.661	7.834	8.016
83	8.206	8.405	8.614	8.834	9.065	9.309
84	9.567	9.839	10.13	10.43	10.76	11.10
85°	11.47	11.87	12.29	12.75	13.23	13.76
86	14.34	14.96	15.64	16.38	17.20	18.10
87	19.11	20.23	21.49	22.93	24.56	26.45
88	28.65	31.26	34.38	38.20	42.98	49.11
89	57.30	68.76	85.95	114.6	171.9	343.8
90°	∞					

TABLE

8

Csc x (x in degrees and minutes)

x	0′	10′	20′	30′	40′	50′
0°	∞	343.8	171.9	114.6	85.95	68.76
1	57.30	49.11	42.98	38.20	34.38	31.26
2	28.65	26.45	24.56	22.93	21.49	20.23
3	19.11	18.10	17.20	16.38	15.64	14.96
4	14.34	13.76	13.23	12.75	12.29	11.87
5°	11.47	11.10	10.76	10.43	10.13	9.839
6	9.567	9.309	9.065	8.834	8.614	8.405
7	8.206	8.016	7.834	7.661	7.496	7.337
8	7.185	7.040	6.900	6.765	6.636	6.512
9	6.392	6.277	6.166	6.059	5.955	5.855
10°	5.759	5.665	5.575	5.487	5.403	5.320
11	5.241	5.164	5.089	5.016	4.945	4.876
12	4.810	4.745	4.682	4.620	4.560	4.502
13	4.445	4.390	4.336	4.284	4.232	4.182
14	4.134	4.086	4.039	3.994	3.950	3.906
15°	3.864	3.822	3.782	3.742	3.703	3.665
16	3.628	3.592	3.556	3.521	3.487	3.453
17	3.420	3.388	3.357	3.326	3.295	3.265
18	3.236	3.207	3.179	3.152	3.124	3.098
19	3.072	3.046	3.021	2.996	2.971	2.947
20°	2.924	2.901	2.878	2.855	2.833	2.812
21	2.790	2.769	2.749	2.729	2.709	2.689
22	2.669	2.650	2.632	2.613	2.595	2.577
23	2.559	2.542	2.525	2.508	2.491	2.475
24	2.459	2.443	2.427	2.411	2.396	2.381
25°	2.366	2.352	2.337	2.323	2.309	2.295
26	2.281	2.268	2.254	2.241	2.228	2.215
27	2.203	2.190	2.178	2.166	2.154	2.142
28	2.130	2.118	2.107	2.096	2.085	2.074
29	2.063	2.052	2.041	2.031	2.020	2.010
30°	2.000	1.990	1.980	1.970	1.961	1.951
31	1.942	1.932	1.923	1.914	1.905	1.896
32	1.887	1.878	1.870	1.861	1.853	1.844
33	1.836	1.828	1.820	1.812	1.804	1.796
34	1.788	1.781	1.773	1.766	1.758	1.751
35°	1.743	1.736	1.729	1.722	1.715	1.708
36	1.701	1.695	1.688	1.681	1.675	1.668
37	1.662	1.655	1.649	1.643	1.636	1.630
38	1.624	1.618	1.612	1.606	1.601	1.595
39	1.589	1.583	1.578	1.572	1.567	1.561
40°	1.556	1.550	1.545	1.540	1.535	1.529
41	1.524	1.519	1.514	1.509	1.504	1.499
42	1.494	1.490	1.485	1.480	1.476	1.471
43	1.466	1.462	1.457	1.453	1.448	1.444
44	1.440	1.435	1.431	1.427	1.423	1.418
45°	1.414	1.410	1.406	1.402	1.398	1.394

x	0′	10′	20′	30′	40′	50′
45°	1.414	1.410	1.406	1.402	1.398	1.394
46	1.390	1.386	1.382	1.379	1.375	1.371
47	1.367	1.364	1.360	1.356	1.353	1.349
48	1.346	1.342	1.339	1.335	1.332	1.328
49	1.325	1.322	1.318	1.315	1.312	1.309
50°	1.305	1.302	1.299	1.296	1.293	1.290
51	1.287	1.284	1.281	1.278	1.275	1.272
52	1.269	1.266	1.263	1.260	1.258	1.255
53	1.252	1.249	1.247	1.244	1.241	1.239
54	1.236	1.233	1.231	1.228	1.226	1.223
55°	1.221	1.218	1.216	1.213	1.211	1.209
56	1.206	1.204	1.202	1.199	1.197	1.195
57	1.192	1.190	1.188	1.186	1.184	1.181
58	1.179	1.177	1.175	1.173	1.171	1.169
59	1.167	1.165	1.163	1.161	1.159	1.157
60°	1.155	1.153	1.151	1.149	1.147	1.145
61	1.143	1.142	1.140	1.138	1.136	1.134
62	1.133	1.131	1.129	1.127	1.126	1.124
63	1.122	1.121	1.119	1.117	1.116	1.114
64	1.113	1.111	1.109	1.108	1.106	1.105
65°	1.103	1.102	1.100	1.099	1.097	1.096
66	1.095	1.093	1.092	1.090	1.089	1.088
67	1.086	1.085	1.084	1.082	1.081	1.080
68	1.079	1.077	1.076	1.075	1.074	1.072
69	1.071	1.070	1.069	1.068	1.066	1.065
70°	1.064	1.063	1.062	1.061	1.060	1.059
71	1.058	1.057	1.056	1.054	1.053	1.052
72	1.051	1.050	1.049	1.048	1.048	1.047
73	1.046	1.045	1.044	1.043	1.042	1.041
74	1.040	1.039	1.039	1.038	1.037	1.036
75°	1.035	1.034	1.034	1.033	1.032	1.031
76	1.031	1.030	1.029	1.028	1.028	1.027
77	1.026	1.026	1.025	1.024	1.024	1.023
78	1.022	1.022	1.021	1.020	1.020	1.019
79	1.019	1.018	1.018	1.017	1.016	1.016
80°	1.015	1.015	1.014	1.014	1.013	1.013
81	1.012	1.012	1.012	1.011	1.011	1.010
82	1.010	1.009	1.009	1.009	1.008	1.008
83	1.008	1.007	1.007	1.006	1.006	1.006
84	1.006	1.005	1.005	1.005	1.004	1.004
85°	1.004	1.004	1.003	1.003	1.003	1.003
86	1.002	1.002	1.002	1.002	1.002	1.002
87	1.001	1.001	1.001	1.001	1.001	1.001
88	1.001	1.001	1.000	1.000	1.000	1.000
89	1.000	1.000	1.000	1.000	1.000	1.000
90°	1.000					

TABLE 9

NATURAL TRIGONOMETRIC FUNCTIONS (in radians)

x	Sin x	Cos x	Tan x	Cot x	Sec x	Csc x
.00	.00000	1.00000	.00000	∞	1.00000	∞
.01	.01000	.99995	.01000	99.9967	1.00005	100.0017
.02	.02000	.99980	.02000	49.9933	1.00020	50.0033
.03	.03000	.99955	.03001	33.3233	1.00045	33.3383
.04	.03999	.99920	.04002	24.9867	1.00080	25.0067
.05	.04998	.99875	.05004	19.9833	1.00125	20.0083
.06	.05996	.99820	.06007	16.6467	1.00180	16.6767
.07	.06994	.99755	.07011	14.2624	1.00246	14.2974
.08	.07991	.99680	.08017	12.4733	1.00321	12.5133
.09	.08988	.99595	.09024	11.0811	1.00406	11.1261
.10	.09983	.99500	.10033	9.9666	1.00502	10.0167
.11	.10978	.99396	.11045	9.0542	1.00608	9.1093
.12	.11971	.99281	.12058	8.2933	1.00724	8.3534
.13	.12963	.99156	.13074	7.6489	1.00851	7.7140
.14	.13954	.99022	.14092	7.0961	1.00988	7.1662
.15	.14944	.98877	.15114	6.6166	1.01136	6.6917
.16	.15932	.98723	.16138	6.1966	1.01294	6.2767
.17	.16918	.98558	.17166	5.8256	1.01463	5.9108
.18	.17903	.98384	.18197	5.4954	1.01642	5.5857
.19	.18886	.98200	.19232	5.1997	1.01833	5.2950
.20	.19867	.98007	.20271	4.9332	1.02034	5.0335
.21	.20846	.97803	.21314	4.6917	1.02246	4.7971
.22	.21823	.97590	.22362	4.4719	1.02470	4.5823
.23	.22798	.97367	.23414	4.2709	1.02705	4.3864
.24	.23770	.97134	.24472	4.0864	1.02951	4.2069
.25	.24740	.96891	.25534	3.9163	1.03209	4.0420
.26	.25708	.96639	.26602	3.7591	1.03478	3.8898
.27	.26673	.96377	.27676	3.6133	1.03759	3.7491
.28	.27636	.96106	.28755	3.4776	1.04052	3.6185
.29	.28595	.95824	.29841	3.3511	1.04358	3.4971
.30	.29552	.95534	.30934	3.2327	1.04675	3.3839
.31	.30506	.95233	.32033	3.1218	1.05005	3.2781
.32	.31457	.94924	.33139	3.0176	1.05348	3.1790
.33	.32404	.94604	.34252	2.9195	1.05704	3.0860
.34	.33349	.94275	.35374	2.8270	1.06072	2.9986
.35	.34290	.93937	.36503	2.7395	1.06454	2.9163
.36	.35227	.93590	.37640	2.6567	1.06849	2.8387
.37	.36162	.93233	.38786	2.5782	1.07258	2.7654
.38	.37092	.92866	.39941	2.5037	1.07682	2.6960
.39	.38019	.92491	.41105	2.4328	1.08119	2.6303
.40	.38942	.92106	.42279	2.3652	1.08570	2.5679

Table 9 (continued) NATURAL TRIGONOMETRIC FUNCTIONS (in radians)

x	Sin x	Cos x	Tan x	Cot x	Sec x	Csc x
.40	.38942	.92106	.42279	2.3652	1.0857	2.5679
.41	.39861	.91712	.43463	2.3008	1.0904	2.5087
.42	.40776	.91309	.44657	2.2393	1.0952	2.4524
.43	.41687	.90897	.45862	2.1804	1.1002	2.3988
.44	.42594	.90475	.47078	2.1241	1.1053	2.3478
.45	.43497	.90045	.48306	2.0702	1.1106	2.2990
.46	.44395	.89605	.49545	2.0184	1.1160	2.2525
.47	.45289	.89157	.50797	1.9686	1.1216	2.2081
.48	.46178	.88699	.52061	1.9208	1.1274	2.1655
.49	.47063	.88233	.53339	1.8748	1.1334	2.1248
.50	.47943	.87758	.54630	1.8305	1.1395	2.0858
.51	.48818	.87274	.55936	1.7878	1.1458	2.0484
.52	.49688	.86782	.57256	1.7465	1.1523	2.0126
.53	.50553	.86281	.58592	1.7067	1.1590	1.9781
.54	.51414	.85771	.59943	1.6683	1.1659	1.9450
.55	.52269	.85252	.61311	1.6310	1.1730	1.9132
.56	.53119	.84726	.62695	1.5950	1.1803	1.8826
.57	.53963	.84190	.64097	1.5601	1.1878	1.8531
.58	.54802	.83646	.65517	1.5263	1.1955	1.8247
.59	.55636	.83094	.66956	1.4935	1.2035	1.7974
.60	.56464	.82534	.68414	1.4617	1.2116	1.7710
.61	.57287	.81965	.69892	1.4308	1.2200	1.7456
.62	.58104	.81388	.71391	1.4007	1.2287	1.7211
.63	.58914	.80803	.72911	1.3715	1.2376	1.6974
.64	.59720	.80210	.74454	1.3431	1.2467	1.6745
.65	.60519	.79608	.76020	1.3154	1.2561	1.6524
.66	.61312	.78999	.77610	1.2885	1.2658	1.6310
.67	.62099	.78382	.79225	1.2622	1.2758	1.6103
.68	.62879	.77757	.80866	1.2366	1.2861	1.5903
.69	.63654	.77125	.82534	1.2116	1.2966	1.5710
.70	.64422	.76484	.84229	1.1872	1.3075	1.5523
.71	.65183	.75836	.85953	1.1634	1.3186	1.5341
.72	.65938	.75181	.87707	1.1402	1.3301	1.5166
.73	.66687	.74517	.89492	1.1174	1.3420	1.4995
.74	.67429	.73847	.91309	1.0952	1.3542	1.4830
.75	.68164	.73169	.93160	1.0734	1.3667	1.4671
.76	.68892	.72484	.95045	1.0521	1.3796	1.4515
.77	.69614	.71791	.96967	1.0313	1.3929	1.4365
.78	.70328	.71091	.98926	1.0109	1.4066	1.4219
.79	.71035	.70385	1.0092	.99084	1.4208	1.4078
.80	.71736	.69671	1.0296	.97121	1.4353	1.3940

Table 9
(continued)

NATURAL TRIGONOMETRIC FUNCTIONS (in radians)

x	Sin x	Cos x	Tan x	Cot x	Sec x	Csc x
.80	.71736	.69671	1.0296	.97121	1.4353	1.3940
.81	.72429	.68950	1.0505	.95197	1.4503	1.3807
.82	.73115	.68222	1.0717	.93309	1.4658	1.3677
.83	.73793	.67488	1.0934	.91455	1.4818	1.3551
.84	.74464	.66746	1.1156	.89635	1.4982	1.3429
.85	.75128	.65998	1.1383	.87848	1.5152	1.3311
.86	.75784	.65244	1.1616	.86091	1.5327	1.3195
.87	.76433	.64483	1.1853	.84365	1.5508	1.3083
.88	.77074	.63715	1.2097	.82668	1.5695	1.2975
.89	.77707	.62941	1.2346	.80998	1.5888	1.2869
.90	.78333	.62161	1.2602	.79355	1.6087	1.2766
.91	.78950	.61375	1.2864	.77738	1.6293	1.2666
.92	.79560	.60582	1.3133	.76146	1.6507	1.2569
.93	.80162	.59783	1.3409	.74578	1.6727	1.2475
.94	.80756	.58979	1.3692	.73034	1.6955	1.2383
.95	.81342	.58168	1.3984	.71511	1.7191	1.2294
.96	.81919	.57352	1.4284	.70010	1.7436	1.2207
.97	.82489	.56530	1.4592	.68531	1.7690	1.2123
.98	.83050	.55702	1.4910	.67071	1.7953	1.2041
.99	.83603	.54869	1.5237	.65631	1.8225	1.1961
1.00	.84147	.54030	1.5574	.64209	1.8508	1.1884
1.01	.84683	.53186	1.5922	.62806	1.8802	1.1809
1.02	.85211	.52337	1.6281	.61420	1.9107	1.1736
1.03	.85730	.51482	1.6652	.60051	1.9424	1.1665
1.04	.86240	.50622	1.7036	.58699	1.9754	1.1595
1.05	.86742	.49757	1.7433	.57362	2.0098	1.1528
1.06	.87236	.48887	1.7844	.56040	2.0455	1.1463
1.07	.87720	.48012	1.8270	.54734	2.0828	1.1400
1.08	.88196	.47133	1.8712	.53441	2.1217	1.1338
1.09	.88663	.46249	1.9171	.52162	2.1622	1.1279
1.10	.89121	.45360	1.9648	.50897	2.2046	1.1221
1.11	.89570	.44466	2.0143	.49644	2.2489	1.1164
1.12	.90010	.43568	2.0660	.48404	2.2952	1.1110
1.13	.90441	.42666	2.1198	.47175	2.3438	1.1057
1.14	.90863	.41759	2.1759	.45959	2.3947	1.1006
1.15	.91276	.40849	2.2345	.44753	2.4481	1.0956
1.16	.91680	.39934	2.2958	.43558	2.5041	1.0907
1.17	.92075	.39015	2.3600	.42373	2.5631	1.0861
1.18	.92461	.38092	2.4273	.41199	2.6252	1.0815
1.19	.92837	.37166	2.4979	.40034	2.6906	1.0772
1.20	.93204	.36236	2.5722	.38878	2.7597	1.0729

Table 9
(continued)

NATURAL TRIGONOMETRIC FUNCTIONS (in radians)

x	Sin x	Cos x	Tan x	Cot x	Sec x	Csc x
1.20	.93204	.36236	2.5722	.38878	2.7597	1.07292
1.21	.93562	.35302	2.6503	.37731	2.8327	1.06881
1.22	.93910	.34365	2.7328	.36593	2.9100	1.06485
1.23	.94249	.33424	2.8198	.35463	2.9919	1.06102
1.24	.94578	.32480	2.9119	.34341	3.0789	1.05732
1.25	.94898	.31532	3.0096	.33227	3.1714	1.05376
1.26	.95209	.30582	3.1133	.32121	3.2699	1.05032
1.27	.95510	.29628	3.2236	.31021	3.3752	1.04701
1.28	.95802	.28672	3.3414	.29928	3.4878	1.04382
1.29	.96084	.27712	3.4672	.28842	3.6085	1.04076
1.30	.96356	.26750	3.6021	.27762	3.7383	1.03782
1.31	.96618	.25785	3.7471	.26687	3.8782	1.03500
1.32	.96872	.24818	3.9033	.25619	4.0294	1.03230
1.33	.97115	.23848	4.0723	.24556	4.1933	1.02971
1.34	.97348	.22875	4.2556	.23498	4.3715	1.02724
1.35	.97572	.21901	4.4552	.22446	4.5661	1.02488
1.36	.97786	.20924	4.6734	.21398	4.7792	1.02264
1.37	.97991	.19945	4.9131	.20354	5.0138	1.02050
1.38	.98185	.18964	5.1774	.19315	5.2731	1.01848
1.39	.98370	.17981	5.4707	.18279	5.5613	1.01657
1.40	.98545	.16997	5.7979	.17248	5.8835	1.01477
1.41	.98710	.16010	6.1654	.16220	6.2459	1.01307
1.42	.98865	.15023	6.5811	.15195	6.6567	1.01148
1.43	.99010	.14033	7.0555	.14173	7.1260	1.00999
1.44	.99146	.13042	7.6018	.13155	7.6673	1.00862
1.45	.99271	.12050	8.2381	.12139	8.2986	1.00734
1.46	.99387	.11057	8.9886	.11125	9.0441	1.00617
1.47	.99492	.10063	9.8874	.10114	9.9378	1.00510
1.48	.99588	.09067	10.9834	.09105	11.0288	1.00414
1.49	.99674	.08071	12.3499	.08097	12.3903	1.00327
1.50	.99749	.07074	14.1014	.07091	14.1368	1.00251
1.51	.99815	.06076	16.4281	.06087	16.4585	1.00185
1.52	.99871	.05077	19.6695	.05084	19.6949	1.00129
1.53	.99917	.04079	24.4984	.04082	24.5188	1.00083
1.54	.99953	.03079	32.4611	.03081	32.4765	1.00047
1.55	.99978	.02079	48.0785	.02080	48.0889	1.00022
1.56	.99994	.01080	92.6205	.01080	92.6259	1.00006
1.57	1.00000	.00080	1255.77	.00080	1255.77	1.00000
1.58	.99996	−.00920	−108.649	−.00920	−108.654	1.00004
1.59	.99982	−.01920	−52.0670	−.01921	−52.0766	1.00018
1.60	.99957	−.02920	−34.2325	−.02921	−34.2471	1.00043

TABLE 10

log sin x (x in degrees and minutes)

[subtract 10 from each entry]

x	0′	10′	20′	30′	40′	50′
0°	–	7.4637	7.7648	7.9408	8.0658	8.1627
1	8.2419	8.3088	8.3668	8.4179	8.4637	8.5050
2	8.5428	8.5776	8.6097	8.6397	8.6677	8.6940
3	8.7188	8.7423	8.7645	8.7857	8.8059	8.8251
4	8.8436	8.8613	8.8783	8.8946	8.9104	8.9256
5°	8.9403	8.9545	8.9682	8.9816	8.9945	9.0070
6	9.0192	9.0311	9.0426	9.0539	9.0648	9.0755
7	9.0859	9.0961	9.1060	9.1157	9.1252	9.1345
8	9.1436	9.1525	9.1612	9.1697	9.1781	9.1863
9	9.1943	9.2022	9.2100	9.2176	9.2251	9.2324
10°	9.2397	9.2468	9.2538	9.2606	9.2674	9.2740
11	9.2806	9.2870	9.2934	9.2997	9.3058	9.3119
12	9.3179	9.3238	9.3296	9.3353	9.3410	9.3466
13	9.3521	9.3575	9.3629	9.3682	9.3734	9.3786
14	9.3837	9.3887	9.3937	9.3986	9.4035	9.4083
15°	9.4130	9.4177	9.4223	9.4269	9.4314	9.4359
16	9.4403	9.4447	9.4491	9.4533	9.4576	9.4618
17	9.4659	9.4700	9.4741	9.4781	9.4821	9.4861
18	9.4900	9.4939	9.4977	9.5015	9.5052	9.5090
19	9.5126	9.5163	9.5199	9.5235	9.5270	9.5306
20°	9.5341	9.5375	9.5409	9.5443	9.5477	9.5510
21	9.5543	9.5576	9.5609	9.5641	9.5673	9.5704
22	9.5736	9.5767	9.5798	9.5828	9.5859	9.5889
23	9.5919	9.5948	9.5978	9.6007	9.6036	9.6065
24	9.6093	9.6121	9.6149	9.6177	9.6205	9.6232
25°	9.6259	9.6286	9.6313	9.6340	9.6366	9.6392
26	9.6418	9.6444	9.6470	9.6495	9.6521	9.6546
27	9.6570	9.6595	9.6620	9.6644	9.6668	9.6692
28	9.6716	9.6740	9.6763	9.6787	9.6810	9.6833
29	9.6856	9.6878	9.6901	9.6923	9.6946	9.6968
30°	9.6990	9.7012	9.7033	9.7055	9.7076	9.7097
31	9.7118	9.7139	9.7160	9.7181	9.7201	9.7222
32	9.7242	9.7262	9.7282	9.7302	9.7322	9.7342
33	9.7361	9.7380	9.7400	9.7419	9.7438	9.7457
34	9.7476	9.7494	9.7513	9.7531	9.7550	9.7568
35°	9.7586	9.7604	9.7622	9.7640	9.7657	9.7675
36	9.7692	9.7710	9.7727	9.7744	9.7761	9.7778
37	9.7795	9.7811	9.7828	9.7844	9.7861	9.7877
38	9.7893	9.7910	9.7926	9.7941	9.7957	9.7973
39	9.7989	9.8004	9.8020	9.8035	9.8050	9.8066
40°	9.8081	9.8096	9.8111	9.8125	9.8140	9.8155
41	9.8169	9.8184	9.8198	9.8213	9.8227	9.8241
42	9.8255	9.8269	9.8283	9.8297	9.8311	9.8324
43	9.8338	9.8351	9.8365	9.8378	9.8391	9.8405
44	9.8418	9.8431	9.8444	9.8457	9.8469	9.8482
45°	9.8495	9.8507	9.8520	9.8532	9.8545	9.8557

Table 10
(continued)

log sin x (x in degrees and minutes)

[subtract 10 from each entry]

x	0′	10′	20′	30′	40′	50′
45°	9.8495	9.8507	9.8520	9.8532	9.8545	9.8557
46	9.8569	9.8582	9.8594	9.8606	9.8618	9.8629
47	9.8641	9.8653	9.8665	9.8676	9.8688	9.8699
48	9.8711	9.8722	9.8733	9.8745	9.8756	9.8767
49	9.8778	9.8789	9.8800	9.8810	9.8821	9.8832
50°	9.8843	9.8853	9.8864	9.8874	9.8884	9.8895
51	9.8905	9.8915	9.8925	9.8935	9.8945	3.8955
52	9.8965	9.8975	9.8985	9.8995	9.9004	9.9014
53	9.9023	9.9033	9.9042	9.9052	9.9061	9.9070
54	9.9080	9.9089	9.9098	9.9107	9.9116	9.9125
55°	9.9134	9.9142	9.9151	9.9160	9.9169	9.9177
56	9.9186	9.9194	9.9203	9.9211	9.9219	9.9228
57	9.9236	9.9244	9.9252	9.9260	9.9268	9.9276
58	9.9284	9.9292	9.9300	9.9308	9.9315	9.9323
59	9.9331	9.9338	9.9346	9.9353	9.9361	9.9368
60°	9.9375	9.9383	9.9390	9.9397	9.9404	9.9411
61	9.9418	9.9425	9.9432	9.9439	9.9446	9.9453
62	9.9459	9.9466	9.9473	9.9479	9.9486	9.9492
63	9.9499	9.9505	9.9512	9.9518	9.9524	9.9530
64	9.9537	9.9543	9.9549	9.9555	9.9561	9.9567
65°	9.9573	9.9579	9.9584	9.9590	9.9596	9.9602
66	9.9607	9.9613	9.9618	9.9624	9.9629	9.9635
67	9.9640	9.9646	9.9651	9.9656	9.9661	9.9667
68	9.9672	9.9677	9.9682	9.9687	9.9692	9.9697
69	9.9702	9.9706	9.9711	9.9716	9.9721	9.9725
70°	9.9730	9.9734	9.9739	9.9743	9.9748	9.9752
71	9.9757	9.9761	9.9765	9.9770	9.9774	9.9778
72	9.9782	9.9786	9.9790	9.9794	9.9798	9.9802
73	9.9806	9.9810	9.9814	9.9817	9.9821	9.9825
74	9.9828	9.9832	9.9836	9.9839	9.9843	9.9846
75°	9.9849	9.9853	9.9856	9.9859	9.9863	9.9866
76	9.9869	9.9872	9.9875	9.9878	9.9881	9.9884
77	9.9887	9.9890	9.9893	9.9896	9.9899	9.9901
78	9.9904	9.9907	9.9909	9.9912	9.9914	9.9917
79	9.9919	9.9922	9.9924	9.9927	9.9929	9.9931
80°	9.9934	9.9936	9.9938	9.9940	9.9942	9.9944
81	9.9946	9.9948	9.9950	9.9952	9.9954	9.9956
82	9.9958	9.9959	9.9961	9.9963	9.9964	9.9966
83	9.9968	9.9969	9.9971	9.9972	9.9973	9.9975
84	9.9976	9.9977	9.9979	9.9980	9.9981	9.9982
85°	9.9983	9.9985	9.9986	9.9987	9.9988	9.9989
86	9.9989	9.9990	9.9991	9.9992	9.9993	9.9993
87	9.9994	9.9995	9.9995	9.9996	9.9996	9.9997
88	9.9997	9.9998	9.9998	9.9999	9.9999	9.9999
89	9.9999	10.0000	10.0000	10.0000	10.0000	10.0000
90°	10.0000					

TABLE 11

log cos x (x in degrees and minutes)

[subtract 10 from each entry]

x	0′	10′	20′	30′	40′	50′
0°	10.0000	10.0000	10.0000	10.0000	10.0000	10.0000
1	9.9999	9.9999	9.9999	9.9999	9.9998	9.9998
2	9.9997	9.9997	9.9996	9.9996	9.9995	9.9995
3	9.9994	9.9993	9.9993	9.9992	9.9991	9.9990
4	9.9989	9.9989	9.9988	9.9987	9.9986	9.9985
5°	9.9983	9.9982	9.9981	9.9980	9.9979	9.9977
6	9.9976	9.9975	9.9973	9.9972	9.9971	9.9969
7	9.9968	9.9966	9.9964	9.9963	9.9961	9.9959
8	9.9958	9.9956	9.9954	9.9952	9.9950	9.9948
9	9.9946	9.9944	9.9942	9.9940	9.9938	9.9936
10°	9.9934	9.9931	9.9929	9.9927	9.9924	9.9922
11	9.9919	9.9917	9.9914	9.9912	9.9909	9.9907
12	9.9904	9.9901	9.9899	9.9896	9.9893	9.9890
13	9.9887	9.9884	9.9881	9.9878	9.9875	9.9872
14	9.9869	9.9866	9.9863	9.9859	9.9856	9.9853
15°	9.9849	9.9846	9.9843	9.9839	9.9836	9.9832
16	9.9828	9.9825	9.9821	9.9817	9.9814	9.9810
17	9.9806	9.9802	9.9798	9.9794	9.9790	9.9786
18	9.9782	9.9778	9.9774	9.9770	9.9765	9.9761
19	9.9757	9.9752	9.9748	9.9743	9.9739	9.9734
20°	9.9730	9.9725	9.9721	9.9716	9.9711	9.9706
21	9.9702	9.9697	9.9692	9.9687	9.9682	9.9677
22	9.9672	9.9667	9.9661	9.9656	9.9651	9.9646
23	9.9640	9.9635	9.9629	9.9624	9.9618	9.9613
24	9.9607	9.9602	9.9596	9.9590	9.9584	9.9579
25°	9.9573	9.9567	9.9561	9.9555	9.9549	9.9543
26	9.9537	9.9530	9.9524	9.9518	9.9512	9.9505
27	9.9499	9.9492	9.9486	9.9479	9.9473	9.9466
28	9.9459	9.9453	9.9446	9.9439	9.9432	9.9425
29	9.9418	9.9411	9.9404	9.9397	9.9390	9.9383
30°	9.9375	9.9368	9.9361	9.9353	9.9346	9.9338
31	9.9331	9.9323	9.9315	9.9308	9.9300	9.9292
32	9.9284	9.9276	9.9268	9.9260	9.9252	9.9244
33	9.9236	9.9228	9.9219	9.9211	9.9203	9.9194
34	9.9186	9.9177	9.9169	9.9160	9.9151	9.9142
35°	9.9134	9.9125	9.9116	9.9107	9.9098	9.9089
36	9.9080	9.9070	9.9061	9.9052	9.9042	9.9033
37	9.9023	9.9014	9.9004	9.8995	9.8985	9.8975
38	9.8965	9.8955	9.8945	9.8935	9.8925	9.8915
39	9.8905	9.8895	9.8884	9.8874	9.8864	9.8853
40°	9.8843	9.8832	9.8821	9.8810	9.8800	9.8789
41	9.8778	9.8767	9.8756	9.8745	9.8733	9.8722
42	9.8711	9.8699	9.8688	9.8676	9.8665	9.8653
43	9.8641	9.8629	9.8618	9.8606	9.8594	9.8582
44	9.8569	9.8557	9.8545	9.8532	9.8520	9.8507
45°	9.8495	9.8482	9.8469	9.8457	9.8444	9.8431

Table 11
(continued)

log cos x (x in degrees and minutes)

[subtract 10 from each entry]

x	0′	10′	20′	30′	40′	50′
45°	9.8495	9.8482	9.8469	9.8457	9.8444	9.8431
46	9.8418	9.8405	9.8391	9.8378	9.8365	9.8351
47	9.8338	9.8324	9.8311	9.8297	9.8283	9.8269
48	9.8255	9.8241	9.8227	9.8213	9.8198	9.8184
49	9.8169	9.8155	9.8140	9.8125	9.8111	9.8096
50°	9.8081	9.8066	9.8050	9.8035	9.8020	9.8004
51	9.7989	9.7973	9.7957	9.7941	9.7926	9.7910
52	9.7893	9.7877	9.7861	9.7844	9.7828	9.7811
53	9.7795	9.7778	9.7761	9.7744	9.7727	9.7710
54	9.7692	9.7675	9.7657	9.7640	9.7622	9.7604
55°	9.7586	9.7568	9.7550	9.7531	9.7513	9.7494
56	9.7476	9.7457	9.7438	9.7419	9.7400	9.7380
57	9.7361	9.7342	9.7322	9.7302	9.7282	9.7262
58	9.7242	9.7222	9.7201	9.7181	9.7160	9.7139
59	9.7118	9.7097	9.7076	9.7055	9.7033	9.7012
60°	9.6990	9.6968	9.6946	9.6923	9.6901	9.6878
61	9.6856	9.6833	9.6810	9.6787	9.6763	9.6740
62	9.6716	9.6692	9.6668	9.6644	9.6620	9.6595
63	9.6570	9.6546	9.6521	9.6495	9.6470	9.6444
64	9.6418	9.6392	9.6366	9.6340	9.6313	9.6286
65°	9.6259	9.6232	9.6205	9.6177	9.6149	9.6121
66	9.6093	9.6065	9.6036	9.6007	9.5978	9.5948
67	9.5919	9.5889	9.5859	9.5828	9.5798	9.5767
68	9.5736	9.5704	9.5673	9.5641	9.5609	9.5576
69	9.5543	9.5510	9.5477	9.5443	9.5409	9.5375
70°	9.5341	9.5306	9.5270	9.5235	9.5199	9.5163
71	9.5126	9.5090	9.5052	9.5015	9.4977	9.4939
72	9.4900	9.4861	9.4821	9.4781	9.4741	9.4700
73	9.4659	9.4618	9.4576	9.4533	9.4491	9.4447
74	9.4403	9.4359	9.4314	9.4269	9.4223	9.4177
75°	9.4130	9.4083	9.4035	9.3986	9.3937	9.3887
76	9.3837	9.3786	9.3734	9.3682	9.3629	9.3575
77	9.3521	9.3466	9.3410	9.3353	9.3296	9.3238
78	9.3179	9.3119	9.3058	9.2997	9.2934	9.2870
79	9.2806	9.2740	9.2674	9.2606	9.2538	9.2468
80°	9.2397	9.2324	9.2251	9.2176	9.2100	9.2022
81	9.1943	9.1863	9.1781	9.1697	9.1612	9.1525
82	9.1436	9.1345	9.1252	9.1157	9.1060	9.0961
83	9.0859	9.0755	9.0648	9.0539	9.0426	9.0311
84	9.0192	9.0070	8.9945	8.9816	8.9682	8.9545
85°	8.9403	8.9256	8.9104	8.8946	8.8783	8.8613
86	8.8436	8.8251	8.8059	8.7857	8.7645	8.7423
87	8.7188	8.6940	8.6677	8.6397	8.6097	8.5776
88	8.5428	8.5050	8.4637	8.4179	8.3668	8.3088
89	8.2419	8.1627	8.0658	7.9408	7.7648	7.4637
90°						

TABLE 12

log tan x (x in degrees and minutes)

[subtract 10 from each entry]

x	0′	10′	20′	30′	40′	50′
0°	—	7.4637	7.7648	7.9409	8.0658	8.1627
1	8.2419	8.3089	8.3669	8.4181	8.4638	8.5053
2	8.5431	8.5779	8.6101	8.6401	8.6682	8.6945
3	8.7194	8.7429	8.7652	8.7865	8.8067	8.8261
4	8.8446	8.8624	8.8795	8.8960	8.9118	8.9272
5°	8.9420	8.9563	8.9701	8.9836	8.9966	9.0093
6	9.0216	9.0336	9.0453	9.0567	9.0678	9.0786
7	9.0891	9.0995	9.1096	9.1194	9.1291	9.1385
8	9.1478	9.1569	9.1658	9.1745	9.1831	9.1915
9	9.1997	9.2078	9.2158	9.2236	9.2313	9.2389
10°	9.2463	9.2536	9.2609	9.2680	9.2750	9.2819
11	9.2887	9.2953	9.3020	9.3085	9.3149	9.3212
12	9.3275	9.3336	9.3397	9.3458	9.3517	9.3576
13	9.3634	9.3691	9.3748	9.3804	9.3859	9.3914
14	9.3968	9.4021	9.4074	9.4127	9.4178	9.4230
15°	9.4281	9.4331	9.4381	9.4430	9.4479	9.4527
16	9.4575	9.4622	9.4669	9.4716	9.4762	9.4808
17	9.4853	9.4898	9.4943	9.4987	9.5031	9.5075
18	9.5118	9.5161	9.5203	9.5245	9.5287	9.5329
19	9.5370	9.5411	9.5451	9.5491	9.5531	9.5571
20°	9.5611	9.5650	9.5689	9.5727	9.5766	9.5804
21	9.5842	9.5879	9.5917	9.5954	9.5991	9.6028
22	9.6064	9.6100	9.6136	9.6172	9.6208	9.6243
23	9.6279	9.6314	9.6348	9.6383	9.6417	9.6452
24	9.6486	9.6520	9.6553	9.6587	9.6620	9.6654
25°	9.6687	9.6720	9.6752	9.6785	9.6817	9.6850
26	9.6882	9.6914	9.6946	9.6977	9.7009	9.7040
27	9.7072	9.7103	9.7134	9.7165	9.7196	9.7226
28	9.7257	9.7287	9.7317	9.7348	9.7378	9.7408
29	9.7438	9.7467	9.7497	9.7526	9.7556	9.7585
30°	9.7614	9.7644	9.7673	9.7701	9.7730	9.7759
31	9.7788	9.7816	9.7845	9.7873	9.7902	9.7930
32	9.7958	9.7986	9.8014	9.8042	9.8070	9.8097
33	9.8125	9.8153	9.8180	9.8208	9.8235	9.8263
34	9.8290	9.8317	9.8344	9.8371	9.8398	9.8425
35°	9.8452	9.8479	9.8506	9.8533	9.8559	9.8586
36	9.8613	9.8639	9.8666	9.8692	9.8718	9.8745
37	9.8771	9.8797	9.8824	9.8850	9.8876	9.8902
38	9.8928	9.8954	9.8980	9.9006	9.9032	9.9058
39	9.9084	9.9110	9.9135	9.9161	9.9187	9.9212
40°	9.9238	9.9264	9.9289	9.9315	9.9341	9.9366
41	9.9392	9.9417	9.9443	9.9468	9.9494	9.9519
42	9.9544	9.9570	9.9595	9.9621	9.9646	9.9671
43	9.9697	9.9722	9.9747	9.9772	9.9798	9.9823
44	9.9848	9.9874	9.9899	9.9924	9.9949	9.9975
45°	10.0000	10.0025	10.0051	10.0076	10.0101	10.0126

Table 12 (continued)

log tan x (x in degrees and minutes)

[subtract 10 from each entry]

x	0′	10′	20′	30′	40′	50′
45°	10.0000	10.0025	10.0051	10.0076	10.0101	10.0126
46	10.0152	10.0177	10.0202	10.0228	10.0253	10.0278
47	10.0303	10.0329	10.0354	10.0379	10.0405	10.0430
48	10.0456	10.0481	10.0506	10.0532	10.0557	10.0583
49	10.0608	10.0634	10.0659	10.0685	10.0711	10.0736
50°	10.0762	10.0788	10.0813	10.0839	10.0865	10.0890
51	10.0916	10.0942	10.0968	10.0994	10.1020	10.1046
52	10.1072	10.1098	10.1124	10.1150	10.1176	10.1203
53	10.1229	10.1255	10.1282	10.1308	10.1334	10.1361
54	10.1387	10.1414	10.1441	10.1467	10.1494	10.1521
55°	10.1548	10.1575	10.1602	10.1629	10.1656	10.1683
56	10.1710	10.1737	10.1765	10.1792	10.1820	10.1847
57	10.1875	10.1903	10.1930	10.1958	10.1986	10.2014
58	10.2042	10.2070	10.2098	10.2127	10.2155	10.2184
59	10.2212	10.2241	10.2270	10.2299	10.2327	10.2356
60°	10.2386	10.2415	10.2444	10.2474	10.2503	10.2533
61	10.2562	10.2592	10.2622	10.2652	10.2683	10.2713
62	10.2743	10.2774	10.2804	10.2835	10.2866	10.2897
63	10.2928	10.2960	10.2991	10.3023	10.3054	10.3086
64	10.3118	10.3150	10.3183	10.3215	10.3248	10.3280
65°	10.3313	10.3346	10.3380	10.3413	10.3447	10.3480
66	10.3514	10.3548	10.3583	10.3617	10.3652	10.3686
67	10.3721	10.3757	10.3792	10.3828	10.3864	10.3900
68	10.3936	10.3972	10.4009	10.4046	10.4083	10.4121
69	10.4158	10.4196	10.4234	10.4273	10.4311	10.4350
70°	10.4389	10.4429	10.4469	10.4509	10.4549	10.4589
71	10.4630	10.4671	10.4713	10.4755	10.4797	10.4839
72	10.4882	10.4925	10.4969	10.5013	10.5057	10.5102
73	10.5147	10.5192	10.5238	10.5284	10.5331	10.5378
74	10.5425	10.5473	10.5521	10.5570	10.5619	10.5669
75°	10.5719	10.5770	10.5822	10.5873	10.5926	10.5979
76	10.6032	10.6086	10.6141	10.6196	10.6252	10.6309
77	10.6366	10.6424	10.6483	10.6542	10.6603	10.6664
78	10.6725	10.6788	10.6851	10.6915	10.6980	10.7047
79	10.7113	10.7181	10.7250	10.7320	10.7391	10.7464
80°	10.7537	10.7611	10.7687	10.7764	10.7842	10.7922
81	10.8003	10.8085	10.8169	10.8255	10.8342	10.8431
82	10.8522	10.8615	10.8709	10.8806	10.8904	10.9005
83	10.9109	10.9214	10.9322	10.9433	10.9547	10.9664
84	10.9784	10.9907	11.0034	11.0164	11.0299	11.0437
85°	11.0580	11.0728	11.0882	11.1040	11.1205	11.1376
86	11.1554	11.1739	11.1933	11.2135	11.2348	11.2571
87	11.2806	11.3055	11.3318	11.3599	11.3899	11.4221
88	11.4569	11.4947	11.5362	11.5819	11.6331	11.6911
89	11.7581	11.8373	11.9342	12.0591	12.2352	12.5363
90°						

TABLE 13

CONVERSION OF RADIANS TO DEGREES, MINUTES AND SECONDS OR FRACTIONS OF DEGREES

Radians	Deg.	Min.	Sec.	Fractions of Degrees
1	57°	17′	44.8″	57.2958°
2	114°	35′	29.6″	114.5916°
3	171°	53′	14.4″	171.8873°
4	229°	10′	59.2″	229.1831°
5	286°	28′	44.0″	286.4789°
6	343°	46′	28.8″	343.7747°
7	401°	4′	13.6″	401.0705°
8	458°	21′	58.4″	458.3662°
9	515°	39′	43.3″	515.6620°
10	572°	57′	28.1″	572.9578°
.1	5°	43′	46.5″	
.2	11°	27′	33.0″	
.3	17°	11′	19.4″	
.4	22°	55′	5.9″	
.5	28°	38′	52.4″	
.6	34°	22′	38.9″	
.7	40°	6′	25.4″	
.8	45°	50′	11.8″	
.9	51°	33′	58.3″	
.01	0°	34′	22.6″	
.02	1°	8′	45.3″	
.03	1°	43′	7.9″	
.04	2°	17′	30.6″	
.05	2°	51′	53.2″	
.06	3°	26′	15.9″	
.07	4°	0′	38.5″	
.08	4°	35′	1.2″	
.09	5°	9′	23.8″	
.001	0°	3′	26.3″	
.002	0°	6′	52.5″	
.003	0°	10′	18.8″	
.004	0°	13′	45.1″	
.005	0°	17′	11.3″	
.006	0°	20′	37.6″	
.007	0°	24′	3.9″	
.008	0°	27′	30.1″	
.009	0°	30′	56.4″	
.0001	0°	0′	20.6″	
.0002	0°	0′	41.3″	
.0003	0°	1′	1.9″	
.0004	0°	1′	22.5″	
.0005	0°	1′	43.1″	
.0006	0°	2′	3.8″	
.0007	0°	2′	24.4″	
.0008	0°	2′	45.0″	
.0009	0°	3′	5.6″	

TABLE 14

CONVERSION OF DEGREES, MINUTES AND SECONDS TO RADIANS

Degrees	Radians
1°	.0174533
2°	.0349066
3°	.0523599
4°	.0698132
5°	.0872665
6°	.1047198
7°	.1221730
8°	.1396263
9°	.1570796
10°	.1745329

Minutes	Radians
1′	.00029089
2′	.00058178
3′	.00087266
4′	.00116355
5′	.00145444
6′	.00174533
7′	.00203622
8′	.00232711
9′	.00261800
10′	.00290888

Seconds	Radians
1″	.0000048481
2″	.0000096963
3″	.0000145444
4″	.0000193925
5″	.0000242407
6″	.0000290888
7″	.0000339370
8″	.0000387851
9″	.0000436332
10″	.0000484814

TABLE

15

NATURAL OR NAPIERIAN LOGARITHMS

$\log_e x$ or $\ln x$

x	0	1	2	3	4	5	6	7	8	9
1.0	.00000	.00995	.01980	.02956	.03922	.04879	.05827	.06766	.07696	.08618
1.1	.09531	.10436	.11333	.12222	.13103	.13976	.14842	.15700	.16551	.17395
1.2	.18232	.19062	.19885	.20701	.21511	.22314	.23111	.23902	.24686	.25464
1.3	.26236	.27003	.27763	.28518	.29267	.30010	.30748	.31481	.32208	.32930
1.4	.33647	.34359	.35066	.35767	.36464	.37156	.37844	.38526	.39204	.39878
1.5	.40547	.41211	.41871	.42527	.43178	.43825	.44469	.45108	.45742	.46373
1.6	.47000	.47623	.48243	.48858	.49470	.50078	.50682	.51282	.51879	.52473
1.7	.53063	.53649	.54232	.54812	.55389	.55962	.56531	.57098	.57661	.58222
1.8	.58779	.59333	.59884	.60432	.60977	.61519	.62058	.62594	.63127	.63658
1.9	.64185	.64710	.65233	.65752	.66269	.66783	.67294	.67803	.68310	.68813
2.0	.69315	.69813	.70310	.70804	.71295	.71784	.72271	.72755	.73237	.73716
2.1	.74194	.74669	.75142	.75612	.76081	.76547	.77011	.77473	.77932	.78390
2.2	.78846	.79299	.79751	.80200	.80648	.81093	.81536	.81978	.82418	.82855
2.3	.83291	.83725	.84157	.84587	.85015	.85442	.85866	.86289	.86710	.87129
2.4	.87547	.87963	.88377	.88789	.89200	.89609	.90016	.90422	.90826	.91228
2.5	.91629	.92028	.92426	.92822	.93216	.93609	.94001	.94391	.94779	.95166
2.6	.95551	.95935	.96317	.96698	.97078	.97456	.97833	.98208	.98582	.98954
2.7	.99325	.99695	1.00063	1.00430	1.00796	1.01160	1.01523	1.01885	1.02245	1.02604
2.8	1.02962	1.03318	1.03674	1.04028	1.04380	1.04732	1.05082	1.05431	1.05779	1.06126
2.9	1.06471	1.06815	1.07158	1.07500	1.07841	1.08181	1.08519	1.08856	1.09192	1.09527
3.0	1.09861	1.10194	1.10526	1.10856	1.11186	1.11514	1.11841	1.12168	1.12493	1.12817
3.1	1.13140	1.13462	1.13783	1.14103	1.14422	1.14740	1.15057	1.15373	1.15688	1.16002
3.2	1.16315	1.16627	1.16938	1.17248	1.17557	1.17865	1.18173	1.18479	1.18784	1.19089
3.3	1.19392	1.19695	1.19996	1.20297	1.20597	1.20896	1.21194	1.21491	1.21788	1.22083
3.4	1.22378	1.22671	1.22964	1.23256	1.23547	1.23837	1.24127	1.24415	1.24703	1.24990
3.5	1.25276	1.25562	1.25846	1.26130	1.26413	1.26695	1.26976	1.27257	1.27536	1.27815
3.6	1.28093	1.28371	1.28647	1.28923	1.29198	1.29473	1.29746	1.30019	1.30291	1.30563
3.7	1.30833	1.31103	1.31372	1.31641	1.31909	1.32176	1.32442	1.32708	1.32972	1.33237
3.8	1.33500	1.33763	1.34025	1.34286	1.34547	1.34807	1.35067	1.35325	1.35584	1.35841
3.9	1.36098	1.36354	1.36609	1.36864	1.37118	1.37372	1.37624	1.37877	1.38128	1.38379
4.0	1.38629	1.38879	1.39128	1.39377	1.39624	1.39872	1.40118	1.40364	1.40610	1.40854
4.1	1.41099	1.41342	1.41585	1.41828	1.42070	1.42311	1.42552	1.42792	1.43031	1.43270
4.2	1.43508	1.43746	1.43984	1.44220	1.44456	1.44692	1.44927	1.45161	1.45395	1.45629
4.3	1.45862	1.46094	1.46326	1.46557	1.46787	1.47018	1.47247	1.47476	1.47705	1.47933
4.4	1.48160	1.48387	1.48614	1.48840	1.49065	1.49290	1.49515	1.49739	1.49962	1.50185
4.5	1.50408	1.50630	1.50851	1.51072	1.51293	1.51513	1.51732	1.51951	1.52170	1.52388
4.6	1.52606	1.52823	1.53039	1.53256	1.53471	1.53687	1.53902	1.54116	1.54330	1.54543
4.7	1.54756	1.54969	1.55181	1.55393	1.55604	1.55814	1.56025	1.56235	1.56444	1.56653
4.8	1.56862	1.57070	1.57277	1.57485	1.57691	1.57898	1.58104	1.58309	1.58515	1.58719
4.9	1.58924	1.59127	1.59331	1.59534	1.59737	1.59939	1.60141	1.60342	1.60543	1.60744

$\ln 10 = 2.30259$ $\quad 4 \ln 10 = 9.21034$ $\quad 7 \ln 10 = 16.11810$

$2 \ln 10 = 4.60517$ $\quad 5 \ln 10 = 11.51293$ $\quad 8 \ln 10 = 18.42068$

$3 \ln 10 = 6.90776$ $\quad 6 \ln 10 = 13.81551$ $\quad 9 \ln 10 = 20.72327$

Table 15
(continued)

NATURAL OR NAPIERIAN LOGARITHMS

$\log_e x$ or $\ln x$

x	0	1	2	3	4	5	6	7	8	9
5.0	1.60944	1.61144	1.61343	1.61542	1.61741	1.61939	1.62137	1.62334	1.62531	1.62728
5.1	1.62924	1.63120	1.63315	1.63511	1.63705	1.63900	1.64094	1.64287	1.64481	1.64673
5.2	1.64866	1.65058	1.65250	1.65441	1.65632	1.65823	1.66013	1.66203	1.66393	1.66582
5.3	1.66771	1.66959	1.67147	1.67335	1.67523	1.67710	1.67896	1.68083	1.68269	1.68455
5.4	1.68640	1.68825	1.69010	1.69194	1.69378	1.69562	1.69745	1.69928	1.70111	1.70293
5.5	1.70475	1.70656	1.70838	1.71019	1.71199	1.71380	1.71560	1.71740	1.71919	1.72098
5.6	1.72277	1.72455	1.72633	1.72811	1.72988	1.73166	1.73342	1.73519	1.73695	1.73871
5.7	1.74047	1.74222	1.74397	1.74572	1.74746	1.74920	1.75094	1.75267	1.75440	1.75613
5.8	1.75786	1.75958	1.76130	1.76302	1.76473	1.76644	1.76815	1.76985	1.77156	1.77326
5.9	1.77495	1.77665	1.77834	1.78002	1.78171	1.78339	1.78507	1.78675	1.78842	1.79009
6.0	1.79176	1.79342	1.79509	1.79675	1.79840	1.80006	1.80171	1.80336	1.80500	1.80665
6.1	1.80829	1.80993	1.81156	1.81319	1.81482	1.81645	1.81808	1.81970	1.82132	1.82294
6.2	1.82455	1.82616	1.82777	1.82938	1.83098	1.83258	1.83418	1.83578	1.83737	1.83896
6.3	1.84055	1.84214	1.84372	1.84530	1.84688	1.84845	1.85003	1.85160	1.85317	1.85473
6.4	1.85630	1.85786	1.85942	1.86097	1.86253	1.86408	1.86563	1.86718	1.86872	1.87026
6.5	1.87180	1.87334	1.87487	1.87641	1.87794	1.87947	1.88099	1.88251	1.88403	1.88555
6.6	1.88707	1.88858	1.89010	1.89160	1.89311	1.89462	1.89612	1.89762	1.89912	1.90061
6.7	1.90211	1.90360	1.90509	1.90658	1.90806	1.90954	1.91102	1.91250	1.91398	1.91545
6.8	1.91692	1.91839	1.91986	1.92132	1.92279	1.92425	1.92571	1.92716	1.92862	1.93007
6.9	1.93152	1.93297	1.93442	1.93586	1.93730	1.93874	1.94018	1.94162	1.94305	1.94448
7.0	1.94591	1.94734	1.94876	1.95019	1.95161	1.95303	1.95445	1.95586	1.95727	1.95869
7.1	1.96009	1.96150	1.96291	1.96431	1.96571	1.96711	1.96851	1.96991	1.97130	1.97269
7.2	1.97408	1.97547	1.97685	1.97824	1.97962	1.98100	1.98238	1.98376	1.98513	1.98650
7.3	1.98787	1.98924	1.99061	1.99198	1.99334	1.99470	1.99606	1.99742	1.99877	2.00013
7.4	2.00148	2.00283	2.00418	2.00553	2.00687	2.00821	2.00956	2.01089	2.01223	2.01357
7.5	2.01490	2.01624	2.01757	2.01890	2.02022	2.02155	2.02287	2.02419	2.02551	2.02683
7.6	2.02815	2.02946	2.03078	2.03209	2.03340	2.03471	2.03601	2.03732	2.03862	2.03992
7.7	2.04122	2.04252	2.04381	2.04511	2.04640	2.04769	2.04898	2.05027	2.05156	2.05284
7.8	2.05412	2.05540	2.05668	2.05796	2.05924	2.06051	2.06179	2.06306	2.06433	2.06560
7.9	2.06686	2.06813	2.06939	2.07065	2.07191	2.07317	2.07443	2.07568	2.07694	2.07819
8.0	2.07944	2.08069	2.08194	2.08318	2.08443	2.08567	2.08691	2.08815	2.08939	2.09063
8.1	2.09186	2.09310	2.09433	2.09556	2.09679	2.09802	2.09924	2.10047	2.10169	2.10291
8.2	2.10413	2.10535	2.10657	2.10779	2.10900	2.11021	2.11142	2.11263	2.11384	2.11505
8.3	2.11626	2.11746	2.11866	2.11986	2.12106	2.12226	2.12346	2.12465	2.12585	2.12704
8.4	2.12823	2.12942	2.13061	2.13180	2.13298	2.13417	2.13535	2.13653	2.13771	2.13889
8.5	2.14007	2.14124	2.14242	2.14359	2.14476	2.14593	2.14710	2.14827	2.14943	2.15060
8.6	2.15176	2.15292	2.15409	2.15524	2.15640	2.15756	2.15871	2.15987	2.16102	2.16217
8.7	2.16332	2.16447	2.16562	2.16677	2.16791	2.16905	2.17020	2.17134	2.17248	2.17361
8.8	2.17475	2.17589	2.17702	2.17816	2.17929	2.18042	2.18155	2.18267	2.18380	2.18493
8.9	2.18605	2.18717	2.18830	2.18942	2.19054	2.19165	2.19277	2.19389	2.19500	2.19611
9.0	2.19722	2.19834	2.19944	2.20055	2.20166	2.20276	2.20387	2.20497	2.20607	2.20717
9.1	2.20827	2.20937	2.21047	2.21157	2.21266	2.21375	2.21485	2.21594	2.21703	2.21812
9.2	2.21920	2.22029	2.22138	2.22246	2.22354	2.22462	2.22570	2.22678	2.22786	2.22894
9.3	2.23001	2.23109	2.23216	2.23324	2.23431	2.23538	2.23645	2.23751	2.23858	2.23965
9.4	2.24071	2.24177	2.24284	2.24390	2.24496	2.24601	2.24707	2.24813	2.24918	2.25024
9.5	2.25129	2.25234	2.25339	2.25444	2.25549	2.25654	2.25759	2.25863	2.25968	2.26072
9.6	2.26176	2.26280	2.26384	2.26488	2.26592	2.26696	2.26799	2.26903	2.27006	2.27109
9.7	2.27213	2.27316	2.27419	2.27521	2.27624	2.27727	2.27829	2.27932	2.28034	2.28136
9.8	2.28238	2.28340	2.28442	2.28544	2.28646	2.28747	2.28849	2.28950	2.29051	2.29152
9.9	2.29253	2.29354	2.29455	2.29556	2.29657	2.29757	2.29858	2.29958	2.30058	2.30158

TABLE 16

EXPONENTIAL FUNCTIONS

e^x

x	0	1	2	3	4	5	6	7	8	9
.0	1.0000	1.0101	1.0202	1.0305	1.0408	1.0513	1.0618	1.0725	1.0833	1.0942
.1	1.1052	1.1163	1.1275	1.1388	1.1503	1.1618	1.1735	1.1853	1.1972	1.2092
.2	1.2214	1.2337	1.2461	1.2586	1.2712	1.2840	1.2969	1.3100	1.3231	1.3364
.3	1.3499	1.3634	1.3771	1.3910	1.4049	1.4191	1.4333	1.4477	1.4623	1.4770
.4	1.4918	1.5068	1.5220	1.5373	1.5527	1.5683	1.5841	1.6000	1.6161	1.6323
.5	1.6487	1.6653	1.6820	1.6989	1.7160	1.7333	1.7507	1.7683	1.7860	1.8040
.6	1.8221	1.8404	1.8589	1.8776	1.8965	1.9155	1.9348	1.9542	1.9739	1.9937
.7	2.0138	2.0340	2.0544	2.0751	2.0959	2.1170	2.1383	2.1598	2.1815	2.2034
.8	2.2255	2.2479	2.2705	2.2933	2.3164	2.3396	2.3632	2.3869	2.4109	2.4351
.9	2.4596	2.4843	2.5093	2.5345	2.5600	2.5857	2.6117	2.6379	2.6645	2.6912
1.0	2.7183	2.7456	2.7732	2.8011	2.8292	2.8577	2.8864	2.9154	2.9447	2.9743
1.1	3.0042	3.0344	3.0649	3.0957	3.1268	3.1582	3.1899	3.2220	3.2544	3.2871
1.2	3.3201	3.3535	3.3872	3.4212	3.4556	3.4903	3.5254	3.5609	3.5966	3.6328
1.3	3.6693	3.7062	3.7434	3.7810	3.8190	3.8574	3.8962	3.9354	3.9749	4.0149
1.4	4.0552	4.0960	4.1371	4.1787	4.2207	4.2631	4.3060	4.3492	4.3929	4.4371
1.5	4.4817	4.5267	4.5722	4.6182	4.6646	4.7115	4.7588	4.8066	4.8550	4.9037
1.6	4.9530	5.0028	5.0531	5.1039	5.1552	5.2070	5.2593	5.3122	5.3656	5.4195
1.7	5.4739	5.5290	5.5845	5.6407	5.6973	5.7546	5.8124	5.8709	5.9299	5.9895
1.8	6.0496	6.1104	6.1719	6.2339	6.2965	6.3598	6.4237	6.4883	6.5535	6.6194
1.9	6.6859	6.7531	6.8210	6.8895	6.9588	7.0287	7.0993	7.1707	7.2427	7.3155
2.0	7.3891	7.4633	7.5383	7.6141	7.6906	7.7679	7.8460	7.9248	8.0045	8.0849
2.1	8.1662	8.2482	8.3311	8.4149	8.4994	8.5849	8.6711	8.7583	8.8463	8.9352
2.2	9.0250	9.1157	9.2073	9.2999	9.3933	9.4877	9.5831	9.6794	9.7767	9.8749
2.3	9.9742	10.074	10.176	10.278	10.381	10.486	10.591	10.697	10.805	10.913
2.4	11.023	11.134	11.246	11.359	11.473	11.588	11.705	11.822	11.941	12.061
2.5	12.182	12.305	12.429	12.554	12.680	12.807	12.936	13.066	13.197	13.330
2.6	13.464	13.599	13.736	13.874	14.013	14.154	14.296	14.440	14.585	14.732
2.7	14.880	15.029	15.180	15.333	15.487	15.643	15.800	15.959	16.119	16.281
2.8	16.445	16.610	16.777	16.945	17.116	17.288	17.462	17.637	17.814	17.993
2.9	18.174	18.357	18.541	18.728	18.916	19.106	19.298	19.492	19.688	19.886
3.0	20.086	20.287	20.491	20.697	20.905	21.115	21.328	21.542	21.758	21.977
3.1	22.198	22.421	22.646	22.874	23.104	23.336	23.571	23.807	24.047	24.288
3.2	24.533	24.779	25.028	25.280	25.534	25.790	26.050	26.311	26.576	26.843
3.3	27.113	27.385	27.660	27.938	28.219	28.503	28.789	29.079	29.371	29.666
3.4	29.964	30.265	30.569	30.877	31.187	31.500	31.817	32.137	32.460	32.786
3.5	33.115	33.448	33.784	34.124	34.467	34.813	35.163	35.517	35.874	36.234
3.6	36.598	36.966	37.338	37.713	38.092	38.475	38.861	39.252	39.646	40.045
3.7	40.447	40.854	41.264	41.679	42.098	42.521	42.948	43.380	43.816	44.256
3.8	44.701	45.150	45.604	46.063	46.525	46.993	47.465	47.942	48.424	48.911
3.9	49.402	49.899	50.400	50.907	51.419	51.935	52.457	52.985	53.517	54.055
4.	54.598	60.340	66.686	73.700	81.451	90.017	99.484	109.95	121.51	134.29
5.	148.41	164.02	181.27	200.34	221.41	244.69	270.43	298.87	330.30	365.04
6.	403.43	445.86	492.75	544.57	601.85	665.14	735.10	812.41	897.85	992.27
7.	1096.6	1212.0	1339.4	1480.3	1636.0	1808.0	1998.2	2208.3	2440.6	2697.3
8.	2981.0	3294.5	3641.0	4023.9	4447.1	4914.8	5431.7	6002.9	6634.2	7332.0
9.	8103.1	8955.3	9897.1	10938	12088	13360	14765	16318	18034	19930
10.	22026									

TABLE 17

EXPONENTIAL FUNCTIONS

e^{-x}

x	0	1	2	3	4	5	6	7	8	9
.0	1.00000	.99005	.98020	.97045	.96079	.95123	.94176	.93239	.92312	.91393
.1	.90484	.89583	.88692	.87810	.86936	.86071	.85214	.84366	.83527	.82696
.2	.81873	.81058	.80252	.79453	.78663	.77880	.77105	.76338	.75578	.74826
.3	.74082	.73345	.72615	.71892	.71177	.70469	.69768	.69073	.68386	.67706
.4	.67032	.66365	.65705	.65051	.64404	.63763	.63128	.62500	.61878	.61263
.5	.60653	.60050	.59452	.58860	.58275	.57695	.57121	.56553	.55990	.55433
.6	.54881	.54335	.53794	.53259	.52729	.52205	.51685	.51171	.50662	.50158
.7	.49659	.49164	.48675	.48191	.47711	.47237	.46767	.46301	.45841	.45384
.8	.44933	.44486	.44043	.43605	.43171	.42741	.42316	.41895	.41478	.41066
.9	.40657	.40252	.39852	.39455	.39063	.38674	.38289	.37908	.37531	.37158
1.0	.36788	.36422	.36060	.35701	.35345	.34994	.34646	.34301	.33960	.33622
1.1	.33287	.32956	.32628	.32303	.31982	.31664	.31349	.31037	.30728	.30422
1.2	.30119	.29820	.29523	.29229	.28938	.28650	.28365	.28083	.27804	.27527
1.3	.27253	.26982	.26714	.26448	.26185	.25924	.25666	.25411	.25158	.24908
1.4	.24660	.24414	.24171	.23931	.23693	.23457	.23224	.22993	.22764	.22537
1.5	.22313	.22091	.21871	.21654	.21438	.21225	.21014	.20805	.20598	.20393
1.6	.20190	.19989	.19790	.19593	.19398	.19205	.19014	.18825	.18637	.18452
1.7	.18268	.18087	.17907	.17728	.17552	.17377	.17204	.17033	.16864	.16696
1.8	.16530	.16365	.16203	.16041	.15882	.15724	.15567	.15412	.15259	.15107
1.9	.14957	.14808	.14661	.14515	.14370	.14227	.14086	.13946	.13807	.13670
2.0	.13534	.13399	.13266	.13134	.13003	.12873	.12745	.12619	.12493	.12369
2.1	.12246	.12124	.12003	.11884	.11765	.11648	.11533	.11418	.11304	.11192
2.2	.11080	.10970	.10861	.10753	.10646	.10540	.10435	.10331	.10228	.10127
2.3	.10026	.09926	.09827	.09730	.09633	.09537	.09442	.09348	.09255	.09163
2.4	.09072	.08982	.08892	.08804	.08716	.08629	.08543	.08458	.08374	.08291
2.5	.08208	.08127	.08046	.07966	.07887	.07808	.07730	.07654	.07577	.07502
2.6	.07427	.07353	.07280	.07208	.07136	.07065	.06995	.06925	.06856	.06788
2.7	.06721	.06654	.06587	.06522	.06457	.06393	.06329	.06266	.06204	.06142
2.8	.06081	.06020	.05961	.05901	.05843	.05784	.05727	.05670	.05613	.05558
2.9	.05502	.05448	.05393	.05340	.05287	.05234	.05182	.05130	.05079	.05029
3.0	.04979	.04929	.04880	.04832	.04783	.04736	.04689	.04642	.04596	.04550
3.1	.04505	.04460	.04416	.04372	.04328	.04285	.04243	.04200	.04159	.04117
3.2	.04076	.04036	.03996	.03956	.03916	.03877	.03839	.03801	.03763	.03725
3.3	.03688	.03652	.03615	.03579	.03544	.03508	.03474	.03439	.03405	.03371
3.4	.03337	.03304	.03271	.03239	.03206	.03175	.03143	.03112	.03081	.03050
3.5	.03020	.02990	.02960	.02930	.02901	.02872	.02844	.02816	.02788	.02760
3.6	.02732	.02705	.02678	.02652	.02625	.02599	.02573	.02548	.02522	.02497
3.7	.02472	.02448	.02423	.02399	.02375	.02352	.02328	.02305	.02282	.02260
3.8	.02237	.02215	.02193	.02171	.02149	.02128	.02107	.02086	.02065	.02045
3.9	.02024	.02004	.01984	.01964	.01945	.01925	.01906	.01887	.01869	.01850
4.	.018316	.016573	.014996	.013569	.012277	.011109	.010052	$.0^{2}90953$	$.0^{2}82297$	$.0^{2}74466$
5.	$.0^{2}67379$	$.0^{2}60967$	$.0^{2}55166$	$.0^{2}49916$	$.0^{2}45166$	$.0^{2}40868$	$.0^{2}36979$	$.0^{2}33460$	$.0^{2}30276$	$.0^{2}27394$
6.	$.0^{2}24788$	$.0^{2}22429$	$.0^{2}20294$	$.0^{2}18363$	$.0^{2}16616$	$.0^{2}15034$	$.0^{2}13604$	$.0^{2}12309$	$.0^{2}11138$	$.0^{2}10078$
7.	$.0^{3}91188$	$.0^{3}82510$	$.0^{3}74659$	$.0^{3}67554$	$.0^{3}61125$	$.0^{3}55308$	$.0^{3}50045$	$.0^{3}45283$	$.0^{3}40973$	$.0^{3}37074$
8.	$.0^{3}33546$	$.0^{3}30354$	$.0^{3}27465$	$.0^{3}24852$	$.0^{3}22487$	$.0^{3}20347$	$.0^{3}18411$	$.0^{3}16659$	$.0^{3}15073$	$.0^{3}13639$
9.	$.0^{3}12341$	$.0^{3}11167$	$.0^{3}10104$	$.0^{4}91424$	$.0^{4}82724$	$.0^{4}74852$	$.0^{4}67729$	$.0^{4}61283$	$.0^{4}55452$	$.0^{4}50175$
10.	$.0^{4}45400$									

TABLE 18a

HYPERBOLIC FUNCTIONS

sinh x

x	0	1	2	3	4	5	6	7	8	9
.0	.0000	.0100	.0200	.0300	.0400	.0500	.0600	.0701	.0801	.0901
.1	.1002	.1102	.1203	.1304	.1405	.1506	.1607	.1708	.1810	.1911
.2	.2013	.2115	.2218	.2320	.2423	.2526	.2629	.2733	.2837	.2941
.3	.3045	.3150	.3255	.3360	.3466	.3572	.3678	.3785	.3892	.4000
.4	.4108	.4216	.4325	.4434	.4543	.4653	.4764	.4875	.4986	.5098
.5	.5211	.5324	.5438	.5552	.5666	.5782	.5897	.6014	.6131	.6248
.6	.6367	.6485	.6605	.6725	.6846	.6967	.7090	.7213	.7336	.7461
.7	.7586	.7712	.7838	.7966	.8094	.8223	.8353	.8484	.8615	.8748
.8	.8881	.9015	.9150	.9286	.9423	.9561	.9700	.9840	.9981	1.0122
.9	1.0265	1.0409	1.0554	1.0700	1.0847	1.0995	1.1144	1.1294	1.1446	1.1598
1.0	1.1752	1.1907	1.2063	1.2220	1.2379	1.2539	1.2700	1.2862	1.3025	1.3190
1.1	1.3356	1.3524	1.3693	1.3863	1.4035	1.4208	1.4382	1.4558	1.4735	1.4914
1.2	1.5095	1.5276	1.5460	1.5645	1.5831	1.6019	1.6209	1.6400	1.6593	1.6788
1.3	1.6984	1.7182	1.7381	1.7583	1.7786	1.7991	1.8198	1.8406	1.8617	1.8829
1.4	1.9043	1.9259	1.9477	1.9697	1.9919	2.0143	2.0369	2.0597	2.0827	2.1059
1.5	2.1293	2.1529	2.1768	2.2008	2.2251	2.2496	2.2743	2.2993	2.3245	2.3499
1.6	2.3756	2.4015	2.4276	2.4540	2.4806	2.5075	2.5346	2.5620	2.5896	2.6175
1.7	2.6456	2.6740	2.7027	2.7317	2.7609	2.7904	2.8202	2.8503	2.8806	2.9112
1.8	2.9422	2.9734	3.0049	3.0367	3.0689	3.1013	3.1340	3.1671	3.2005	3.2341
1.9	3.2682	3.3025	3.3372	3.3722	3.4075	3.4432	3.4792	3.5156	3.5523	3.5894
2.0	3.6269	3.6647	3.7028	3.7414	3.7803	3.8196	3.8593	3.8993	3.9398	3.9806
2.1	4.0219	4.0635	4.1056	4.1480	4.1909	4.2342	4.2779	4.3221	4.3666	4.4116
2.2	4.4571	4.5030	4.5494	4.5962	4.6434	4.6912	4.7394	4.7880	4.8372	4.8868
2.3	4.9370	4.9876	5.0387	5.0903	5.1425	5.1951	5.2483	5.3020	5.3562	5.4109
2.4	5.4662	5.5221	5.5785	5.6354	5.6929	5.7510	5.8097	5.8689	5.9288	5.9892
2.5	6.0502	6.1118	6.1741	6.2369	6.3004	6.3645	6.4293	6.4946	6.5607	6.6274
2.6	6.6947	6.7628	6.8315	6.9008	6.9709	7.0417	7.1132	7.1854	7.2583	7.3319
2.7	7.4063	7.4814	7.5572	7.6338	7.7112	7.7894	7.8683	7.9480	8.0285	8.1098
2.8	8.1919	8.2749	8.3586	8.4432	8.5287	8.6150	8.7021	8.7902	8.8791	8.9689
2.9	9.0596	9.1512	9.2437	9.3371	9.4315	9.5268	9.6231	9.7203	9.8185	9.9177

Table 18a (continued)

HYPERBOLIC FUNCTIONS

sinh x

x	0	1	2	3	4	5	6	7	8	9
3.0	10.018	10.119	10.221	10.324	10.429	10.534	10.640	10.748	10.856	10.966
3.1	11.076	11.188	11.301	11.415	11.530	11.647	11.764	11.883	12.003	12.124
3.2	12.246	12.369	12.494	12.620	12.747	12.876	13.006	13.137	13.269	13.403
3.3	13.538	13.674	13.812	13.951	14.092	14.234	14.377	14.522	14.668	14.816
3.4	14.965	15.116	15.268	15.422	15.577	15.734	15.893	16.053	16.215	16.378
3.5	16.543	16.709	16.877	17.047	17.219	17.392	17.567	17.744	17.923	18.103
3.6	18.285	18.470	18.655	18.843	19.033	19.224	19.418	19.613	19.811	20.010
3.7	20.211	20.415	20.620	20.828	21.037	21.249	21.463	21.679	21.897	22.117
3.8	22.339	22.564	22.791	23.020	23.252	23.486	23.722	23.961	24.202	24.445
3.9	24.691	24.939	25.190	25.444	25.700	25.958	26.219	26.483	26.749	27.018
4.0	27.290	27.564	27.842	28.122	28.404	28.690	28.979	29.270	29.564	29.862
4.1	30.162	30.465	30.772	31.081	31.393	31.709	32.028	32.350	32.675	33.004
4.2	33.336	33.671	34.009	34.351	34.697	35.046	35.398	35.754	36.113	36.476
4.3	36.843	37.214	37.588	37.965	38.347	38.733	39.122	39.515	39.913	40.314
4.4	40.719	41.129	41.542	41.960	42.382	42.808	43.238	43.673	44.112	44.555
4.5	45.003	45.455	45.912	46.374	46.840	47.311	47.787	48.267	48.752	49.242
4.6	49.737	50.237	50.742	51.252	51.767	52.288	52.813	53.344	53.880	54.422
4.7	54.969	55.522	56.080	56.643	57.213	57.788	58.369	58.995	59.548	60.147
4.8	60.751	61.362	61.979	62.601	63.231	63.866	64.508	65.157	65.812	66.473
4.9	67.141	67.816	68.498	69.186	69.882	70.584	71.293	72.010	72.734	73.465
5.0	74.203	74.949	75.702	76.463	77.232	78.008	78.792	79.584	80.384	81.192
5.1	82.008	82.832	83.665	84.506	85.355	86.213	87.079	87.955	88.839	89.732
5.2	90.633	91.544	92.464	93.394	94.332	95.281	96.238	97.205	98.182	99.169
5.3	100.17	101.17	102.19	103.22	104.25	105.30	106.36	107.43	108.51	109.60
5.4	110.70	111.81	112.94	114.07	115.22	116.38	117.55	118.73	119.92	121.13
5.5	122.34	123.57	124.82	126.07	127.34	128.62	129.91	131.22	132.53	133.87
5.6	135.21	136.57	137.94	139.33	140.73	142.14	143.57	145.02	146.47	147.95
5.7	149.43	150.93	152.45	153.98	155.53	157.09	158.67	160.27	161.88	163.51
5.8	165.15	166.81	168.48	170.18	171.89	173.62	175.36	177.12	178.90	180.70
5.9	182.52	184.35	186.20	188.08	189.97	191.88	193.80	195.75	197.72	199.71

TABLE 18b

HYPERBOLIC FUNCTIONS
cosh x

x	0	1	2	3	4	5	6	7	8	9
.0	1.0000	1.0001	1.0002	1.0005	1.0008	1.0013	1.0018	1.0025	1.0032	1.0041
.1	1.0050	1.0061	1.0072	1.0085	1.0098	1.0113	1.0128	1.0145	1.0162	1.0181
.2	1.0201	1.0221	1.0243	1.0266	1.0289	1.0314	1.0340	1.0367	1.0395	1.0423
.3	1.0453	1.0484	1.0516	1.0549	1.0584	1.0619	1.0655	1.0692	1.0731	1.0770
.4	1.0811	1.0852	1.0895	1.0939	1.0984	1.1030	1.1077	1.1125	1.1174	1.1225
.5	1.1276	1.1329	1.1383	1.1438	1.1494	1.1551	1.1609	1.1669	1.1730	1.1792
.6	1.1855	1.1919	1.1984	1.2051	1.2119	1.2188	1.2258	1.2330	1.2402	1.2476
.7	1.2552	1.2628	1.2706	1.2785	1.2865	1.2947	1.3030	1.3114	1.3199	1.3286
.8	1.3374	1.3464	1.3555	1.3647	1.3740	1.3835	1.3932	1.4029	1.4128	1.4229
.9	1.4331	1.4434	1.4539	1.4645	1.4753	1.4862	1.4973	1.5085	1.5199	1.5314
1.0	1.5431	1.5549	1.5669	1.5790	1.5913	1.6038	1.6164	1.6292	1.6421	1.6552
1.1	1.6685	1.6820	1.6956	1.7093	1.7233	1.7374	1.7517	1.7662	1.7808	1.7957
1.2	1.8107	1.8258	1.8412	1.8568	1.8725	1.8884	1.9045	1.9208	1.9373	1.9540
1.3	1.9709	1.9880	2.0053	2.0228	2.0404	2.0583	2.0764	2.0947	2.1132	2.1320
1.4	2.1509	2.1700	2.1894	2.2090	2.2288	2.2488	2.2691	2.2896	2.3103	2.3312
1.5	2.3524	2.3738	2.3955	2.4174	2.4395	2.4619	2.4845	2.5073	2.5305	2.5538
1.6	2.5775	2.6013	2.6255	2.6499	2.6746	2.6995	2.7247	2.7502	2.7760	2.8020
1.7	2.8283	2.8549	2.8818	2.9090	2.9364	2.9642	2.9922	3.0206	3.0492	3.0782
1.8	3.1075	3.1371	3.1669	3.1972	3.2277	3.2585	3.2897	3.3212	3.3530	3.3852
1.9	3.4177	3.4506	3.4838	3.5173	3.5512	3.5855	3.6201	3.6551	3.6904	3.7261
2.0	3.7622	3.7987	3.8355	3.8727	3.9103	3.9483	3.9867	4.0255	4.0647	4.1043
2.1	4.1443	4.1847	4.2256	4.2669	4.3085	4.3507	4.3932	4.4362	4.4797	4.5236
2.2	4.5679	4.6127	4.6580	4.7037	4.7499	4.7966	4.8437	4.8914	4.9395	4.9881
2.3	5.0372	5.0868	5.1370	5.1876	5.2388	5.2905	5.3427	5.3954	5.4487	5.5026
2.4	5.5569	5.6119	5.6674	5.7235	5.7801	5.8373	5.8951	5.9535	6.0125	6.0721
2.5	6.1323	6.1931	6.2545	6.3166	6.3793	6.4426	6.5066	6.5712	6.6365	6.7024
2.6	6.7690	6.8363	6.9043	6.9729	7.0423	7.1123	7.1831	7.2546	7.3268	7.3998
2.7	7.4735	7.5479	7.6231	7.6991	7.7758	7.8533	7.9316	8.0106	8.0905	8.1712
2.8	8.2527	8.3351	8.4182	8.5022	8.5871	8.6728	8.7594	8.8469	8.9352	9.0244
2.9	9.1146	9.2056	9.2976	9.3905	9.4844	9.5791	9.6749	9.7716	9.8693	9.9680

Table 18b
(continued)

HYPERBOLIC FUNCTIONS

cosh x

x	0	1	2	3	4	5	6	7	8	9
3.0	10.068	10.168	10.270	10.373	10.476	10.581	10.687	10.794	10.902	11.011
3.1	11.121	11.233	11.345	11.459	11.574	11.689	11.806	11.925	12.044	12.165
3.2	12.287	12.410	12.534	12.660	12.786	12.915	13.044	13.175	13.307	13.440
3.3	13.575	13.711	13.848	13.987	14.127	14.269	14.412	14.556	14.702	14.850
3.4	14.999	15.149	15.301	15.455	15.610	15.766	15.924	16.084	16.245	16.408
3.5	16.573	16.739	16.907	17.077	17.248	17.421	17.596	17.772	17.951	18.131
3.6	18.313	18.497	18.682	18.870	19.059	19.250	19.444	19.639	19.836	20.035
3.7	20.236	20.439	20.644	20.852	21.061	21.272	21.486	21.702	21.919	22.139
3.8	22.362	22.586	22.813	23.042	23.273	23.507	23.743	23.982	24.222	24.466
3.9	24.711	24.959	25.210	25.463	25.719	25.977	26.238	26.502	26.768	27.037
4.0	27.308	27.583	27.860	28.139	28.422	28.707	28.996	29.287	29.581	29.878
4.1	30.178	30.482	30.788	31.097	31.409	31.725	32.044	32.365	32.691	33.019
4.2	33.351	33.686	34.024	34.366	34.711	35.060	35.412	35.768	36.127	36.490
4.3	36.857	37.227	37.601	37.979	38.360	38.746	39.135	39.528	39.925	40.326
4.4	40.732	41.141	41.554	41.972	42.393	42.819	43.250	43.684	44.123	44.566
4.5	45.014	45.466	45.923	46.385	46.851	47.321	47.797	48.277	48.762	49.252
4.6	49.747	50.247	50.752	51.262	51.777	52.297	52.823	53.354	53.890	54.431
4.7	54.978	55.531	56.089	56.652	57.221	57.796	58.377	58.964	59.556	60.155
4.8	60.759	61.370	61.987	62.609	63.239	63.874	64.516	65.164	65.819	66.481
4.9	67.149	67.823	68.505	69.193	69.889	70.591	71.300	72.017	72.741	73.472
5.0	74.210	74.956	75.709	76.470	77.238	78.014	78.798	79.590	80.390	81.198
5.1	82.014	82.838	83.671	84.512	85.361	86.219	87.085	87.960	88.844	89.737
5.2	90.639	91.550	92.470	93.399	94.338	95.286	96.243	97.211	98.188	99.174
5.3	100.17	101.18	102.19	103.22	104.26	105.31	106.67	107.43	108.51	109.60
5.4	110.71	111.82	112.94	114.08	115.22	116.38	117.55	118.73	119.93	121.13
5.5	122.35	123.58	124.82	126.07	127.34	128.62	129.91	131.22	132.54	133.87
5.6	135.22	136.57	137.95	139.33	140.73	142.15	143.58	145.02	146.48	147.95
5.7	149.44	150.94	152.45	153.99	155.53	157.10	158.68	160.27	161.88	163.51
5.8	165.15	166.81	168.49	170.18	171.89	173.62	175.36	177.13	178.91	180.70
5.9	182.52	184.35	186.21	188.08	189.97	191.88	193.81	195.75	197.72	199.71

TABLE 18c

HYPERBOLIC FUNCTIONS

tanh x

x	0	1	2	3	4	5	6	7	8	9
.0	.00000	.01000	.02000	.02999	.03998	.04996	.05993	.06989	.07983	.08976
.1	.09967	.10956	.11943	.12927	.13909	.14889	.15865	.16838	.17808	.18775
.2	.19738	.20697	.21652	.22603	.23550	.24492	.25430	.26362	.27291	.28213
.3	.29131	.30044	.30951	.31852	.32748	.33638	.34521	.35399	.36271	.37136
.4	.37995	.38847	.39693	.40532	.41364	.42190	.43008	.43820	.44624	.45422
.5	.46212	.46995	.47770	.48538	.49299	.50052	.50798	.51536	.52267	.52990
.6	.53705	.54413	.55113	.55805	.56490	.57167	.57836	.58498	.59152	.59798
.7	.60437	.61068	.61691	.62307	.62915	.63515	.64108	.64693	.65271	.65841
.8	.66404	.66959	.67507	.68048	.68581	.69107	.69626	.70137	.70642	.71139
.9	.71630	.72113	.72590	.73059	.73522	.73978	.74428	.74870	.75307	.75736
1.0	.76159	.76576	.76987	.77391	.77789	.78181	.78566	.78946	.79320	.79688
1.1	.80050	.80406	.80757	.81102	.81441	.81775	.82104	.82427	.82745	.83058
1.2	.83365	.83668	.83965	.84258	.84546	.84828	.85106	.85380	.85648	.85913
1.3	.86172	.86428	.86678	.86925	.87167	.87405	.87639	.87869	.88095	.88317
1.4	.88535	.88749	.88960	.89167	.89370	.89569	.89765	.89958	.90147	.90332
1.5	.90515	.90694	.90870	.91042	.91212	.91379	.91542	.91703	.91860	.92015
1.6	.92167	.92316	.92462	.92606	.92747	.92886	.93022	.93155	.93286	.93415
1.7	.93541	.93665	.93786	.93906	.94023	.94138	.94250	.94361	.94470	.94576
1.8	.94681	.94783	.94884	.94983	.95080	.95175	.95268	.95359	.95449	.95537
1.9	.95624	.95709	.95792	.95873	.95953	.96032	.96109	.96185	.96259	.96331
2.0	.96403	.96473	.96541	.96609	.96675	.96740	.96803	.96865	.96926	.96986
2.1	.97045	.97103	.97159	.97215	.97269	.97323	.97375	.97426	.97477	.97526
2.2	.97574	.97622	.97668	.97714	.97759	.97803	.97846	.97888	.97929	.97970
2.3	.98010	.98049	.98087	.98124	.98161	.98197	.98233	.98267	.98301	.98335
2.4	.98367	.98400	.98431	.98462	.98492	.98522	.98551	.98579	.98607	.98635
2.5	.98661	.98688	.98714	.98739	.98764	.98788	.98812	.98835	.98858	.98881
2.6	.98903	.98924	.98946	.98966	.98987	.99007	.99026	.99045	.99064	.99083
2.7	.99101	.99118	.99136	.99153	.99170	.99186	.99202	.99218	.99233	.99248
2.8	.99263	.99278	.99292	.99306	.99320	.99333	.99346	.99359	.99372	.99384
2.9	.99396	.99408	.99420	.99431	.99443	.99454	.99464	.99475	.99485	.99496

Table 18c
(continued)

HYPERBOLIC FUNCTIONS

tanh x

x	0	1	2	3	4	5	6	7	8	9
3.0	.99505	.99515	.99525	.99534	.99543	.99552	.99561	.99570	.99578	.99587
3.1	.99595	.99603	.99611	.99618	.99626	.99633	.99641	.99648	.99655	.99662
3.2	.99668	.99675	.99681	.99688	.99694	.99700	.99706	.99712	.99717	.99723
3.3	.99728	.99734	.99739	.99744	.99749	.99754	.99759	.99764	.99768	.99773
3.4	.99777	.99782	.99786	.99790	.99795	.99799	.99803	.99807	.99810	.99814
3.5	.99818	.99821	.99825	.99828	.99832	.99835	.99838	.99842	.99845	.99848
3.6	.99851	.99853	.99857	.99859	.99862	.99865	.99868	.99870	.99873	.99875
3.7	.99878	.99880	.99883	.99885	.99887	.99889	.99892	.99894	.99896	.99898
3.8	.99900	.99902	.99904	.99906	.99908	.99909	.99911	.99913	.99915	.99916
3.9	.99918	.99920	.99921	.99923	.99924	.99926	.99927	.99929	.99930	.99932
4.0	.99933	.99934	.99936	.99937	.99938	.99939	.99941	.99942	.99943	.99944
4.1	.99945	.99946	.99947	.99948	.99949	.99950	.99951	.99952	.99953	.99954
4.2	.99955	.99956	.99957	.99958	.99958	.99959	.99960	.99961	.99962	.99962
4.3	.99963	.99964	.99965	.99966	.99966	.99967	.99967	.99968	.99969	.99969
4.4	.99970	.99970	.99971	.99972	.99972	.99973	.99973	.99974	.99974	.99975
4.5	.99975	.99976	.99976	.99977	.99977	.99978	.99978	.99979	.99979	.99979
4.6	.99980	.99980	.99981	.99981	.99981	.99982	.99982	.99982	.99983	.99983
4.7	.99983	.99984	.99984	.99984	.99985	.99985	.99985	.99986	.99986	.99986
4.8	.99986	.99987	.99987	.99987	.99987	.99988	.99988	.99988	.99988	.99989
4.9	.99989	.99989	.99990	.99990	.99990	.99990	.99990	.99990	.99991	.99991
5.0	.99991	.99991	.99991	.99991	.99992	.99992	.99992	.99992	.99992	.99992
5.1	.99993	.99993	.99993	.99993	.99993	.99993	.99993	.99994	.99994	.99994
5.2	.99994	.99994	.99994	.99994	.99994	.99994	.99995	.99995	.99995	.99995
5.3	.99995	.99995	.99995	.99995	.99995	.99995	.99996	.99996	.99996	.99996
5.4	.99996	.99996	.99996	.99996	.99996	.99996	.99996	.99996	.99997	.99997
5.5	.99997	.99997	.99997	.99997	.99997	.99997	.99997	.99997	.99997	.99997
5.6	.99997	.99997	.99997	.99997	.99997	.99998	.99998	.99998	.99998	.99998
5.7	.99998	.99998	.99998	.99998	.99998	.99998	.99998	.99998	.99998	.99998
5.8	.99998	.99998	.99998	.99998	.99998	.99998	.99998	.99998	.99998	.99998
5.9	.99998	.99999	.99999	.99999	.99999	.99999	.99999	.99999	.99999	.99999

TABLE 19

FACTORIAL n

$$n! = 1 \cdot 2 \cdot 3 \cdot \cdots \cdot n$$

n	$n!$
0	1 (by definition)
1	1
2	2
3	6
4	24
5	120
6	720
7	5040
8	40,320
9	362,880
10	3,628,800
11	39,916,800
12	479,001,600
13	6,227,020,800
14	87,178,291,200
15	1,307,674,368,000
16	20,922,789,888,000
17	355,687,428,096,000
18	6,402,373,705,728,000
19	121,645,100,408,832,000
20	2,432,902,008,176,640,000
21	51,090,942,171,709,440,000
22	1,124,000,727,777,607,680,000
23	25,852,016,738,884,976,640,000
24	620,448,401,733,239,439,360,000
25	15,511,210,043,330,985,984,000,000
26	403,291,461,126,605,635,584,000,000
27	10,888,869,450,418,352,160,768,000,000
28	304,888,344,611,713,860,501,504,000,000
29	8,841,761,993,739,701,954,543,616,000,000
30	265,252,859,812,191,058,636,308,480,000,000
31	8.22284×10^{33}
32	2.63131×10^{35}
33	8.68332×10^{36}
34	2.95233×10^{38}
35	1.03331×10^{40}
36	3.71993×10^{41}
37	1.37638×10^{43}
38	5.23023×10^{44}
39	2.03979×10^{46}

n	$n!$
40	8.15915×10^{47}
41	3.34525×10^{49}
42	1.40501×10^{51}
43	6.04153×10^{52}
44	2.65827×10^{54}
45	1.19622×10^{56}
46	5.50262×10^{57}
47	2.58623×10^{59}
48	1.24139×10^{61}
49	6.08282×10^{62}
50	3.04141×10^{64}
51	1.55112×10^{66}
52	8.06582×10^{67}
53	4.27488×10^{69}
54	2.30844×10^{71}
55	1.26964×10^{73}
56	7.10999×10^{74}
57	4.05269×10^{76}
58	2.35056×10^{78}
59	1.38683×10^{80}
60	8.32099×10^{81}
61	5.07580×10^{83}
62	3.14700×10^{85}
63	1.98261×10^{87}
64	1.26887×10^{89}
65	8.24765×10^{90}
66	5.44345×10^{92}
67	3.64711×10^{94}
68	2.48004×10^{96}
69	1.71122×10^{98}
70	1.19786×10^{100}
71	8.50479×10^{101}
72	6.12345×10^{103}
73	4.47012×10^{105}
74	3.30789×10^{107}
75	2.48091×10^{109}
76	1.88549×10^{111}
77	1.45183×10^{113}
78	1.13243×10^{115}
79	8.94618×10^{116}

n	$n!$
80	7.15695×10^{118}
81	5.79713×10^{120}
82	4.75364×10^{122}
83	3.94552×10^{124}
84	3.31424×10^{126}
85	2.81710×10^{128}
86	2.42271×10^{130}
87	2.10776×10^{132}
88	1.85483×10^{134}
89	1.65080×10^{136}
90	1.48572×10^{138}
91	1.35200×10^{140}
92	1.24384×10^{142}
93	1.15677×10^{144}
94	1.08737×10^{146}
95	1.03300×10^{148}
96	9.91678×10^{149}
97	9.61928×10^{151}
98	9.42689×10^{153}
99	9.33262×10^{155}
100	9.33262×10^{157}

TABLE 20

GAMMA FUNCTION

$$\Gamma(x) = \int_0^\infty t^{x-1}e^{-t}\,dt \quad \text{for } 1 \leqq x \leqq 2$$

[For other values use the formula $\Gamma(x+1) = x\,\Gamma(x)$]

x	$\Gamma(x)$	x	$\Gamma(x)$
1.00	1.00000	1.50	.88623
1.01	.99433	1.51	.88659
1.02	.98884	1.52	.88704
1.03	.98355	1.53	.88757
1.04	.97844	1.54	.88818
1.05	.97350	1.55	.88887
1.06	.96874	1.56	.88964
1.07	.96415	1.57	.89049
1.08	.95973	1.58	.89142
1.09	.95546	1.59	.89243
1.10	.95135	1.60	.89352
1.11	.94740	1.61	.89468
1.12	.94359	1.62	.89592
1.13	.93993	1.63	.89724
1.14	.93642	1.64	.89864
1.15	.93304	1.65	.90012
1.16	.92980	1.66	.90167
1.17	.92670	1.67	.90330
1.18	.92373	1.68	.90500
1.19	.92089	1.69	.90678
1.20	.91817	1.70	.90864
1.21	.91558	1.71	.91057
1.22	.91311	1.72	.91258
1.23	.91075	1.73	.91467
1.24	.90852	1.74	.91683
1.25	.90640	1.75	.91906
1.26	.90440	1.76	.92137
1.27	.90250	1.77	.92376
1.28	.90072	1.78	.92623
1.29	.89904	1.79	.92877
1.30	.89747	1.80	.93138
1.31	.89600	1.81	.93408
1.32	.89464	1.82	.93685
1.33	.89338	1.83	.93969
1.34	.89222	1.84	.94261
1.35	.89115	1.85	.94561
1.36	.89018	1.86	.94869
1.37	.88931	1.87	.95184
1.38	.88854	1.88	.95507
1.39	.88785	1.89	.95838
1.40	.88726	1.90	.96177
1.41	.88676	1.91	.96523
1.42	.88636	1.92	.96877
1.43	.88604	1.93	.97240
1.44	.88581	1.94	.97610
1.45	.88566	1.95	.97988
1.46	.88560	1.96	.98374
1.47	.88563	1.97	.98768
1.48	.88575	1.98	.99171
1.49	.88595	1.99	.99581
1.50	.88623	2.00	1.00000

TABLE

21

BINOMIAL COEFFICIENTS

$$\binom{n}{k} = \frac{n!}{k!\,(n-k)!} = \frac{n(n-1)\cdots(n-k+1)}{k!} = \binom{n}{n-k}, \quad 0! = 1$$

Note that each number is the sum of two numbers in the row above; one of these numbers is in the same column and the other is in the preceding column [e.g. $56 = 35 + 21$]. The arrangement is often called *Pascal's triangle* [see 3.6, page 4].

n \ k	0	1	2	3	4	5	6	7	8	9
1	1	1								
2	1	2	1							
3	1	3	3	1						
4	1	4	6	4	1					
5	1	5	10	10	5	1				
6	1	6	15	20	15	6	1			
7	1	7	21	35	35	21	7	1		
8	1	8	28	56	70	56	28	8	1	
9	1	9	36	84	126	126	84	36	9	1
10	1	10	45	120	210	252	210	120	45	10
11	1	11	55	165	330	462	462	330	165	55
12	1	12	66	220	495	792	924	792	495	220
13	1	13	78	286	715	1287	1716	1716	1287	715
14	1	14	91	364	1001	2002	3003	3432	3003	2002
15	1	15	105	455	1365	3003	5005	6435	6435	5005
16	1	16	120	560	1820	4368	8008	11440	12870	11440
17	1	17	136	680	2380	6188	12376	19448	24310	24310
18	1	18	153	816	3060	8568	18564	31824	43758	48620
19	1	19	171	969	3876	11628	27132	50388	75582	92378
20	1	20	190	1140	4845	15504	38760	77520	125970	167960
21	1	21	210	1330	5985	20349	54264	116280	203490	293930
22	1	22	231	1540	7315	26334	74613	170544	319770	497420
23	1	23	253	1771	8855	33649	100947	245157	490314	817190
24	1	24	276	2024	10626	42504	134596	346104	735471	1307504
25	1	25	300	2300	12650	53130	177100	480700	1081575	2042975
26	1	26	325	2600	14950	65780	230230	657800	1562275	3124550
27	1	27	351	2925	17550	80730	296010	888030	2220075	4686825
28	1	28	378	3276	20475	98280	376740	1184040	3108105	6906900
29	1	29	406	3654	23751	118755	475020	1560780	4292145	10015005
30	1	30	435	4060	27405	142506	593775	2035800	5852925	14307150

Table 21 (continued)

BINOMIAL COEFFICIENTS

$$\binom{n}{k} = \frac{n!}{k!\,(n-k)!} = \frac{n(n-1)\cdots(n-k+1)}{k!} = \binom{n}{n-k}, \quad 0! = 1$$

n \ k	10	11	12	13	14	15
10	1					
11	11	1				
12	66	12	1			
13	286	78	13	1		
14	1001	364	91	14	1	
15	3003	1365	455	105	15	1
16	8008	4368	1820	560	120	16
17	19448	12376	6188	2380	680	136
18	43758	31824	18564	8568	3060	816
19	92378	75582	50388	27132	11628	3876
20	184756	167960	125970	77520	38760	15504
21	352716	352716	293930	203490	116280	54264
22	646646	705432	646646	497420	319770	170544
23	1144066	1352078	1352078	1144066	817190	490314
24	1961256	2496144	2704156	2496144	1961256	1307504
25	3268760	4457400	5200300	5200300	4457400	3268760
26	5311735	7726160	9657700	10400600	9657700	7726160
27	8436285	13037895	17383860	20058300	20058300	17383860
28	13123110	21474180	30421755	37442160	40116600	37442160
29	20030010	34597290	51895935	67863915	77558760	77558760
30	30045015	54627300	86493225	119759850	145422675	155117520

For $k > 15$ use the fact that $\binom{n}{k} = \binom{n}{n-k}$.

TABLE 22

SQUARES, CUBES, ROOTS AND RECIPROCALS

n	n^2	n^3	$\sqrt{n}$	$\sqrt{10n}$	$\sqrt[3]{n}$	$\sqrt[3]{10n}$	$\sqrt[3]{100n}$	$1/n$
1	1	1	1.000 000	3.162 278	1.000 000	2.154 435	4.641 589	1.000 000
2	4	8	1.414 214	4.472 136	1.259 921	2.714 418	5.848 035	.500 000
3	9	27	1.732 051	5.477 226	1.442 250	3.107 233	6.694 330	.333 333
4	16	64	2.000 000	6.324 555	1.587 401	3.419 952	7.368 063	.250 000
5	25	125	2.236 068	7.071 068	1.709 976	3.684 031	7.937 005	.200 000
6	36	216	2.449 490	7.745 967	1.817 121	3.914 868	8.434 327	.166 667
7	49	343	2.645 751	8.366 600	1.912 931	4.121 285	8.879 040	.142 857
8	64	512	2.828 427	8.944 272	2.000 000	4.308 869	9.283 178	.125 000
9	81	729	3.000 000	9.486 833	2.080 084	4.481 405	9.654 894	.111 111
10	100	1 000	3.162 278	10.000 00	2.154 435	4.641 589	10.000 00	.100 000
11	121	1 331	3.316 625	10.488 09	2.223 980	4.791 420	10.322 80	.090 909
12	144	1 728	3.464 102	10.954 45	2.289 428	4.932 424	10.626 59	.083 333
13	169	2 197	3.605 551	11.401 75	2.351 335	5.065 797	10.913 93	.076 923
14	196	2 744	3.741 657	11.832 16	2.410 142	5.192 494	11.186 89	.071 429
15	225	3 375	3.872 983	12.247 45	2.466 212	5.313 293	11.447 14	.066 667
16	256	4 096	4.000 000	12.649 11	2.519 842	5.428 835	11.696 07	.062 500
17	289	4 913	4.123 106	13.038 40	2.571 282	5.539 658	11.934 83	.058 824
18	324	5 832	4.242 641	13.416 41	2.620 741	5.646 216	12.164 40	.055 556
19	361	6 859	4.358 899	13.784 05	2.668 402	5.748 897	12.385 62	.052 632
20	400	8 000	4.472 136	14.142 14	2.714 418	5.848 035	12.599 21	.050 000
21	441	9 261	4.582 576	14.491 38	2.758 924	5.943 922	12.805 79	.047 619
22	484	10 648	4.690 416	14.832 40	2.802 039	6.036 811	13.005 91	.045 455
23	529	12 167	4.795 832	15.165 75	2.843 867	6.126 926	13.200 06	.043 478
24	576	13 824	4.898 979	15.491 93	2.884 499	6.214 465	13.388 66	.041 667
25	625	15 625	5.000 000	15.811 39	2.924 018	6.299 605	13.572 09	.040 000
26	676	17 576	5.099 020	16.124 52	2.962 496	6.382 504	13.750 69	.038 462
27	729	19 683	5.196 152	16.431 68	3.000 000	6.463 304	13.924 77	.037 037
28	784	21 952	5.291 503	16.733 20	3.036 589	6.542 133	14.094 60	.035 714
29	841	24 389	5.385 165	17.029 39	3.072 317	6.619 106	14.260 43	.034 483
30	900	27 000	5.477 226	17.320 51	3.107 233	6.694 330	14.422 50	.033 333
31	961	29 791	5.567 764	17.606 82	3.141 381	6.767 899	14.581 00	.032 258
32	1 024	32 768	5.656 854	17.888 54	3.174 802	6.839 904	14.736 13	.031 250
33	1 089	35 937	5.744 563	18.165 90	3.207 534	6.910 423	14.888 06	.030 303
34	1 156	39 304	5.830 952	18.439 09	3.239 612	6.979 532	15.036 95	.029 412
35	1 225	42 875	5.916 080	18.708 29	3.271 066	7.047 299	15.182 94	.028 571
36	1 296	46 656	6.000 000	18.973 67	3.301 927	7.113 787	15.326 19	.027 778
37	1 369	50 653	6.082 763	19.235 38	3.332 222	7.179 054	15.466 80	.027 027
38	1 444	54 872	6.164 414	19.493 59	3.361 975	7.243 156	15.604 91	.026 316
39	1 521	59 319	6.244 998	19.748 42	3.391 211	7.306 144	15.740 61	.025 641
40	1 600	64 000	6.324 555	20.000 00	3.419 952	7.368 063	15.874 01	.025 000
41	1 681	68 921	6.403 124	20.248 46	3.448 217	7.428 959	16.005 21	.024 390
42	1 764	74 088	6.480 741	20.493 90	3.476 027	7.488 872	16.134 29	.023 810
43	1 849	79 507	6.557 439	20.736 44	3.503 398	7.547 842	16.261 33	.023 256
44	1 936	85 184	6.633 250	20.976 18	3.530 348	7.605 905	16.386 43	.022 727
45	2 025	91 125	6.708 204	21.213 20	3.556 893	7.663 094	16.509 64	.022 222
46	2 116	97 336	6.782 330	21.447 61	3.583 048	7.719 443	16.631 03	.021 739
47	2 209	103 823	6.855 655	21.679 48	3.608 826	7.774 980	16.750 69	.021 277
48	2 304	110 592	6.928 203	21.908 90	3.634 241	7.829 735	16.868 65	.020 833
49	2 401	117 649	7.000 000	22.135 94	3.659 306	7.883 735	16.984 99	.020 408
50	2 500	125 000	7.071 068	22.360 68	3.684 031	7.937 005	17.099 76	.020 000

Table 22 (continued)

SQUARES, CUBES, ROOTS AND RECIPROCALS

n	n^2	n^3	$\sqrt{n}$	$\sqrt{10n}$	$\sqrt[3]{n}$	$\sqrt[3]{10n}$	$\sqrt[3]{100n}$	$1/n$
50	2 500	125 000	7.071 068	22.360 68	3.684 031	7.937 005	17.099 76	.020 000
51	2 601	132 651	7.141 428	22.583 18	3.708 430	7.989 570	17.213 01	.019 608
52	2 704	140 608	7.211 103	22.803 51	3.732 511	8.041 452	17.324 78	.019 231
53	2 809	148 877	7.280 110	23.021 73	3.756 286	8.092 672	17.435 13	.018 868
54	2 916	157 464	7.348 469	23.237 90	3.779 763	8.143 253	17.544 11	.018 519
55	3 025	166 375	7.416 198	23.452 08	3.802 952	8.193 213	17.651 74	.018 182
56	3 136	175 616	7.483 315	23.664 32	3.825 862	8.242 571	17.758 08	.017 857
57	3 249	185 193	7.549 834	23.874 67	3.848 501	8.291 344	17.863 16	.017 544
58	3 364	195 112	7.615 773	24.083 19	3.870 877	8.339 551	17.967 02	.017 241
59	3 481	205 379	7.681 146	24.289 92	3.892 996	8.387 207	18.069 69	.016 949
60	3 600	216 000	7.745 967	24.494 90	3.914 868	8.434 327	18.171 21	.016 667
61	3 721	226 981	7.810 250	24.698 18	3.936 497	8.480 926	18.271 60	.016 393
62	3 844	238 328	7.874 008	24.899 80	3.957 892	8.527 019	18.370 91	.016 129
63	3 969	250 047	7.937 254	25.099 80	3.979 057	8.572 619	18.469 15	.015 873
64	4 096	262 144	8.000 000	25.298 22	4.000 000	8.617 739	18.566 36	.015 625
65	4 225	274 625	8.062 258	25.495 10	4.020 726	8.662 391	18.662 56	.015 385
66	4 356	287 496	8.124 038	25.690 47	4.041 240	8.706 588	18.757 77	.015 152
67	4 489	300 763	8.185 353	25.884 36	4.061 548	8.750 340	18.852 04	.014 925
68	4 624	314 432	8.246 211	26.076 81	4.081 655	8.793 659	18.945 36	.014 706
69	4 761	328 509	8.306 624	26.267 85	4.101 566	8.836 556	19.037 78	.014 493
70	4 900	343 000	8.366 600	26.457 51	4.121 285	8.879 040	19.129 31	.014 286
71	5 041	357 911	8.426 150	26.645 83	4.140 818	8.921 121	19.219 97	.014 085
72	5 184	373 248	8.485 281	26.832 82	4.160 168	8.962 809	19.309 79	.013 889
73	5 329	389 017	8.544 004	27.018 51	4.179 339	9.004 113	19.398 77	.013 699
74	5 476	405 224	8.602 325	27.202 94	4.198 336	9.045 042	19.486 95	.013 514
75	5 625	421 875	8.660 254	27.386 13	4.217 163	9.085 603	19.574 34	.013 333
76	5 776	438 976	8.717 798	27.568 10	4.235 824	9.125 805	19.660 95	.013 158
77	5 929	456 533	8.774 964	27.748 87	4.254 321	9.165 656	19.746 81	.012 987
78	6 084	474 552	8.831 761	27.928 48	4.272 659	9.205 164	19.831 92	.012 821
79	6 241	493 039	8.888 194	28.106 94	4.290 840	9.244 335	19.916 32	.012 658
80	6 400	512 000	8.944 272	28.284 27	4.308 869	9.283 178	20.000 00	.012 500
81	6 561	531 441	9.000 000	28.460 50	4.326 749	9.321 698	20.082 99	.012 346
82	6 724	551 368	9.055 385	28.635 64	4.344 481	9.359 902	20.165 30	.012 195
83	6 889	571 787	9.110 434	28.809 72	4.362 071	9.397 796	20.246 94	.012 048
84	7 056	592 704	9.165 151	28.982 75	4.379 519	9.435 388	20.327 93	.011 905
85	7 225	614 125	9.219 544	29.154 76	4.396 830	9.472 682	20.408 28	.011 765
86	7 396	636 056	9.273 618	29.325 76	4.414 005	9.509 685	20.488 00	.011 628
87	7 569	658 503	9.327 379	29.495 76	4.431 048	9.546 403	20.567 10	.011 494
88	7 744	681 472	9.380 832	29.664 79	4.447 960	9.582 840	20.645 60	.011 364
89	7 921	704 969	9.433 981	29.832 87	4.464 745	9.619 002	20.723 51	.011 236
90	8 100	729 000	9.486 833	30.000 00	4.481 405	9.654 894	20.800 84	.011 111
91	8 281	753 571	9.539 392	30.166 21	4.497 941	9.690 521	20.877 59	.010 989
92	8 464	778 688	9.591 663	30.331 50	4.514 357	9.725 888	20.953 79	.010 870
93	8 649	804 357	9.643 651	30.495 90	4.530 655	9.761 000	21.029 44	.010 753
94	8 836	830 584	9.695 360	30.659 42	4.546 836	9.795 861	21.104 54	.010 638
95	9 025	857 375	9.746 794	30.822 07	4.562 903	9.830 476	21.179 12	.010 526
96	9 216	884 736	9.797 959	30.983 87	4.578 857	9.864 848	21.253 17	.010 417
97	9 409	912 673	9.848 858	31.144 82	4.594 701	9.898 983	21.326 71	.010 309
98	9 604	941 192	9.899 495	31.304 95	4.610 436	9.932 884	21.399 75	.010 204
99	9 801	970 299	9.949 874	31.464 27	4.626 065	9.966 555	21.472 29	.010 101
100	10 000	1 000 000	10.00 000	31.622 78	4.641 589	10.00 000	21.544 35	.010 000

TABLE

23

COMPOUND AMOUNT: $(1+r)^n$

If a principal P is deposited at interest rate r (in decimals) compounded annually, then at the end of n years the accumulated amount $A = P(1+r)^n$.

n \ r	1%	$1\frac{1}{4}$%	$1\frac{1}{2}$%	2%	$2\frac{1}{2}$%	3%	4%	5%	6%
1	1.0100	1.0125	1.0150	1.0200	1.0250	1.0300	1.0400	1.0500	1.0600
2	1.0201	1.0252	1.0302	1.0404	1.0506	1.0609	1.0816	1.1025	1.1236
3	1.0303	1.0380	1.0457	1.0612	1.0769	1.0927	1.1249	1.1576	1.1910
4	1.0406	1.0509	1.0614	1.0824	1.1038	1.1255	1.1699	1.2155	1.2635
5	1.0510	1.0641	1.0773	1.1041	1.1314	1.1593	1.2167	1.2763	1.3382
6	1.0615	1.0774	1.0934	1.1262	1.1597	1.1941	1.2653	1.3401	1.4185
7	1.0721	1.0909	1.1098	1.1487	1.1887	1.2299	1.3159	1.4071	1.5036
8	1.0829	1.1045	1.1265	1.1717	1.2184	1.2668	1.3688	1.4775	1.5938
9	1.0937	1.1183	1.1434	1.1951	1.2489	1.3048	1.4233	1.5513	1.6895
10	1.1046	1.1323	1.1605	1.2190	1.2801	1.3439	1.4802	1.6289	1.7908
11	1.1157	1.1464	1.1779	1.2434	1.3121	1.3842	1.5395	1.7103	1.8983
12	1.1268	1.1608	1.1956	1.2682	1.3449	1.4258	1.6010	1.7959	2.0122
13	1.1381	1.1753	1.2136	1.2936	1.3785	1.4685	1.6651	1.8856	2.1329
14	1.1495	1.1900	1.2318	1.3195	1.4130	1.5126	1.7317	1.9799	2.2609
15	1.1610	1.2048	1.2502	1.3459	1.4483	1.5580	1.8009	2.0789	2.3966
16	1.1726	1.2199	1.2690	1.3728	1.4845	1.6047	1.8730	2.1829	2.5404
17	1.1843	1.2351	1.2880	1.4002	1.5216	1.6528	1.9479	2.2920	2.6928
18	1.1961	1.2506	1.3073	1.4282	1.5597	1.7024	2.0258	2.4066	2.8543
19	1.2081	1.2662	1.3270	1.4568	1.5987	1.7535	2.1068	2.5270	3.0256
20	1.2202	1.2820	1.3469	1.4859	1.6386	1.8061	2.1911	2.6533	3.2071
21	1.2324	1.2981	1.3671	1.5157	1.6796	1.8603	2.2788	2.7860	3.3996
22	1.2447	1.3143	1.3876	1.5460	1.7216	1.9161	2.3699	2.9253	3.6035
23	1.2572	1.3307	1.4084	1.5769	1.7646	1.9736	2.4647	3.0715	3.8197
24	1.2697	1.3474	1.4295	1.6084	1.8087	2.0328	2.5633	3.2251	4.0489
25	1.2824	1.3642	1.4509	1.6406	1.8539	2.0938	2.6658	3.3864	4.2919
26	1.2953	1.3812	1.4727	1.6734	1.9003	2.1566	2.7725	3.5557	4.5494
27	1.3082	1.3985	1.4948	1.7069	1.9478	2.2213	2.8834	3.7335	4.8223
28	1.3213	1.4160	1.5172	1.7410	1.9965	2.2879	2.9987	3.9201	5.1117
29	1.3345	1.4337	1.5400	1.7758	2.0464	2.3566	3.1187	4.1161	5.4184
30	1.3478	1.4516	1.5631	1.8114	2.0976	2.4273	3.2434	4.3219	5.7435
31	1.3613	1.4698	1.5865	1.8476	2.1500	2.5001	3.3731	4.5380	6.0881
32	1.3749	1.4881	1.6103	1.8845	2.2038	2.5751	3.5081	4.7649	6.4534
33	1.3887	1.5067	1.6345	1.9222	2.2589	2.6523	3.6484	5.0032	6.8406
34	1.4026	1.5256	1.6590	1.9607	2.3153	2.7319	3.7943	5.2533	7.2510
35	1.4166	1.5446	1.6839	1.9999	2.3732	2.8139	3.9461	5.5160	7.6861
36	1.4308	1.5639	1.7091	2.0399	2.4325	2.8983	4.1039	5.7918	8.1473
37	1.4451	1.5835	1.7348	2.0807	2.4933	2.9852	4.2681	6.0814	8.6361
38	1.4595	1.6033	1.7608	2.1223	2.5557	3.0748	4.4388	6.3855	9.1543
39	1.4741	1.6233	1.7872	2.1647	2.6196	3.1670	4.6164	6.7048	9.7035
40	1.4889	1.6436	1.8140	2.2080	2.6851	3.2620	4.8010	7.0400	10.2857
41	1.5038	1.6642	1.8412	2.2522	2.7522	3.3599	4.9931	7.3920	10.9029
42	1.5188	1.6850	1.8688	2.2972	2.8210	3.4607	5.1928	7.7616	11.5570
43	1.5340	1.7060	1.8969	2.3432	2.8915	3.5645	5.4005	8.1497	12.2505
44	1.5493	1.7274	1.9253	2.3901	2.9638	3.6715	5.6165	8.5572	12.9855
45	1.5648	1.7489	1.9542	2.4379	3.0379	3.7816	5.8412	8.9850	13.7646
46	1.5805	1.7708	1.9835	2.4866	3.1139	3.8950	6.0748	9.4343	14.5905
47	1.5963	1.7929	2.0133	2.5363	3.1917	4.0119	6.3178	9.9060	15.4659
48	1.6122	1.8154	2.0435	2.5871	3.2715	4.1323	6.5705	10.4013	16.3939
49	1.6283	1.8380	2.0741	2.6388	3.3533	4.2562	6.8333	10.9213	17.3775
50	1.6446	1.8610	2.1052	2.6916	3.4371	4.3839	7.1067	11.4674	18.4202

TABLE 24

PRESENT VALUE OF AN AMOUNT: $(1+r)^{-n}$

The present value P which will amount to A in n years at an interest rate of r (in decimals) compounded annually is $P = A(1+r)^n$.

n \ r	1%	$1\frac{1}{4}$%	$1\frac{1}{2}$%	2%	$2\frac{1}{2}$%	3%	4%	5%	6%
1	.99010	.98765	.98522	.98039	.97561	.97087	.96154	.95238	.94340
2	.98030	.97546	.97066	.96117	.95181	.94260	.92456	.90703	.89000
3	.97059	.96342	.95632	.94232	.92860	.91514	.88900	.86384	.83962
4	.96098	.95152	.94218	.92385	.90595	.88849	.85480	.82270	.79209
5	.95147	.93978	.92826	.90573	.88385	.86261	.82193	.78353	.74726
6	.94205	.92817	.91454	.88797	.86230	.83748	.79031	.74622	.70496
7	.93272	.91672	.90103	.87056	.84127	.81309	.75992	.71068	.66506
8	.92348	.90540	.88771	.85349	.82075	.78941	.73069	.67684	.62741
9	.91434	.89422	.87459	.83676	.80073	.76642	.70259	.64461	.59190
10	.90529	.88318	.86167	.82035	.78120	.74409	.67556	.61391	.55839
11	.89632	.87228	.84893	.80426	.76214	.72242	.64958	.58468	.52679
12	.88745	.86151	.83639	.78849	.74356	.70138	.62460	.55684	.49697
13	.87866	.85087	.82403	.77303	.72542	.68095	.60057	.53032	.46884
14	.86996	.84037	.81185	.75788	.70773	.66112	.57748	.50507	.44230
15	.86135	.82999	.79985	.74301	.69047	.64186	.55526	.48102	.41727
16	.85282	.81975	.78803	.72845	.67362	.62317	.53391	.45811	.39365
17	.84438	.80963	.77639	.71416	.65720	.60502	.51337	.43630	.37136
18	.83602	.79963	.76491	.70016	.64117	.58739	.49363	.41552	.35034
19	.82774	.78976	.75361	.68643	.62553	.57029	.47464	.39573	.33051
20	.81954	.78001	.74247	.67297	.61027	.55368	.45639	.37689	.31180
21	.81143	.77038	.73150	.65978	.59539	.53755	.43883	.35894	.29416
22	.80340	.76087	.72069	.64684	.58086	.52189	.42196	.34185	.27751
23	.79544	.75147	.71004	.63416	.56670	.50669	.40573	.32557	.26180
24	.78757	.74220	.69954	.62172	.55288	.49193	.39012	.31007	.24698
25	.77977	.73303	.68921	.60953	.53939	.47761	.37512	.29530	.23300
26	.77205	.72398	.67902	.59758	.52623	.46369	.36069	.28124	.21981
27	.76440	.71505	.66899	.58586	.51340	.45019	.34682	.26785	.20737
28	.75684	.70622	.65910	.57437	.50088	.43708	.33348	.25509	.19563
29	.74934	.69750	.64936	.56311	.48866	.42435	.32065	.24295	.18456
30	.74192	.68889	.63976	.55207	.47674	.41199	.30832	.23138	.17411
31	.73458	.68038	.63031	.54125	.46511	.39999	.29646	.22036	.16425
32	.72730	.67198	.62099	.53063	.45377	.38834	.28506	.20987	.15496
33	.72010	.66369	.61182	.52023	.44270	.37703	.27409	.19987	.14619
34	.71297	.65549	.60277	.51003	.43191	.36604	.26355	.19035	.13791
35	.70591	.64740	.59387	.50003	.42137	.35538	.25342	.18129	.13011
36	.69892	.63941	.58509	.49022	.41109	.34503	.24367	.17266	.12274
37	.69200	.63152	.57644	.48061	.40107	.33498	.23430	.16444	.11579
38	.68515	.62372	.56792	.47119	.39128	.32523	.22529	.15661	.10924
39	.67837	.61602	.55953	.46195	.38174	.31575	.21662	.14915	.10306
40	.67165	.60841	.55126	.45289	.37243	.30656	.20829	.14205	.09722
41	.66500	.60090	.54312	.44401	.36335	.29763	.20028	.13528	.09172
42	.65842	.59348	.53509	.43530	.35448	.28896	.19257	.12884	.08653
43	.65190	.58616	.52718	.42677	.34584	.28054	.18517	.12270	.08163
44	.64545	.57892	.51939	.41840	.33740	.27237	.17805	.11686	.07701
45	.63905	.57177	.51171	.41020	.32917	.26444	.17120	.11130	.07265
46	.63273	.56471	.50415	.40215	.32115	.25674	.16461	.10600	.06854
47	.62646	.55774	.49670	.39427	.31331	.24926	.15828	.10095	.06466
48	.62026	.55086	.48936	.38654	.30567	.24200	.15219	.09614	.06100
49	.61412	.54406	.48213	.37896	.29822	.23495	.14634	.09156	.05755
50	.60804	.53734	.47500	.37153	.29094	.22811	.14071	.08720	.05429

TABLE 25

AMOUNT OF AN ANNUITY: $\dfrac{(1+r)^n - 1}{r}$

If a principal P is deposited at the end of each year at interest rate r (in decimals) compounded annually, then at the end of n years the accumulated amount is $P\left[\dfrac{(1+r)^n - 1}{r}\right]$. The process is often called an *annuity*.

n \ r	1%	$1\frac{1}{4}$%	$1\frac{1}{2}$%	2%	$2\frac{1}{2}$%	3%	4%	5%	6%
1	1.0000	1.0000	1.0000	1.0000	1.0000	1.0000	1.0000	1.0000	1.0000
2	2.0100	2.0125	2.0150	2.0200	2.0250	2.0300	2.0400	2.0500	2.0600
3	3.0301	3.0377	3.0452	3.0604	3.0756	3.0909	3.1216	3.1525	3.1836
4	4.0604	4.0756	4.0909	4.1216	4.1525	4.1836	4.2465	4.3101	4.3746
5	5.1010	5.1266	5.1523	5.2040	5.2563	5.3091	5.4163	5.5256	5.6371
6	6.1520	6.1907	6.2296	6.3081	6.3877	6.4684	6.6330	6.8019	6.9753
7	7.2135	7.2680	7.3230	7.4343	7.5474	7.6625	7.8983	8.1420	8.3938
8	8.2857	8.3589	8.4328	8.5830	8.7361	8.8923	9.2142	9.5491	9.8975
9	9.3685	9.4634	9.5593	9.7546	9.9545	10.1591	10.5828	11.0266	11.4913
10	10.4622	10.5817	10.7027	10.9497	11.2034	11.4639	12.0061	12.5779	13.1808
11	11.5668	11.7139	11.8633	12.1687	12.4835	12.8078	13.4864	14.2068	14.9716
12	12.6825	12.8604	13.0412	13.4121	13.7956	14.1920	15.0258	15.9171	16.8699
13	13.8093	14.0211	14.2368	14.6803	15.1404	15.6178	16.6268	17.7130	18.8821
14	14.9474	15.1964	15.4504	15.9739	16.5190	17.0863	18.2919	19.5986	21.0151
15	16.0969	16.3863	16.6821	17.2934	17.9319	18.5989	20.0236	21.5786	23.2760
16	17.2579	17.5912	17.9324	18.6393	19.3802	20.1569	21.8245	23.6575	25.6725
17	18.4304	18.8111	19.2014	20.0121	20.8647	21.7616	23.6975	25.8404	28.2129
18	19.6147	20.0462	20.4894	21.4123	22.3863	23.4144	25.6454	28.1324	30.9057
19	20.8109	21.2968	21.7967	22.8406	23.9460	25.1169	27.6712	30.5390	33.7600
20	22.0190	22.5630	23.1237	24.2974	25.5447	26.8704	29.7781	33.0660	36.7856
21	23.2392	23.8450	24.4705	25.7833	27.1833	28.6765	31.9692	35.7193	39.9927
22	24.4716	25.1431	25.8376	27.2990	28.8629	30.5368	34.2480	38.5052	43.3923
23	25.7163	26.4574	27.2251	28.8450	30.5844	32.4529	36.6179	41.4305	46.9958
24	26.9735	27.7881	28.6335	30.4219	32.3490	34.4265	39.0826	44.5020	50.8156
25	28.2432	29.1354	30.0630	32.0303	34.1578	36.4593	41.6459	47.7271	54.8645
26	29.5256	30.4996	31.5140	33.6709	36.0117	38.5530	44.3117	51.1135	59.1564
27	30.8209	31.8809	32.9867	35.3443	37.9120	40.7096	47.0842	54.6691	63.7058
28	32.1291	33.2794	34.4815	37.0512	39.8598	42.9309	49.9676	58.4026	68.5281
29	33.4504	34.6954	35.9987	38.7922	41.8563	45.2189	52.9663	62.3227	73.6398
30	34.7849	36.1291	37.5387	40.5681	43.9027	47.5754	56.0849	66.4388	79.0582
31	36.1327	37.5807	39.1018	42.3794	46.0003	50.0027	59.3283	70.7608	84.8017
32	37.4941	39.0504	40.6883	44.2270	48.1503	52.5028	62.7015	75.2988	90.8898
33	38.8690	40.5386	42.2986	46.1116	50.3540	55.0778	66.2095	80.0638	97.3432
34	40.2577	42.0453	43.9331	48.0338	52.6129	57.7302	69.8579	85.0670	104.1838
35	41.6603	43.5709	45.5921	49.9945	54.9282	60.4621	73.6522	90.3203	111.4348
36	43.0769	45.1155	47.2760	51.9944	57.3014	63.2759	77.5983	95.8363	119.1209
37	44.5076	46.6794	48.9851	54.0343	59.7339	66.1742	81.7022	101.6281	127.2681
38	45.9527	48.2629	50.7199	56.1149	62.2273	69.1594	85.9703	107.7095	135.9042
39	47.4123	49.8662	52.4807	58.2372	64.7830	72.2342	90.4091	114.0950	145.0585
40	48.8864	51.4896	54.2679	60.4020	67.4026	75.4013	95.0255	120.7998	154.7620
41	50.3752	53.1332	56.0819	62.6100	70.0876	78.6633	99.8265	127.8398	165.0477
42	51.8790	54.7973	57.9231	64.8622	72.8398	82.0232	104.8196	135.2318	175.9505
43	53.3978	56.4823	59.7920	67.1595	75.6608	85.4839	110.0124	142.9933	187.5076
44	54.9318	58.1883	61.6889	69.5027	78.5523	89.0484	115.4129	151.1430	199.7580
45	56.4811	59.9157	63.6142	71.8927	81.5161	92.7199	121.0294	159.7002	212.7435
46	58.0459	61.6646	65.5684	74.3306	84.5540	96.5015	126.8706	168.6852	226.5081
47	59.6263	63.4354	67.5519	76.8172	87.6679	100.3965	132.9454	178.1194	241.0986
48	61.2226	65.2284	69.5652	79.3535	90.8596	104.4084	139.2632	188.0254	256.5645
49	62.8348	67.0437	71.6087	81.9406	94.1311	108.5406	145.8337	198.4267	272.9584
50	64.4632	68.8818	73.6828	84.5794	97.4843	112.7969	152.6671	209.3480	290.3359

TABLE 26

PRESENT VALUE OF AN ANNUITY: $\dfrac{1-(1+r)^{-n}}{r}$

An annuity in which the yearly payment at the end of each of n years is A at an interest rate r (in decimals) compounded annually has present value $A\left[\dfrac{1-(1+r)^{-n}}{r}\right]$.

n \ r	1%	$1\frac{1}{4}$%	$1\frac{1}{2}$%	2%	$2\frac{1}{2}$%	3%	4%	5%	6%
1	0.9901	0.9877	0.9852	0.9804	0.9756	0.9709	0.9615	0.9524	0.9434
2	1.9704	1.9631	1.9559	1.9416	1.9274	1.9135	1.8861	1.8594	1.8334
3	2.9410	2.9265	2.9122	2.8839	2.8560	2.8286	2.7751	2.7232	2.6730
4	3.9020	3.8781	3.8544	3.8077	3.7620	3.7171	3.6299	3.5460	3.4651
5	4.8534	4.8178	4.7826	4.7135	4.6458	4.5797	4.4518	4.3295	4.2124
6	5.7955	5.7460	5.6972	5.6014	5.5081	5.4172	5.2421	5.0757	4.9173
7	6.7282	6.6627	6.5982	6.4720	6.3494	6.2303	6.0021	5.7864	5.5824
8	7.6517	7.5681	7.4859	7.3255	7.1701	7.0197	6.7327	6.4632	6.2098
9	8.5660	8.4623	8.3605	8.1622	7.9709	7.7861	7.4353	7.1078	6.8017
10	9.4713	9.3455	9.2222	8.9826	8.7521	8.5302	8.1109	7.7217	7.3601
11	10.3676	10.2178	10.0711	9.7868	9.5142	9.2526	8.7605	8.3064	7.8869
12	11.2551	11.0793	10.9075	10.5753	10.2578	9.9540	9.3851	8.8633	8.3838
13	12.1337	11.9302	11.7315	11.3484	10.9832	10.6350	9.9856	9.3936	8.8527
14	13.0037	12.7706	12.5434	12.1062	11.6909	11.2961	10.5631	9.8986	9.2950
15	13.8651	13.6005	13.3432	12.8493	12.3814	11.9379	11.1184	10.3797	9.7122
16	14.7179	14.4203	14.1313	13.5777	13.0550	12.5611	11.6523	10.8378	10.1059
17	15.5623	15.2299	14.9076	14.2919	13.7122	13.1661	12.1657	11.2741	10.4773
18	16.3983	16.0295	15.6726	14.9920	14.3534	13.7535	12.6593	11.6896	10.8276
19	17.2260	16.8193	16.4262	15.6785	14.9789	14.3238	13.1339	12.0853	11.1581
20	18.0456	17.5993	17.1686	16.3514	15.5892	14.8775	13.5903	12.4622	11.4699
21	18.8570	18.3697	17.9001	17.0112	16.1845	15.4150	14.0292	12.8212	11.7641
22	19.6604	19.1306	18.6208	17.6580	16.7654	15.9369	14.4511	13.1630	12.0416
23	20.4558	19.8820	19.3309	18.2922	17.3321	16.4436	14.8568	13.4886	12.3034
24	21.2434	20.6242	20.0304	18.9139	17.8850	16.9355	15.2470	13.7986	12.5504
25	22.0232	21.3573	20.7196	19.5235	18.4244	17.4131	15.6221	14.0939	12.7834
26	22.7952	22.0813	21.3986	20.1210	18.9506	17.8768	15.9828	14.3752	13.0032
27	23.5596	22.7963	22.0676	20.7069	19.4640	18.3270	16.3296	14.6430	13.2105
28	24.3164	23.5025	22.7267	21.2813	19.9649	18.7641	16.6631	14.8981	13.4062
29	25.0658	24.2000	23.3761	21.8444	20.4535	19.1885	16.9837	15.1411	13.5907
30	25.8077	24.8889	24.0158	22.3965	20.9303	19.6004	17.2920	15.3725	13.7648
31	26.5423	25.5693	24.6461	22.9377	21.3954	20.0004	17.5885	15.5928	13.9291
32	27.2696	26.2413	25.2671	23.4683	21.8492	20.3888	17.8736	15.8027	14.0840
33	27.9897	26.9050	25.8790	23.9886	22.2919	20.7658	18.1476	16.0025	14.2302
34	28.7027	27.5605	26.4817	24.4986	22.7238	21.1318	18.4112	16.1929	14.3681
35	29.4086	28.2079	27.0756	24.9986	23.1452	21.4872	18.6646	16.3742	14.4982
36	30.1075	28.8473	27.6607	25.4888	23.5563	21.8323	18.9083	16.5469	14.6210
37	30.7995	29.4788	28.2371	25.9695	23.9573	22.1672	19.1426	16.7113	14.7368
38	31.4847	30.1025	28.8051	26.4406	24.3486	22.4925	19.3679	16.8679	14.8460
39	32.1630	30.7185	29.3646	26.9026	24.7303	22.8082	19.5845	17.0170	14.9491
40	32.8347	31.3269	29.9158	27.3555	25.1028	23.1148	19.7928	17.1591	15.0463
41	33.4997	31.9278	30.4590	27.7995	25.4661	23.4124	19.9931	17.2944	15.1380
42	34.1581	32.5213	30.9941	28.2348	25.8206	23.7014	20.1856	17.4232	15.2245
43	34.8100	33.1075	31.5212	28.6616	26.1664	23.9819	20.3708	17.5459	15.3062
44	35.4555	33.6864	32.0406	29.0800	26.5038	24.2543	20.5488	17.6628	15.3832
45	36.0945	34.2582	32.5523	29.4902	26.8330	24.5187	20.7200	17.7741	15.4558
46	36.7272	34.8229	33.0565	29.8923	27.1542	24.7754	20.8847	17.8801	15.5244
47	37.3537	35.3806	33.5532	30.2866	27.4675	25.0247	21.0429	17.9810	15.5890
48	37.9740	35.9315	34.0426	30.6731	27.7732	25.2667	21.1951	18.0772	15.6500
49	38.5881	36.4755	34.5247	31.0521	28.0714	25.5017	21.3415	18.1687	15.7076
50	39.1961	37.0129	34.9997	31.4236	28.3623	25.7298	21.4822	18.2559	15.7619

TABLE 27

BESSEL FUNCTIONS $J_0(x)$

x	0	1	2	3	4	5	6	7	8	9
0.	1.0000	.9975	.9900	.9776	.9604	.9385	.9120	.8812	.8463	.8075
1.	.7652	.7196	.6711	.6201	.5669	.5118	.4554	.3980	.3400	.2818
2.	.2239	.1666	.1104	.0555	.0025	−.0484	−.0968	−.1424	−.1850	−.2243
3.	−.2601	−.2921	−.3202	−.3443	−.3643	−.3801	−.3918	−.3992	−.4026	−.4018
4.	−.3971	−.3887	−.3766	−.3610	−.3423	−.3205	−.2961	−.2693	−.2404	−.2097
5.	−.1776	−.1443	−.1103	−.0758	−.0412	−.0068	.0270	.0599	.0917	.1220
6.	.1506	.1773	.2017	.2238	.2433	.2601	.2740	.2851	.2931	.2981
7.	.3001	.2991	.2951	.2882	.2786	.2663	.2516	.2346	.2154	.1944
8.	.1717	.1475	.1222	.0960	.0692	.0419	.0146	−.0125	−.0392	−.0653
9.	−.0903	−.1142	−.1367	−.1577	−.1768	−.1939	−.2090	−.2218	−.2323	−.2403

TABLE 28

BESSEL FUNCTIONS $J_1(x)$

x	0	1	2	3	4	5	6	7	8	9
0.	.0000	.0499	.0995	.1483	.1960	.2423	.2867	.3290	.3688	.4059
1.	.4401	.4709	.4983	.5220	.5419	.5579	.5699	.5778	.5815	.5812
2.	.5767	.5683	.5560	.5399	.5202	.4971	.4708	.4416	.4097	.3754
3.	.3391	.3009	.2613	.2207	.1792	.1374	.0955	.0538	.0128	−.0272
4.	−.0660	−.1033	−.1386	−.1719	−.2028	−.2311	−.2566	−.2791	−.2985	−.3147
5.	−.3276	−.3371	−.3432	−.3460	−.3453	−.3414	−.3343	−.3241	−.3110	−.2951
6.	−.2767	−.2559	−.2329	−.2081	−.1816	−.1538	−.1250	−.0953	−.0652	−.0349
7.	−.0047	.0252	.0543	.0826	.1096	.1352	.1592	.1813	.2014	.2192
8.	.2346	.2476	.2580	.2657	.2708	.2731	.2728	.2697	.2641	.2559
9.	.2453	.2324	.2174	.2004	.1816	.1613	.1395	.1166	.0928	.0684

TABLE 29

BESSEL FUNCTIONS $Y_0(x)$

x	0	1	2	3	4	5	6	7	8	9
0.	$-\infty$	−1.5342	−1.0811	−.8073	−.6060	−.4445	−.3085	−.1907	−.0868	.0056
1.	.0883	.1622	.2281	.2865	.3379	.3824	.4204	.4520	.4774	.4968
2.	.5104	.5183	.5208	.5181	.5104	.4981	.4813	.4605	.4359	.4079
3.	.3769	.3431	.3071	.2691	.2296	.1890	.1477	.1061	.0645	.0234
4.	−.0169	−.0561	−.0938	−.1296	−.1633	−.1947	−.2235	−.2494	−.2723	−.2921
5.	−.3085	−.3216	−.3313	−.3374	−.3402	−.3395	−.3354	−.3282	−.3177	−.3044
6.	−.2882	−.2694	−.2483	−.2251	−.1999	−.1732	−.1452	−.1162	−.0864	−.0563
7.	−.0259	.0042	.0339	.0628	.0907	.1173	.1424	.1658	.1872	.2065
8.	.2235	.2381	.2501	.2595	.2662	.2702	.2715	.2700	.2659	.2592
9.	.2499	.2383	.2245	.2086	.1907	.1712	.1502	.1279	.1045	.0804

TABLE 30

BESSEL FUNCTIONS $Y_1(x)$

x	0	1	2	3	4	5	6	7	8	9
0.	$-\infty$	−6.4590	−3.3238	−2.2931	−1.7809	−1.4715	−1.2604	−1.1032	−.9781	−.8731
1.	−.7812	−.6981	−.6211	−.5485	−.4791	−.4123	−.3476	−.2847	−.2237	−.1644
2.	−.1070	−.0517	.0015	.0523	.1005	.1459	.1884	.2276	.2635	.2959
3.	.3247	.3496	.3707	.3879	.4010	.4102	.4154	.4167	.4141	.4078
4.	.3979	.3846	.3680	.3484	.3260	.3010	.2737	.2445	.2136	.1812
5.	.1479	.1137	.0792	.0445	.0101	−.0238	−.0568	−.0887	−.1192	−.1481
6.	−.1750	−.1998	−.2223	−.2422	−.2596	−.2741	−.2857	−.2945	−.3002	−.3029
7.	−.3027	−.2995	−.2934	−.2846	−.2731	−.2591	−.2428	−.2243	−.2039	−.1817
8.	−.1581	−.1331	−.1072	−.0806	−.0535	−.0262	.0011	.0280	.0544	.0799
9.	.1043	.1275	.1491	.1691	.1871	.2032	.2171	.2287	.2379	.2447

TABLE 31

BESSEL FUNCTIONS

$I_0(x)$

x	0	1	2	3	4	5	6	7	8	9
0.	1.000	1.003	1.010	1.023	1.040	1.063	1.092	1.126	1.167	1.213
1.	1.266	1.326	1.394	1.469	1.553	1.647	1.750	1.864	1.990	2.128
2.	2.280	2.446	2.629	2.830	3.049	3.290	3.553	3.842	4.157	4.503
3.	4.881	5.294	5.747	6.243	6.785	7.378	8.028	8.739	9.517	10.37
4.	11.30	12.32	13.44	14.67	16.01	17.48	19.09	20.86	22.79	24.91
5.	27.24	29.79	32.58	35.65	39.01	42.69	46.74	51.17	56.04	61.38
6.	67.23	73.66	80.72	88.46	96.96	106.3	116.5	127.8	140.1	153.7
7.	168.6	185.0	202.9	222.7	244.3	268.2	294.3	323.1	354.7	389.4
8.	427.6	469.5	515.6	566.3	621.9	683.2	750.5	824.4	905.8	995.2
9.	1094	1202	1321	1451	1595	1753	1927	2119	2329	2561

TABLE 32

BESSEL FUNCTIONS

$I_1(x)$

x	0	1	2	3	4	5	6	7	8	9
0.	.0000	.0501	.1005	.1517	.2040	.2579	.3137	.3719	.4329	.4971
1.	.5652	.6375	.7147	.7973	.8861	.9817	1.085	1.196	1.317	1.448
2.	1.591	1.745	1.914	2.098	2.298	2.517	2.755	3.016	3.301	3.613
3.	3.953	4.326	4.734	5.181	5.670	6.206	6.793	7.436	8.140	8.913
4.	9.759	10.69	11.71	12.82	14.05	15.39	16.86	18.48	20.25	22.20
5.	24.34	26.68	29.25	32.08	35.18	38.59	42.33	46.44	50.95	55.90
6.	61.34	67.32	73.89	81.10	89.03	97.74	107.3	117.8	129.4	142.1
7.	156.0	171.4	188.3	206.8	227.2	249.6	274.2	301.3	331.1	363.9
8.	399.9	439.5	483.0	531.0	583.7	641.6	705.4	775.5	852.7	937.5
9.	1031	1134	1247	1371	1508	1658	1824	2006	2207	2428

TABLE 33

BESSEL FUNCTIONS

$K_0(x)$

x	0	1	2	3	4	5	6	7	8	9
0.	∞	2.4271	1.7527	1.3725	1.1145	.9244	.7775	.6605	.5653	.4867
1.	.4210	.3656	.3185	.2782	.2437	.2138	.1880	.1655	.1459	.1288
2.	.1139	.1008	.08927	.07914	.07022	.06235	.05540	.04926	.04382	.03901
3.	.03474	.03095	.02759	.02461	.02196	.01960	.01750	.01563	.01397	.01248
4.	.01116	$.0^{2}9980$	$.0^{2}8927$	$.0^{2}7988$	$.0^{2}7149$	$.0^{2}6400$	$.0^{2}5730$	$.0^{2}5132$	$.0^{2}4597$	$.0^{2}4119$
5.	$.0^{2}3691$	$.0^{2}3308$	$.0^{2}2966$	$.0^{2}2659$	$.0^{2}2385$	$.0^{2}2139$	$.0^{2}1918$	$.0^{2}1721$	$.0^{2}1544$	$.0^{2}1386$
6.	$.0^{2}1244$	$.0^{2}1117$	$.0^{2}1003$	$.0^{3}9001$	$.0^{3}8083$	$.0^{3}7259$	$.0^{3}6520$	$.0^{3}5857$	$.0^{3}5262$	$.0^{3}4728$
7.	$.0^{3}4248$	$.0^{3}3817$	$.0^{3}3431$	$.0^{3}3084$	$.0^{3}2772$	$.0^{3}2492$	$.0^{3}2240$	$.0^{3}2014$	$.0^{3}1811$	$.0^{3}1629$
8.	$.0^{3}1465$	$.0^{3}1317$	$.0^{3}1185$	$.0^{3}1066$	$.0^{4}9588$	$.0^{4}8626$	$.0^{4}7761$	$.0^{4}6983$	$.0^{4}6283$	$.0^{4}5654$
9.	$.0^{4}5088$	$.0^{4}4579$	$.0^{4}4121$	$.0^{4}3710$	$.0^{4}3339$	$.0^{4}3006$	$.0^{4}2706$	$.0^{4}2436$	$.0^{4}2193$	$.0^{4}1975$

TABLE 34

BESSEL FUNCTIONS

$K_1(x)$

x	0	1	2	3	4	5	6	7	8	9
0.	∞	9.8538	4.7760	3.0560	2.1844	1.6564	1.3028	1.0503	.8618	.7165
1.	.6019	.5098	.4346	.3725	.3208	.2774	.2406	.2094	.1826	.1597
2.	.1399	.1227	.1079	.09498	.08372	.07389	.06528	.05774	.05111	.04529
3.	.04016	.03563	.03164	.02812	.02500	.02224	.01979	.01763	.01571	.01400
4.	.01248	.01114	$.0^{2}9938$	$.0^{2}8872$	$.0^{2}7923$	$.0^{2}7078$	$.0^{2}6325$	$.0^{2}5654$	$.0^{2}5055$	$.0^{2}4521$
5.	$.0^{2}4045$	$.0^{2}3619$	$.0^{2}3239$	$.0^{2}2900$	$.0^{2}2597$	$.0^{2}2326$	$.0^{2}2083$	$.0^{2}1866$	$.0^{2}1673$	$.0^{2}1499$
6.	$.0^{2}1344$	$.0^{2}1205$	$.0^{2}1081$	$.0^{3}9691$	$.0^{3}8693$	$.0^{3}7799$	$.0^{3}6998$	$.0^{3}6280$	$.0^{3}5636$	$.0^{3}5059$
7.	$.0^{3}4542$	$.0^{3}4078$	$.0^{3}3662$	$.0^{3}3288$	$.0^{3}2953$	$.0^{3}2653$	$.0^{3}2383$	$.0^{3}2141$	$.0^{3}1924$	$.0^{3}1729$
8.	$.0^{3}1554$	$.0^{3}1396$	$.0^{3}1255$	$.0^{3}1128$	$.0^{3}1014$	$.0^{4}9120$	$.0^{4}8200$	$.0^{4}7374$	$.0^{4}6631$	$.0^{4}5964$
9.	$.0^{4}5364$	$.0^{4}4825$	$.0^{4}4340$	$.0^{4}3904$	$.0^{4}3512$	$.0^{4}3160$	$.0^{4}2843$	$.0^{4}2559$	$.0^{4}2302$	$.0^{4}2072$

TABLE 35

BESSEL FUNCTIONS Ber (x)

x	0	1	2	3	4	5	6	7	8	9
0.	1.0000	1.0000	1.0000	.9999	.9996	.9990	.9980	.9962	.9936	.9898
1.	.9844	.9771	.9676	.9554	.9401	.9211	.8979	.8700	.8367	.7975
2.	.7517	.6987	.6377	.5680	.4890	.4000	.3001	.1887	.06511	−.07137
3.	−.2214	−.3855	−.5644	−.7584	−.9680	−1.1936	−1.4353	−1.6933	−1.9674	−2.2576
4.	−2.5634	−2.8843	−3.2195	−3.5679	−3.9283	−4.2991	−4.6784	−5.0639	−5.4531	−5.8429
5.	−6.2301	−6.6107	−6.9803	−7.3344	−7.6674	−7.9736	−8.2466	−8.4794	−8.6644	−8.7937
6.	−8.8583	−8.8491	−8.7561	−8.5688	−8.2762	−7.8669	−7.3287	−6.6492	−5.8155	−4.8146
7.	−3.6329	−2.2571	−.6737	1.1308	3.1695	5.4550	7.9994	10.814	13.909	17.293
8.	20.974	24.957	29.245	33.840	38.738	43.936	49.423	55.187	61.210	67.469
9.	73.936	80.576	87.350	94.208	101.10	107.95	114.70	121.26	127.54	133.43

TABLE 36

BESSEL FUNCTIONS Bei (x)

x	0	1	2	3	4	5	6	7	8	9
0.	.0000	$.0^{2}2500$	.01000	.02250	.04000	.06249	.08998	.1224	.1599	.2023
1.	.2496	.3017	.3587	.4204	.4867	.5576	.6327	.7120	.7953	.8821
2.	.9723	1.0654	1.1610	1.2585	1.3575	1.4572	1.5569	1.6557	1.7529	1.8472
3.	1.9376	2.0228	2.1016	2.1723	2.2334	2.2832	2.3199	2.3413	2.3454	2.3300
4.	2.2927	2.2309	2.1422	2.0236	1.8726	1.6860	1.4610	1.1946	.8837	.5251
5.	.1160	−.3467	−.8658	−1.4443	−2.0845	−2.7890	−3.5597	−4.3986	−5.3068	−6.2854
6.	−7.3347	−8.4545	−9.6437	−10.901	−12.223	−13.607	−15.047	−16.538	−18.074	−19.644
7.	−21.239	−22.848	−24.456	−26.049	−27.609	−29.116	−30.548	−31.882	−33.092	−34.147
8.	−35.017	−35.667	−36.061	−36.159	−35.920	−35.298	−34.246	−32.714	−30.651	−28.003
9.	−24.713	−20.724	−15.976	−10.412	−3.9693	3.4106	11.787	21.218	31.758	43.459

TABLE 37

BESSEL FUNCTIONS

Ker (x)

x	0	1	2	3	4	5	6	7	8	9
0.	∞	2.4205	1.7331	1.3372	1.0626	.8559	.6931	.5614	.4529	.3625
1.	.2867	.2228	.1689	.1235	.08513	.05293	.02603	$.0^{2}3691$	−.01470	−.02966
2.	−.04166	−.05111	−.05834	−.06367	−.06737	−.06969	−.07083	−.07097	−.07030	−.06894
3.	−.06703	−.06468	−.06198	−.05903	−.05590	−.05264	−.04932	−.04597	−.04265	−.03937
4.	−.03618	−.03308	−.03011	−.02726	−.02456	−.02200	−.01960	−.01734	−.01525	−.01330
5.	−.01151	$-.0^{2}9865$	$-.0^{2}8359$	$-.0^{2}6989$	$-.0^{2}5749$	$-.0^{2}4632$	$-.0^{2}3632$	$-.0^{2}2740$	$-.0^{2}1952$	$-.0^{2}1258$
6.	$-.0^{3}6530$	$-.0^{3}1295$	$.0^{3}3191$	$.0^{3}6991$	$.0^{2}1017$	$.0^{2}1278$	$.0^{2}1488$	$.0^{2}1653$	$.0^{2}1777$	$.0^{2}1866$
7.	$.0^{2}1922$	$.0^{2}1951$	$.0^{2}1956$	$.0^{2}1940$	$.0^{2}1907$	$.0^{2}1860$	$.0^{2}1800$	$.0^{2}1731$	$.0^{2}1655$	$.0^{2}1572$
8.	$.0^{2}1486$	$.0^{2}1397$	$.0^{2}1306$	$.0^{2}1216$	$.0^{2}1126$	$.0^{2}1037$	$.0^{3}9511$	$.0^{3}8675$	$.0^{3}7871$	$.0^{3}7102$
9.	$.0^{3}6372$	$.0^{3}5681$	$.0^{3}5030$	$.0^{3}4422$	$.0^{3}3855$	$.0^{3}3330$	$.0^{3}2846$	$.0^{3}2402$	$.0^{3}1996$	$.0^{3}1628$

TABLE 38

BESSEL FUNCTIONS

Kei (x)

x	0	1	2	3	4	5	6	7	8	9
0.	−.7854	−.7769	−.7581	−.7331	−.7038	−.6716	−.6374	−.6022	−.5664	−.5305
1.	−.4950	−.4601	−.4262	−.3933	−.3617	−.3314	−.3026	−.2752	−.2494	−.2251
2.	−.2024	−.1812	−.1614	−.1431	−.1262	−.1107	−.09644	−.08342	−.07157	−.06083
3.	−.05112	−.04240	−.03458	−.02762	−.02145	−.01600	−.01123	$-.0^{2}7077$	$-.0^{2}3487$	$-.0^{3}4108$
4.	$.0^{2}2198$	$.0^{2}4386$	$.0^{2}6194$	$.0^{2}7661$	$.0^{2}8826$	$.0^{2}9721$	.01038	.01083	.01110	.01121
5.	.01119	.01105	.01082	.01051	.01014	$.0^{2}9716$	$.0^{2}9255$	$.0^{2}8766$	$.0^{2}8258$	$.0^{2}7739$
6.	$.0^{2}7216$	$.0^{2}6696$	$.0^{2}6183$	$.0^{2}5681$	$.0^{2}5194$	$.0^{2}4724$	$.0^{2}4274$	$.0^{2}3846$	$.0^{2}3440$	$.0^{2}3058$
7.	$.0^{2}2700$	$.0^{2}2366$	$.0^{2}2057$	$.0^{2}1770$	$.0^{2}1507$	$.0^{2}1267$	$.0^{2}1048$	$.0^{3}8498$	$.0^{3}6714$	$.0^{3}5117$
8.	$.0^{3}3696$	$.0^{3}2440$	$.0^{3}1339$	$.0^{4}3809$	$-.0^{4}4449$	$-.0^{3}1149$	$-.0^{3}1742$	$-.0^{3}2233$	$-.0^{3}2632$	$-.0^{3}2949$
9.	$-.0^{3}3192$	$-.0^{3}3368$	$-.0^{3}3486$	$-.0^{3}3552$	$-.0^{3}3574$	$-.0^{3}3557$	$-.0^{3}3508$	$-.0^{3}3430$	$-.0^{3}3329$	$-.0^{3}3210$

TABLE 39

VALUES FOR APPROXIMATE ZEROS OF BESSEL FUNCTIONS

The following table lists the first few positive roots of various equations. Note that for all cases listed the successive large roots differ approximately by $\pi = 3.14159\ldots$.

	$n=0$	$n=1$	$n=2$	$n=3$	$n=4$	$n=5$	$n=6$
$J_n(x)=0$	2.4048	3.8317	5.1356	6.3802	7.5883	8.7715	9.9361
	5.5201	7.0156	8.4172	9.7610	11.0647	12.3386	13.5893
	8.6537	10.1735	11.6198	13.0152	14.3725	15.7002	17.0038
	11.7915	13.3237	14.7960	16.2235	17.6160	18.9801	20.3208
	14.9309	16.4706	17.9598	19.4094	20.8269	22.2178	23.5861
	18.0711	19.6159	21.1170	22.5827	24.0190	25.4303	26.8202
$Y_n(x)=0$	0.8936	2.1971	3.3842	4.5270	5.6452	6.7472	7.8377
	3.9577	5.4297	6.7938	8.0976	9.3616	10.5972	11.8110
	7.0861	8.5960	10.0235	11.3965	12.7301	14.0338	15.3136
	10.2223	11.7492	13.2100	14.6231	15.9996	17.3471	18.6707
	13.3611	14.8974	16.3790	17.8185	19.2244	20.6029	21.9583
	16.5009	18.0434	19.5390	20.9973	22.4248	23.8265	25.2062
$J'_n(x)=0$	0.0000	1.8412	3.0542	4.2012	5.3176	6.4156	7.5013
	3.8317	5.3314	6.7061	8.0152	9.2824	10.5199	11.7349
	7.0156	8.5363	9.9695	11.3459	12.6819	13.9872	15.2682
	10.1735	11.7060	13.1704	14.5859	15.9641	17.3128	18.6374
	13.3237	14.8636	16.3475	17.7888	19.1960	20.5755	21.9317
	16.4706	18.0155	19.5129	20.9725	22.4010	23.8036	25.1839
$Y'_n(x)=0$	2.1971	3.6830	5.0026	6.2536	7.4649	8.6496	9.8148
	5.4297	6.9415	8.3507	9.6988	11.0052	12.2809	13.5328
	8.5960	10.1234	11.5742	12.9724	14.3317	15.6608	16.9655
	11.7492	13.2858	14.7609	16.1905	17.5844	18.9497	20.2913
	14.8974	16.4401	17.9313	19.3824	20.8011	22.1928	23.5619
	18.0434	19.5902	21.0929	22.5598	23.9970	25.4091	26.7995

TABLE 40

EXPONENTIAL, SINE AND COSINE INTEGRALS

$$Ei(x) = \int_x^\infty \frac{e^{-u}}{u}\,du, \quad Si(x) = \int_0^x \frac{\sin u}{u}\,du, \quad Ci(x) = \int_x^\infty \frac{\cos u}{u}\,du$$

x	$Ei(x)$	$Si(x)$	$Ci(x)$
.0	∞	.0000	∞
.5	.5598	.4931	.1778
1.0	.2194	.9461	−.3374
1.5	.1000	1.3247	−.4704
2.0	.04890	1.6054	−.4230
2.5	.02491	1.7785	−.2859
3.0	.01305	1.8487	−.1196
3.5	$.0^{2}6970$	1.8331	.0321
4.0	$.0^{2}3779$	1.7582	.1410
4.5	$.0^{2}2073$	1.6541	.1935
5.0	$.0^{2}1148$	1.5499	.1900
5.5	$.0^{3}6409$	1.4687	.1421
6.0	$.0^{3}3601$	1.4247	.0681
6.5	$.0^{3}2034$	1.4218	−.0111
7.0	$.0^{3}1155$	1.4546	−.0767
7.5	$.0^{4}6583$	1.5107	−.1156
8.0	$.0^{4}3767$	1.5742	−.1224
8.5	$.0^{4}2162$	1.6296	−.09943
9.0	$.0^{4}1245$	1.6650	−.05535
9.5	$.0^{5}7185$	1.6745	$-.0^{2}2678$
10.0	$.0^{5}4157$	1.6583	.04546

TABLE 41

LEGENDRE POLYNOMIALS $P_n(x)$

$[P_0(x) = 1,\ P_1(x) = x]$

x	$P_2(x)$	$P_3(x)$	$P_4(x)$	$P_5(x)$
.00	−.5000	.0000	.3750	.0000
.05	−.4963	−.0747	.3657	.0927
.10	−.4850	−.1475	.3379	.1788
.15	−.4663	−.2166	.2928	.2523
.20	−.4400	−.2800	.2320	.3075
.25	−.4063	−.3359	.1577	.3397
.30	−.3650	−.3825	.0729	.3454
.35	−.3163	−.4178	−.0187	.3225
.40	−.2600	−.4400	−.1130	.2706
.45	−.1963	−.4472	−.2050	.1917
.50	−.1250	−.4375	−.2891	.0898
.55	−.0463	−.4091	−.3590	−.0282
.60	.0400	−.3600	−.4080	−.1526
.65	.1338	−.2884	−.4284	−.2705
.70	.2350	−.1925	−.4121	−.3652
.75	.3438	−.0703	−.3501	−.4164
.80	.4600	.0800	−.2330	−.3995
.85	.5838	.2603	−.0506	−.2857
.90	.7150	.4725	.2079	−.0411
.95	.8538	.7184	.5541	.3727
1.00	1.0000	1.0000	1.0000	1.0000

TABLE 42

LEGENDRE POLYNOMIALS $P_n(\cos\theta)$

$[P_0(\cos\theta) = 1]$

θ	$P_1(\cos\theta)$	$P_2(\cos\theta)$	$P_3(\cos\theta)$	$P_4(\cos\theta)$	$P_5(\cos\theta)$
0°	1.0000	1.0000	1.0000	1.0000	1.0000
5°	.9962	.9886	.9773	.9623	.9437
10°	.9848	.9548	.9106	.8532	.7840
15°	.9659	.8995	.8042	.6847	.5471
20°	.9397	.8245	.6649	.4750	.2715
25°	.9063	.7321	.5016	.2465	.0009
30°	.8660	.6250	.3248	.0234	−.2233
35°	.8192	.5065	.1454	−.1714	−.3691
40°	.7660	.3802	−.0252	−.3190	−.4197
45°	.7071	.2500	−.1768	−.4063	−.3757
50°	.6428	.1198	−.3002	−.4275	−.2545
55°	.5736	−.0065	−.3886	−.3852	−.0868
60°	.5000	−.1250	−.4375	−.2891	.0898
65°	.4226	−.2321	−.4452	−.1552	.2381
70°	.3420	−.3245	−.4130	−.0038	.3281
75°	.2588	−.3995	−.3449	.1434	.3427
80°	.1737	−.4548	−.2474	.2659	.2810
85°	.0872	−.4886	−.1291	.3468	.1577
90°	.0000	−.5000	.0000	.3750	.0000

TABLE 43

COMPLETE ELLIPTIC INTEGRALS OF FIRST AND SECOND KINDS

$$K = \int_0^{\pi/2} \frac{d\theta}{\sqrt{1 - k^2 \sin^2\theta}}, \quad E = \int_0^{\pi/2} \sqrt{1 - k^2 \sin^2\theta}\, d\theta, \quad k = \sin\psi$$

ψ	K	E
0°	1.5708	1.5708
1	1.5709	1.5707
2	1.5713	1.5703
3	1.5719	1.5697
4	1.5727	1.5689
5	1.5738	1.5678
6	1.5751	1.5665
7	1.5767	1.5649
8	1.5785	1.5632
9	1.5805	1.5611
10	1.5828	1.5589
11	1.5854	1.5564
12	1.5882	1.5537
13	1.5913	1.5507
14	1.5946	1.5476
15	1.5981	1.5442
16	1.6020	1.5405
17	1.6061	1.5367
18	1.6105	1.5326
19	1.6151	1.5283
20	1.6200	1.5238
21	1.6252	1.5191
22	1.6307	1.5141
23	1.6365	1.5090
24	1.6426	1.5037
25	1.6490	1.4981
26	1.6557	1.4924
27	1.6627	1.4864
28	1.6701	1.4803
29	1.6777	1.4740
30	1.6858	1.4675

ψ	K	E
30°	1.6858	1.4675
31	1.6941	1.4608
32	1.7028	1.4539
33	1.7119	1.4469
34	1.7214	1.4397
35	1.7312	1.4323
36	1.7415	1.4248
37	1.7522	1.4171
38	1.7633	1.4092
39	1.7748	1.4013
40	1.7868	1.3931
41	1.7992	1.3849
42	1.8122	1.3765
43	1.8256	1.3680
44	1.8396	1.3594
45	1.8541	1.3506
46	1.8691	1.3418
47	1.8848	1.3329
48	1.9011	1.3238
49	1.9180	1.3147
50	1.9356	1.3055
51	1.9539	1.2963
52	1.9729	1.2870
53	1.9927	1.2776
54	2.0133	1.2681
55	2.0347	1.2587
56	2.0571	1.2492
57	2.0804	1.2397
58	2.1047	1.2301
59	2.1300	1.2206
60	2.1565	1.2111

ψ	K	E
60°	2.1565	1.2111
61	2.1842	1.2015
62	2.2132	1.1920
63	2.2435	1.1826
64	2.2754	1.1732
65	2.3088	1.1638
66	2.3439	1.1545
67	2.3809	1.1453
68	2.4198	1.1362
69	2.4610	1.1272
70	2.5046	1.1184
71	2.5507	1.1096
72	2.5998	1.1011
73	2.6521	1.0927
74	2.7081	1.0844
75	2.7681	1.0764
76	2.8327	1.0686
77	2.9026	1.0611
78	2.9786	1.0538
79	3.0617	1.0468
80	3.1534	1.0401
81	3.2553	1.0338
82	3.3699	1.0278
83	3.5004	1.0223
84	3.6519	1.0172
85	3.8317	1.0127
86	4.0528	1.0086
87	4.3387	1.0053
88	4.7427	1.0026
89	5.4349	1.0008
90	∞	1.0000

TABLE 44

INCOMPLETE ELLIPTIC INTEGRAL OF THE FIRST KIND

$$F(k, \phi) = \int_0^{\phi} \frac{d\theta}{\sqrt{1 - k^2 \sin^2 \theta}}, \quad k = \sin \psi$$

ϕ \ ψ	0°	10°	20°	30°	40°	50°	60°	70°	80°	90°
0°	0.0000	0.0000	0.0000	0.0000	0.0000	0.0000	0.0000	0.0000	0.0000	0.0000
10°	0.1745	0.1746	0.1746	0.1748	0.1749	0.1751	0.1752	0.1753	0.1754	0.1754
20°	0.3491	0.3493	0.3499	0.3508	0.3520	0.3533	0.3545	0.3555	0.3561	0.3564
30°	0.5236	0.5243	0.5263	0.5294	0.5334	0.5379	0.5422	0.5459	0.5484	0.5493
40°	0.6981	0.6997	0.7043	0.7116	0.7213	0.7323	0.7436	0.7535	0.7604	0.7629
50°	0.8727	0.8756	0.8842	0.8982	0.9173	0.9401	0.9647	0.9876	1.0044	1.0107
60°	1.0472	1.0519	1.0660	1.0896	1.1226	1.1643	1.2126	1.2619	1.3014	1.3170
70°	1.2217	1.2286	1.2495	1.2853	1.3372	1.4068	1.4944	1.5959	1.6918	1.7354
80°	1.3963	1.4056	1.4344	1.4846	1.5597	1.6660	1.8125	2.0119	2.2653	2.4362
90°	1.5708	1.5828	1.6200	1.6858	1.7868	1.9356	2.1565	2.5046	3.1534	∞

TABLE 45

INCOMPLETE ELLIPTIC INTEGRAL OF THE SECOND KIND

$$E(k, \phi) = \int_0^{\phi} \sqrt{1 - k^2 \sin^2 \theta}\, d\theta, \quad k = \sin \psi$$

ϕ \ ψ	0°	10°	20°	30°	40°	50°	60°	70°	80°	90°
0°	0.0000	0.0000	0.0000	0.0000	0.0000	0.0000	0.0000	0.0000	0.0000	0.0000
10°	0.1745	0.1745	0.1744	0.1743	0.1742	0.1740	0.1739	0.1738	0.1737	0.1736
20°	0.3491	0.3489	0.3483	0.3473	0.3462	0.3450	0.3438	0.3429	0.3422	0.3420
30°	0.5236	0.5229	0.5209	0.5179	0.5141	0.5100	0.5061	0.5029	0.5007	0.5000
40°	0.6981	0.6966	0.6921	0.6851	0.6763	0.6667	0.6575	0.6497	0.6446	0.6428
50°	0.8727	0.8698	0.8614	0.8483	0.8317	0.8134	0.7954	0.7801	0.7697	0.7660
60°	1.0472	1.0426	1.0290	1.0076	0.9801	0.9493	0.9184	0.8914	0.8728	0.8660
70°	1.2217	1.2149	1.1949	1.1632	1.1221	1.0750	1.0266	0.9830	0.9514	0.9397
80°	1.3963	1.3870	1.3597	1.3161	1.2590	1.1926	1.1225	1.0565	1.0054	0.9848
90°	1.5708	1.5589	1.5238	1.4675	1.3931	1.3055	1.2111	1.1184	1.0401	1.0000

TABLE 46

ORDINATES OF THE STANDARD NORMAL CURVE

$$y = \frac{1}{\sqrt{2\pi}} e^{-x^2/2}$$

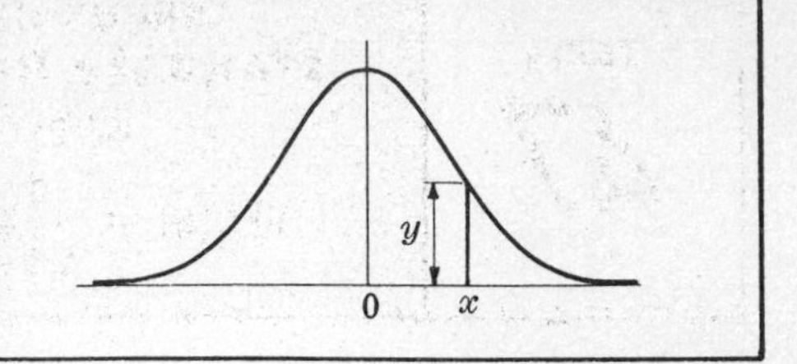

x	0	1	2	3	4	5	6	7	8	9
0.0	.3989	.3989	.3989	.3988	.3986	.3984	.3982	.3980	.3977	.3973
0.1	.3970	.3965	.3961	.3956	.3951	.3945	.3939	.3932	.3925	.3918
0.2	.3910	.3902	.3894	.3885	.3876	.3867	.3857	.3847	.3836	.3825
0.3	.3814	.3802	.3790	.3778	.3765	.3752	.3739	.3725	.3712	.3697
0.4	.3683	.3668	.3653	.3637	.3621	.3605	.3589	.3572	.3555	.3538
0.5	.3521	.3503	.3485	.3467	.3448	.3429	.3410	.3391	.3372	.3352
0.6	.3332	.3312	.3292	.3271	.3251	.3230	.3209	.3187	.3166	.3144
0.7	.3123	.3101	.3079	.3056	.3034	.3011	.2989	.2966	.2943	.2920
0.8	.2897	.2874	.2850	.2827	.2803	.2780	.2756	.2732	.2709	.2685
0.9	.2661	.2637	.2613	.2589	.2565	.2541	.2516	.2492	.2468	.2444
1.0	.2420	.2396	.2371	.2347	.2323	.2299	.2275	.2251	.2227	.2203
1.1	.2179	.2155	.2131	.2107	.2083	.2059	.2036	.2012	.1989	.1965
1.2	.1942	.1919	.1895	.1872	.1849	.1826	.1804	.1781	.1758	.1736
1.3	.1714	.1691	.1669	.1647	.1626	.1604	.1582	.1561	.1539	.1518
1.4	.1497	.1476	.1456	.1435	.1415	.1394	.1374	.1354	.1334	.1315
1.5	.1295	.1276	.1257	.1238	.1219	.1200	.1182	.1163	.1145	.1127
1.6	.1109	.1092	.1074	.1057	.1040	.1023	.1006	.0989	.0973	.0957
1.7	.0940	.0925	.0909	.0893	.0878	.0863	.0848	.0833	.0818	.0804
1.8	.0790	.0775	.0761	.0748	.0734	.0721	.0707	.0694	.0681	.0669
1.9	.0656	.0644	.0632	.0620	.0608	.0596	.0584	.0573	.0562	.0551
2.0	.0540	.0529	.0519	.0508	.0498	.0488	.0478	.0468	.0459	.0449
2.1	.0440	.0431	.0422	.0413	.0404	.0396	.0387	.0379	.0371	.0363
2.2	.0355	.0347	.0339	.0332	.0325	.0317	.0310	.0303	.0297	.0290
2.3	.0283	.0277	.0270	.0264	.0258	.0252	.0246	.0241	.0235	.0229
2.4	.0224	.0219	.0213	.0208	.0203	.0198	.0194	.0189	.0184	.0180
2.5	.0175	.0171	.0167	.0163	.0158	.0154	.0151	.0147	.0143	.0139
2.6	.0136	.0132	.0129	.0126	.0122	.0119	.0116	.0113	.0110	.0107
2.7	.0104	.0101	.0099	.0096	.0093	.0091	.0088	.0086	.0084	.0081
2.8	.0079	.0077	.0075	.0073	.0071	.0069	.0067	.0065	.0063	.0061
2.9	.0060	.0058	.0056	.0055	.0053	.0051	.0050	.0048	.0047	.0046
3.0	.0044	.0043	.0042	.0040	.0039	.0038	.0037	.0036	.0035	.0034
3.1	.0033	.0032	.0031	.0030	.0029	.0028	.0027	.0026	.0025	.0025
3.2	.0024	.0023	.0022	.0022	.0021	.0020	.0020	.0019	.0018	.0018
3.3	.0017	.0017	.0016	.0016	.0015	.0015	.0014	.0014	.0013	.0013
3.4	.0012	.0012	.0012	.0011	.0011	.0010	.0010	.0010	.0009	.0009
3.5	.0009	.0008	.0008	.0008	.0008	.0007	.0007	.0007	.0007	.0006
3.6	.0006	.0006	.0006	.0005	.0005	.0005	.0005	.0005	.0005	.0004
3.7	.0004	.0004	.0004	.0004	.0004	.0004	.0003	.0003	.0003	.0003
3.8	.0003	.0003	.0003	.0003	.0003	.0002	.0002	.0002	.0002	.0002
3.9	.0002	.0002	.0002	.0002	.0002	.0002	.0002	.0002	.0001	.0001

TABLE 47

AREAS UNDER THE STANDARD NORMAL CURVE

from $-\infty$ to x

$$\text{erf}(x) = \frac{1}{\sqrt{2\pi}} \int_{-\infty}^{x} e^{-t^2/2}\, dt$$

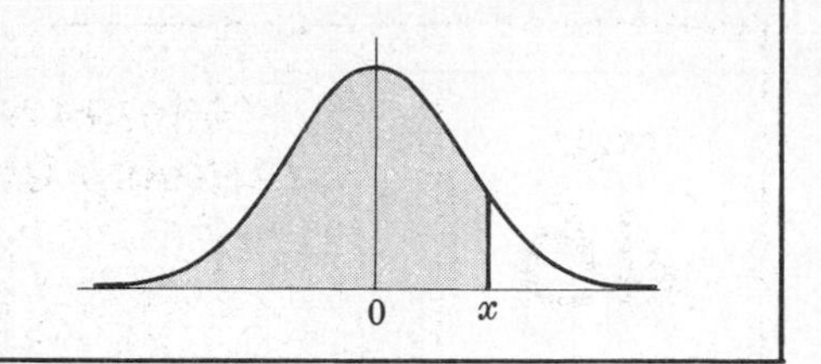

x	0	1	2	3	4	5	6	7	8	9
0.0	.5000	.5040	.5080	.5120	.5160	.5199	.5239	.5279	.5319	.5359
0.1	.5398	.5438	.5478	.5517	.5557	.5596	.5636	.5675	.5714	.5754
0.2	.5793	.5832	.5871	.5910	.5948	.5987	.6026	.6064	.6103	.6141
0.3	.6179	.6217	.6255	.6293	.6331	.6368	.6406	.6443	.6480	.6517
0.4	.6554	.6591	.6628	.6664	.6700	.6736	.6772	.6808	.6844	.6879
0.5	.6915	.6950	.6985	.7019	.7054	.7088	.7123	.7157	.7190	.7224
0.6	.7258	.7291	.7324	.7357	.7389	.7422	.7454	.7486	.7518	.7549
0.7	.7580	.7612	.7642	.7673	.7704	.7734	.7764	.7794	.7823	.7852
0.8	.7881	.7910	.7939	.7967	.7996	.8023	.8051	.8078	.8106	.8133
0.9	.8159	.8186	.8212	.8238	.8264	.8289	.8315	.8340	.8365	.8389
1.0	.8413	.8438	.8461	.8485	.8508	.8531	.8554	.8577	.8599	.8621
1.1	.8643	.8665	.8686	.8708	.8729	.8749	.8770	.8790	.8810	.8830
1.2	.8849	.8869	.8888	.8907	.8925	.8944	.8962	.8980	.8997	.9015
1.3	.9032	.9049	.9066	.9082	.9099	.9115	.9131	.9147	.9162	.9177
1.4	.9192	.9207	.9222	.9236	.9251	.9265	.9279	.9292	.9306	.9319
1.5	.9332	.9345	.9357	.9370	.9382	.9394	.9406	.9418	.9429	.9441
1.6	.9452	.9463	.9474	.9484	.9495	.9505	.9515	.9525	.9535	.9545
1.7	.9554	.9564	.9573	.9582	.9591	.9599	.9608	.9616	.9625	.9633
1.8	.9641	.9649	.9656	.9664	.9671	.9678	.9686	.9693	.9699	.9706
1.9	.9713	.9719	.9726	.9732	.9738	.9744	.9750	.9756	.9761	.9767
2.0	.9772	.9778	.9783	.9788	.9793	.9798	.9803	.9808	.9812	.9817
2.1	.9821	.9826	.9830	.9834	.9838	.9842	.9846	.9850	.9854	.9857
2.2	.9861	.9864	.9868	.9871	.9875	.9878	.9881	.9884	.9887	.9890
2.3	.9893	.9896	.9898	.9901	.9904	.9906	.9909	.9911	.9913	.9916
2.4	.9918	.9920	.9922	.9925	.9927	.9929	.9931	.9932	.9934	.9936
2.5	.9938	.9940	.9941	.9943	.9945	.9946	.9948	.9949	.9951	.9952
2.6	.9953	.9955	.9956	.9957	.9959	.9960	.9961	.9962	.9963	.9964
2.7	.9965	.9966	.9967	.9968	.9969	.9970	.9971	.9972	.9973	.9974
2.8	.9974	.9975	.9976	.9977	.9977	.9978	.9979	.9979	.9980	.9981
2.9	.9981	.9982	.9982	.9983	.9984	.9984	.9985	.9985	.9986	.9986
3.0	.9987	.9987	.9987	.9988	.9988	.9989	.9989	.9989	.9990	.9990
3.1	.9990	.9991	.9991	.9991	.9992	.9992	.9992	.9992	.9993	.9993
3.2	.9993	.9993	.9994	.9994	.9994	.9994	.9994	.9995	.9995	.9995
3.3	.9995	.9995	.9995	.9996	.9996	.9996	.9996	.9996	.9996	.9997
3.4	.9997	.9997	.9997	.9997	.9997	.9997	.9997	.9997	.9997	.9998
3.5	.9998	.9998	.9998	.9998	.9998	.9998	.9998	.9998	.9998	.9998
3.6	.9998	.9998	.9999	.9999	.9999	.9999	.9999	.9999	.9999	.9999
3.7	.9999	.9999	.9999	.9999	.9999	.9999	.9999	.9999	.9999	.9999
3.8	.9999	.9999	.9999	.9999	.9999	.9999	.9999	.9999	.9999	.9999
3.9	1.0000	1.0000	1.0000	1.0000	1.0000	1.0000	1.0000	1.0000	1.0000	1.0000

TABLE 48

PERCENTILE VALUES (t_p) FOR STUDENT'S t DISTRIBUTION

with n degrees of freedom

(shaded area = p)

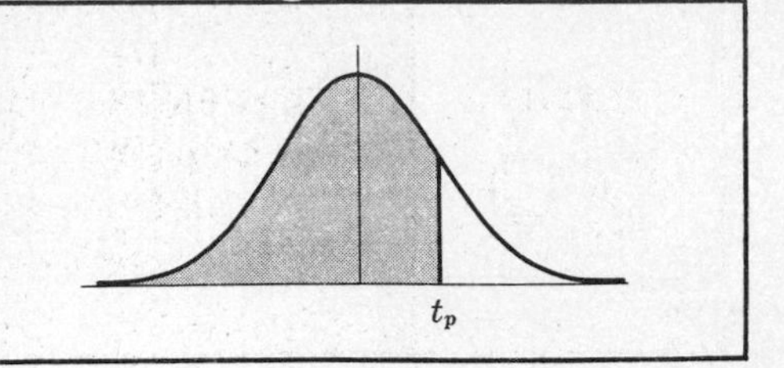

n	$t_{.995}$	$t_{.99}$	$t_{.975}$	$t_{.95}$	$t_{.90}$	$t_{.80}$	$t_{.75}$	$t_{.70}$	$t_{.60}$	$t_{.55}$
1	63.66	31.82	12.71	6.31	3.08	1.376	1.000	.727	.325	.158
2	9.92	6.96	4.30	2.92	1.89	1.061	.816	.617	.289	.142
3	5.84	4.54	3.18	2.35	1.64	.978	.765	.584	.277	.137
4	4.60	3.75	2.78	2.13	1.53	.941	.741	.569	.271	.134
5	4.03	3.36	2.57	2.02	1.48	.920	.727	.559	.267	.132
6	3.71	3.14	2.45	1.94	1.44	.906	.718	.553	.265	.131
7	3.50	3.00	2.36	1.90	1.42	.896	.711	.549	.263	.130
8	3.36	2.90	2.31	1.86	1.40	.889	.706	.546	.262	.130
9	3.25	2.82	2.26	1.83	1.38	.883	.703	.543	.261	.129
10	3.17	2.76	2.23	1.81	1.37	.879	.700	.542	.260	.129
11	3.11	2.72	2.20	1.80	1.36	.876	.697	.540	.260	.129
12	3.06	2.68	2.18	1.78	1.36	.873	.695	.539	.259	.128
13	3.01	2.65	2.16	1.77	1.35	.870	.694	.538	.259	.128
14	2.98	2.62	2.14	1.76	1.34	.868	.692	.537	.258	.128
15	2.95	2.60	2.13	1.75	1.34	.866	.691	.536	.258	.128
16	2.92	2.58	2.12	1.75	1.34	.865	.690	.535	.258	.128
17	2.90	2.57	2.11	1.74	1.33	.863	.689	.534	.257	.128
18	2.88	2.55	2.10	1.73	1.33	.862	.688	.534	.257	.127
19	2.86	2.54	2.09	1.73	1.33	.861	.688	.533	.257	.127
20	2.84	2.53	2.09	1.72	1.32	.860	.687	.533	.257	.127
21	2.83	2.52	2.08	1.72	1.32	.859	.686	.532	.257	.127
22	2.82	2.51	2.07	1.72	1.32	.858	.686	.532	.256	.127
23	2.81	2.50	2.07	1.71	1.32	.858	.685	.532	.256	.127
24	2.80	2.49	2.06	1.71	1.32	.857	.685	.531	.256	.127
25	2.79	2.48	2.06	1.71	1.32	.856	.684	.531	.256	.127
26	2.78	2.48	2.06	1.71	1.32	.856	.684	.531	.256	.127
27	2.77	2.47	2.05	1.70	1.31	.855	.684	.531	.256	.127
28	2.76	2.47	2.05	1.70	1.31	.855	.683	.530	.256	.127
29	2.76	2.46	2.04	1.70	1.31	.854	.683	.530	.256	.127
30	2.75	2.46	2.04	1.70	1.31	.854	.683	.530	.256	.127
40	2.70	2.42	2.02	1.68	1.30	.851	.681	.529	.255	.126
60	2.66	2.39	2.00	1.67	1.30	.848	.679	.527	.254	.126
120	2.62	2.36	1.98	1.66	1.29	.845	.677	.526	.254	.126
∞	2.58	2.33	1.96	1.645	1.28	.842	.674	.524	.253	.126

Source: R. A. Fisher and F. Yates, *Statistical Tables for Biological, Agricultural and Medical Research* (6th edition, 1963), Table III, Oliver and Boyd Ltd., Edinburgh, by permission of the authors and publishers.

TABLE

49

PERCENTILE VALUES (χ^2_p) FOR THE CHI-SQUARE DISTRIBUTION

with n degrees of freedom

(shaded area = p)

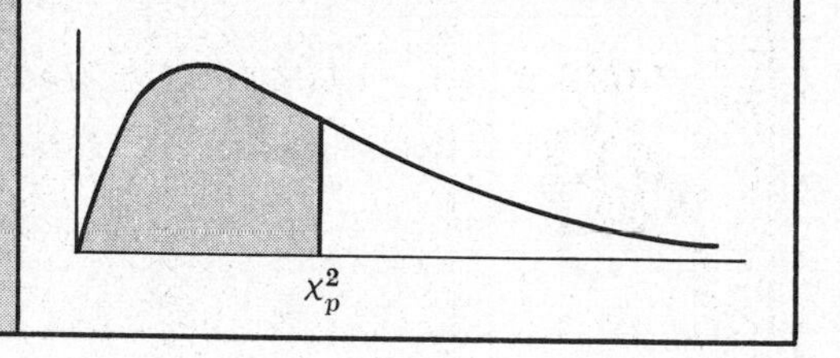

n	$\chi^2_{.995}$	$\chi^2_{.99}$	$\chi^2_{.975}$	$\chi^2_{.95}$	$\chi^2_{.90}$	$\chi^2_{.75}$	$\chi^2_{.50}$	$\chi^2_{.25}$	$\chi^2_{.10}$	$\chi^2_{.05}$	$\chi^2_{.025}$	$\chi^2_{.01}$	$\chi^2_{.005}$
1	7.88	6.63	5.02	3.84	2.71	1.32	.455	.102	.0158	.0039	.0010	.0002	.0000
2	10.6	9.21	7.38	5.99	4.61	2.77	1.39	.575	.211	.103	.0506	.0201	.0100
3	12.8	11.3	9.35	7.81	6.25	4.11	2.37	1.21	.584	.352	.216	.115	.072
4	14.9	13.3	11.1	9.49	7.78	5.39	3.36	1.92	1.06	.711	.484	.297	.207
5	16.7	15.1	12.8	11.1	9.24	6.63	4.35	2.67	1.61	1.15	.831	.554	.412
6	18.5	16.8	14.4	12.6	10.6	7.84	5.35	3.45	2.20	1.64	1.24	.872	.676
7	20.3	18.5	16.0	14.1	12.0	9.04	6.35	4.25	2.83	2.17	1.69	1.24	.989
8	22.0	20.1	17.5	15.5	13.4	10.2	7.34	5.07	3.49	2.73	2.18	1.65	1.34
9	23.6	21.7	19.0	16.9	14.7	11.4	8.34	5.90	4.17	3.33	2.70	2.09	1.73
10	25.2	23.2	20.5	18.3	16.0	12.5	9.34	6.74	4.87	3.94	3.25	2.56	2.16
11	26.8	24.7	21.9	19.7	17.3	13.7	10.3	7.58	5.58	4.57	3.82	3.05	2.60
12	28.3	26.2	23.3	21.0	18.5	14.8	11.3	8.44	6.30	5.23	4.40	3.57	3.07
13	29.8	27.7	24.7	22.4	19.8	16.0	12.3	9.30	7.04	5.89	5.01	4.11	3.57
14	31.3	29.1	26.1	23.7	21.1	17.1	13.3	10.2	7.79	6.57	5.63	4.66	4.07
15	32.8	30.6	27.5	25.0	22.3	18.2	14.3	11.0	8.55	7.26	6.26	5.23	4.60
16	34.3	32.0	28.8	26.3	23.5	19.4	15.3	11.9	9.31	7.96	6.91	5.81	5.14
17	35.7	33.4	30.2	27.6	24.8	20.5	16.3	12.8	10.1	8.67	7.56	6.41	5.70
18	37.2	34.8	31.5	28.9	26.0	21.6	17.3	13.7	10.9	9.39	8.23	7.01	6.26
19	38.6	36.2	32.9	30.1	27.2	22.7	18.3	14.6	11.7	10.1	8.91	7.63	6.84
20	40.0	37.6	34.2	31.4	28.4	23.8	19.3	15.5	12.4	10.9	9.59	8.26	7.43
21	41.4	38.9	35.5	32.7	29.6	24.9	20.3	16.3	13.2	11.6	10.3	8.90	8.03
22	42.8	40.3	36.8	33.9	30.8	26.0	21.3	17.2	14.0	12.3	11.0	9.54	8.64
23	44.2	41.6	38.1	35.2	32.0	27.1	22.3	18.1	14.8	13.1	11.7	10.2	9.26
24	45.6	43.0	39.4	36.4	33.2	28.2	23.3	19.0	15.7	13.8	12.4	10.9	9.89
25	46.9	44.3	40.6	37.7	34.4	29.3	24.3	19.9	16.5	14.6	13.1	11.5	10.5
26	48.3	45.6	41.9	38.9	35.6	30.4	25.3	20.8	17.3	15.4	13.8	12.2	11.2
27	49.6	47.0	43.2	40.1	36.7	31.5	26.3	21.7	18.1	16.2	14.6	12.9	11.8
28	51.0	48.3	44.5	41.3	37.9	32.6	27.3	22.7	18.9	16.9	15.3	13.6	12.5
29	52.3	49.6	45.7	42.6	39.1	33.7	28.3	23.6	19.8	17.7	16.0	14.3	13.1
30	53.7	50.9	47.0	43.8	40.3	34.8	29.3	24.5	20.6	18.5	16.8	15.0	13.8
40	66.8	63.7	59.3	55.8	51.8	45.6	39.3	33.7	29.1	26.5	24.4	22.2	20.7
50	79.5	76.2	71.4	67.5	63.2	56.3	49.3	42.9	37.7	34.8	32.4	29.7	28.0
60	92.0	88.4	83.3	79.1	74.4	67.0	59.3	52.3	46.5	43.2	40.5	37.5	35.5
70	104.2	100.4	95.0	90.5	85.5	77.6	69.3	61.7	55.3	51.7	48.8	45.4	43.3
80	116.3	112.3	106.6	101.9	96.6	88.1	79.3	71.1	64.3	60.4	57.2	53.5	51.2
90	128.3	124.1	118.1	113.1	107.6	98.6	89.3	80.6	73.3	69.1	65.6	61.8	59.2
100	140.2	135.8	129.6	124.3	118.5	109.1	99.3	90.1	82.4	77.9	74.2	70.1	67.3

Source: Catherine M. Thompson, *Table of percentage points of the χ^2 distribution*, Biometrika, Vol. 32 (1941), by permission of the author and publisher.

TABLE

50

95th PERCENTILE VALUES FOR THE F DISTRIBUTION

n_1 = degrees of freedom for numerator

n_2 = degrees of freedom for denominator

(shaded area = .95)

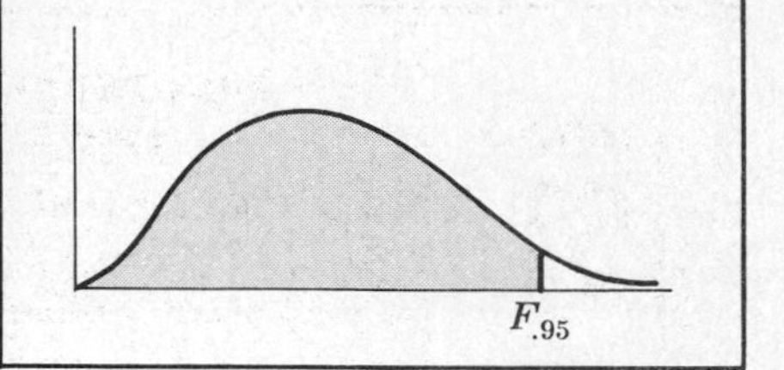

n_2 \ n_1	1	2	3	4	5	6	8	12	16	20	30	40	50	100	∞
1	161.4	199.5	215.7	224.6	230.2	234.0	238.9	243.9	246.3	248.0	250.1	251.1	252.2	253.0	254.3
2	18.51	19.00	19.16	19.25	19.30	19.33	19.37	19.41	19.43	19.45	19.46	19.46	19.47	19.49	19.50
3	10.13	9.55	9.28	9.12	9.01	8.94	8.85	8.74	8.69	8.66	8.62	8.60	8.58	8.56	8.53
4	7.71	6.94	6.59	6.39	6.26	6.16	6.04	5.91	5.84	5.80	5.75	5.71	5.70	5.66	5.63
5	6.61	5.79	5.41	5.19	5.05	4.95	4.82	4.68	4.60	4.56	4.50	4.46	4.44	4.40	4.36
6	5.99	5.14	4.76	4.53	4.39	4.28	4.15	4.00	3.92	3.87	3.81	3.77	3.75	3.71	3.67
7	5.59	4.74	4.35	4.12	3.97	3.87	3.73	3.57	3.49	3.44	3.38	3.34	3.32	3.28	3.23
8	5.32	4.46	4.07	3.84	3.69	3.58	3.44	3.28	3.20	3.15	3.08	3.05	3.03	2.98	2.93
9	5.12	4.26	3.86	3.63	3.48	3.37	3.23	3.07	2.98	2.93	2.86	2.82	2.80	2.76	2.71
10	4.96	4.10	3.71	3.48	3.33	3.22	3.07	2.91	2.82	2.77	2.70	2.67	2.64	2.59	2.54
11	4.84	3.98	3.59	3.36	3.20	3.09	2.95	2.79	2.70	2.65	2.57	2.53	2.50	2.45	2.40
12	4.75	3.89	3.49	3.26	3.11	3.00	2.85	2.69	2.60	2.54	2.46	2.42	2.40	2.35	2.30
13	4.67	3.81	3.41	3.18	3.03	2.92	2.77	2.60	2.51	2.46	2.38	2.34	2.32	2.26	2.21
14	4.60	3.74	3.34	3.11	2.96	2.85	2.70	2.53	2.44	2.39	2.31	2.27	2.24	2.19	2.13
15	4.54	3.68	3.29	3.06	2.90	2.79	2.64	2.48	2.39	2.33	2.25	2.21	2.18	2.12	2.07
16	4.49	3.63	3.24	3.01	2.85	2.74	2.59	2.42	2.33	2.28	2.20	2.16	2.13	2.07	2.01
17	4.45	3.59	3.20	2.96	2.81	2.70	2.55	2.38	2.29	2.23	2.15	2.11	2.08	2.02	1.96
18	4.41	3.55	3.16	2.93	2.77	2.66	2.51	2.34	2.25	2.19	2.11	2.07	2.04	1.98	1.92
19	4.38	3.52	3.13	2.90	2.74	2.63	2.48	2.31	2.21	2.15	2.07	2.02	2.00	1.94	1.88
20	4.35	3.49	3.10	2.87	2.71	2.60	2.45	2.28	2.18	2.12	2.04	1.99	1.96	1.90	1.84
22	4.30	3.44	3.05	2.82	2.66	2.55	2.40	2.23	2.13	2.07	1.98	1.93	1.91	1.84	1.78
24	4.26	3.40	3.01	2.78	2.62	2.51	2.36	2.18	2.09	2.03	1.94	1.89	1.86	1.80	1.73
26	4.23	3.37	2.98	2.74	2.59	2.47	2.32	2.15	2.05	1.99	1.90	1.85	1.82	1.76	1.69
28	4.20	3.34	2.95	2.71	2.56	2.45	2.29	2.12	2.02	1.96	1.87	1.81	1.78	1.72	1.65
30	4.17	3.32	2.92	2.69	2.53	2.42	2.27	2.09	1.99	1.93	1.84	1.79	1.76	1.69	1.62
40	4.08	3.23	2.84	2.61	2.45	2.34	2.18	2.00	1.90	1.84	1.74	1.69	1.66	1.59	1.51
50	4.03	3.18	2.79	2.56	2.40	2.29	2.13	1.95	1.85	1.78	1.69	1.63	1.60	1.52	1.44
60	4.00	3.15	2.76	2.53	2.37	2.25	2.10	1.92	1.81	1.75	1.65	1.59	1.56	1.48	1.39
70	3.98	3.13	2.74	2.50	2.35	2.23	2.07	1.89	1.79	1.72	1.62	1.56	1.53	1.45	1.35
80	3.96	3.11	2.72	2.48	2.33	2.21	2.05	1.88	1.77	1.70	1.60	1.54	1.51	1.42	1.32
100	3.94	3.09	2.70	2.46	2.30	2.19	2.03	1.85	1.75	1.68	1.57	1.51	1.48	1.39	1.28
150	3.91	3.06	2.67	2.43	2.27	2.16	2.00	1.82	1.71	1.64	1.54	1.47	1.44	1.34	1.22
200	3.89	3.04	2.65	2.41	2.26	2.14	1.98	1.80	1.69	1.62	1.52	1.45	1.42	1.32	1.19
400	3.86	3.02	2.62	2.39	2.23	2.12	1.96	1.78	1.67	1.60	1.49	1.42	1.38	1.28	1.13
∞	3.84	2.99	2.60	2.37	2.21	2.09	1.94	1.75	1.64	1.57	1.46	1.40	1.32	1.24	1.00

Source: G. W. Snedecor and W. G. Cochran, *Statistical Methods* (6th edition, 1967), Iowa State University Press, Ames, Iowa, by permission of the authors and publisher.

TABLE

51

99th PERCENTILE VALUES FOR THE F DISTRIBUTION

n_1 = degrees of freedom for numerator
n_2 = degrees of freedom for denominator
(shaded area = .99)

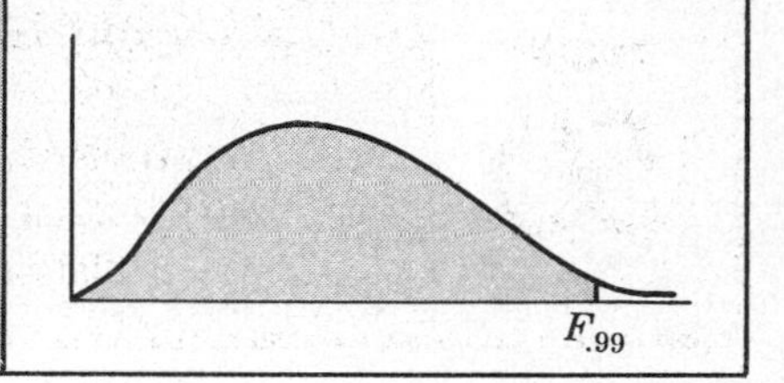

n_2 \ n_1	1	2	3	4	5	6	8	12	16	20	30	40	50	100	∞
1	4052	4999	5403	5625	5764	5859	5981	6106	6169	6208	6258	6286	6302	6334	6366
2	98.49	99.01	99.17	99.25	99.30	99.33	99.36	99.42	99.44	99.45	99.47	99.48	99.48	99.49	99.50
3	34.12	30.81	29.46	28.71	28.24	27.41	27.49	27.05	28.63	26.69	26.50	26.41	26.35	26.23	26.12
4	21.20	18.00	16.69	15.98	15.52	15.21	14.80	14.37	14.15	14.02	13.83	13.74	13.69	13.57	13.46
5	16.26	13.27	12.06	11.39	10.97	10.67	10.27	9.89	9.68	9.55	9.38	9.29	9.24	9.13	9.02
6	13.74	10.92	9.78	9.15	8.75	8.47	8.10	7.72	7.52	7.39	7.23	7.14	7.09	6.99	6.88
7	12.25	9.55	8.45	7.85	7.46	7.19	6.84	6.47	6.27	6.15	5.98	5.90	5.85	5.75	5.65
8	11.26	8.65	7.59	7.01	6.63	6.37	6.03	5.67	5.48	5.36	5.20	5.11	5.06	4.96	4.86
9	10.56	8.02	6.99	6.42	6.06	5.80	5.47	5.11	4.92	4.80	4.64	4.56	4.51	4.41	4.31
10	10.04	7.56	6.55	5.99	5.64	5.39	5.06	4.71	4.52	4.41	4.25	4.17	4.12	4.01	3.91
11	9.05	7.20	6.22	5.67	5.32	5.07	4.74	4.40	4.21	4.10	3.94	3.86	3.80	3.70	3.60
12	9.33	6.93	5.95	5.41	5.06	4.82	4.50	4.16	3.98	3.86	3.70	3.61	3.56	3.46	3.36
13	9.07	6.70	5.74	5.20	4.86	4.62	4.30	3.96	3.78	3.67	3.51	3.42	3.37	3.27	3.16
14	8.86	6.51	5.56	5.03	4.69	4.46	4.14	3.80	3.62	3.51	3.34	3.26	3.21	3.11	3.00
15	8.68	6.36	5.42	4.89	4.56	4.32	4.00	3.67	3.48	3.36	3.20	3.12	3.07	2.97	2.87
16	8.53	6.23	5.29	4.77	4.44	4.20	3.89	3.55	3.37	3.25	3.10	3.01	2.96	2.86	2.75
17	8.40	6.11	5.18	4.67	4.34	4.10	3.79	3.45	3.27	3.16	3.00	2.92	2.86	2.76	2.65
18	8.28	6.01	5.09	4.58	4.25	4.01	3.71	3.37	3.19	3.07	2.91	2.83	2.78	2.68	2.57
19	8.18	5.93	5.01	4.50	4.17	3.94	3.63	3.30	3.12	3.00	2.84	2.76	2.70	2.60	2.49
20	8.10	5.85	4.94	4.43	4.10	3.87	3.56	3.23	3.05	2.94	2.77	2.69	2.63	2.53	2.42
22	7.94	5.72	4.82	4.31	3.99	3.76	3.45	3.12	2.94	2.83	2.67	2.58	2.53	2.42	2.31
24	7.82	5.61	4.72	4.22	3.90	3.67	3.36	3.03	2.85	2.74	2.58	2.49	2.44	2.33	2.21
26	7.72	5.53	4.64	4.14	3.82	3.59	3.29	2.96	2.77	2.66	2.50	2.41	2.36	2.25	2.13
28	7.64	5.45	4.57	4.07	3.76	3.53	3.23	2.90	2.71	2.60	2.44	2.35	2.30	2.18	2.06
30	7.56	5.39	4.51	4.02	3.70	3.47	3.17	2.84	2.66	2.55	2.38	2.29	2.24	2.13	2.01
40	7.31	5.18	4.31	3.83	3.51	3.29	2.99	2.66	2.49	2.37	2.20	2.11	2.05	1.94	1.81
50	7.17	5.06	4.20	3.72	3.41	3.18	2.88	2.56	2.39	2.26	2.10	2.00	1.94	1.82	1.68
60	7.08	4.98	4.13	3.65	3.34	3.12	2.82	2.50	2.32	2.20	2.03	1.93	1.87	1.74	1.60
70	7.01	4.92	4.08	3.60	3.29	3.07	2.77	2.45	2.28	2.15	1.98	1.88	1.82	1.69	1.53
80	6.96	4.88	4.04	3.56	3.25	3.04	2.74	2.41	2.24	2.11	1.94	1.84	1.78	1.65	1.49
100	6.90	4.82	3.98	3.51	3.20	2.99	2.69	2.36	2.19	2.06	1.89	1.79	1.73	1.59	1.43
150	6.81	4.75	3.91	3.44	3.14	2.92	2.62	2.30	2.12	2.00	1.83	1.72	1.66	1.51	1.33
200	6.76	4.71	3.88	3.41	3.11	2.90	2.60	2.28	2.09	1.97	1.79	1.69	1.62	1.48	1.28
400	6.70	4.66	3.83	3.36	3.06	2.85	2.55	2.23	2.04	1.92	1.74	1.64	1.57	1.42	1.19
∞	6.64	4.60	3.78	3.32	3.02	2.80	2.51	2.18	1.99	1.87	1.69	1.59	1.52	1.36	1.00

Source: G. W. Snedecor and W. G. Cochran, *Statistical Methods* (6th edition, 1967), Iowa State University Press, Ames, Iowa, by permission of the authors and publisher.

TABLE 52

RANDOM NUMBERS

51772	74640	42331	29044	46621	62898	93582	04186	19640	87056
24033	23491	83587	06568	21960	21387	76105	10863	97453	90581
45939	60173	52078	25424	11645	55870	56974	37428	93507	94271
30586	02133	75797	45406	31041	86707	12973	17169	88116	42187
03585	79353	81938	82322	96799	85659	36081	50884	14070	74950
64937	03355	95863	20790	65304	55189	00745	65253	11822	15804
15630	64759	51135	98527	62586	41889	25439	88036	24034	67283
09448	56301	57683	30277	94623	85418	68829	06652	41982	49159
21631	91157	77331	60710	52290	16835	48653	71590	16159	14676
91097	17480	29414	06829	87843	28195	27279	47152	35683	47280
50532	25496	95652	42457	73547	76552	50020	24819	52984	76168
07136	40876	79971	54195	25708	51817	36732	72484	94923	75936
27989	64728	10744	08396	56242	90985	28868	99431	50995	20507
85184	73949	36601	46253	00477	25234	09908	36574	72139	70185
54398	21154	97810	36764	32869	11785	55261	59009	38714	38723
65544	34371	09591	07839	58892	92843	72828	91341	84821	63886
08263	65952	85762	64236	39238	18776	84303	99247	46149	03229
39817	67906	48236	16057	81812	15815	63700	85915	19219	45943
62257	04077	79443	95203	02479	30763	92486	54083	23631	05825
53298	90276	62545	21944	16530	03878	07516	95715	02526	33537

Index of Special Symbols and Notations

The following list shows special symbols and notations used in this book together with pages on which they are defined or first appear. Cases where a symbol has more than one meaning will be clear from the context.

Symbols

Symbol	Meaning
$\mathrm{Ber}_n(x)$, $\mathrm{Bei}_n(x)$	140
$B(m,n)$	beta function, 103
B_n	Bernoulli numbers, 114
$C(x)$	Fresnel cosine integral, 184
$Ci(x)$	cosine integral, 184
e	natural base of logarithms, 1
$\mathbf{e}_1, \mathbf{e}_2, \mathbf{e}_3$	unit vectors in curvilinear coordinates, 124
$\mathrm{erf}(x)$	error function, 183
$\mathrm{erfc}(x)$	complementary error function, 183
$E = E(k, \pi/2)$	complete elliptic integral of second kind, 179
$E(k, \phi)$	incomplete elliptic integral of second kind, 179
$Ei(x)$	exponential integral, 183
E_n	Euler numbers, 114
$F(a, b; c; x)$	hypergeometric function, 160
$F(k, \phi)$	incomplete elliptic integral of first kind, 179
$\mathcal{F}, \mathcal{F}^{-1}$	Fourier transform and inverse Fourier transform, 175, 176
h_1, h_2, h_3	scale factors in curvilinear coordinates, 124
$H_n(x)$	Hermite polynomials, 151
$H_n^{(1)}(x)$, $H_n^{(2)}(x)$	Hankel functions of first and second kind, 138
i	imaginary unit, 21
$\mathbf{i}, \mathbf{j}, \mathbf{k}$	unit vectors in rectangular coordinates, 117
$I_n(x)$	modified Bessel function of first kind, 138
$J_n(x)$	Bessel function of first kind, 136
$K = F(k, \pi/2)$	complete elliptic integral of first kind, 179
$\mathrm{Ker}_n(x)$, $\mathrm{Kei}_n(x)$	140
$K_n(x)$	modified Bessel function of second kind, 139
$\ln x$ or $\log_e x$	natural logarithm of x, 24
$\log x$ or $\log_{10} x$	common logarithm of x, 23
$L_n(x)$	Laguerre polynomials, 153
$L_n^m(x)$	associated Laguerre polynomials, 155
$\mathcal{L}, \mathcal{L}^{-1}$	Laplace transform and inverse Laplace transform, 161
$P_n(x)$	Legendre polynomials, 146
$P_n^m(x)$	associated Legendre functions of first kind, 149
$Q_n(x)$	Legendre functions of second kind, 148
$Q_n^m(x)$	associated Legendre functions of second kind, 150
r	cylindrical coordinate, 49
r	polar coordinate, 22, 36
r	spherical coordinate, 50
$S(x)$	Fresnel sine integral, 184
$Si(x)$	sine integral, 183
$T_n(x)$	Chebyshev polynomials of first kind, 157
$U_n(x)$	Chebyshev polynomials of second kind, 158
$Y_n(x)$	Bessel function of second kind, 136

Greek Symbols

Notations

INDEX